AF758184

Leafy Medicinal Herbs

Botany, Chemistry, Postharvest Technology and Uses

FSC
www.fsc.org
MIX
Paper from
responsible sources
FSC® C013604

Leafy Medicinal Herbs

Botany, Chemistry, Postharvest Technology and Uses

Edited by

Dawn C.P. Ambrose

ICAR Central Institute of Agricultural Engineering, Regional Centre, Coimbatore, India

Annamalai Manickavasagan

College of Agricultural and Marine Sciences, Sultan Qaboos University, Oman

Ravindra Naik

ICAR Central Institute of Agricultural Engineering, Regional Centre, Coimbatore, India

CABI

CABI is a trading name of CAB International

CABI
Nosworthy Way
Wallingford
Oxfordshire OX10 8DE
UK

Tel: +44 (0)1491 832111
Fax: +44 (0)1491 833508
E-mail: info@cabi.org
Website: www.cabi.org

CABI
745 Atlantic Avenue
8th Floor
Boston, MA 02111
USA

Tel: +1 (617)682-9015
E-mail: cabi-nao@cabi.org

© CAB International 2016. All rights reserved. No part of this publication may be reproduced in any form or by any means, electronically, mechanically, by photocopying, recording or otherwise, without the prior permission of the copyright owners.

A catalogue record for this book is available from the British Library, London, UK.

Library of Congress Cataloging-in-Publication Data

Names: Ambrose, Dawn C. P., editor.
Title: Leafy medicinal herbs : botany, chemistry, postharvest technology and uses / editors: Dawn C.P. Ambrose, A. Manickavasagan, Ravindra Naik.
Description: Boston, MA : CABI, [2016] | Includes bibliographical references and index.
Identifiers: LCCN 2016002562 (print) | LCCN 2016004050 (ebook) | ISBN 9781780645599 (hbk : alk. paper) | ISBN 9781780645605 (ePDF) | ISBN 9781780647555 (ePub)
Subjects: LCSH: Herbs--Therapeutic use. | Medicinal plants.
Classification: LCC RM666.H33 L43 2016 (print) | LCC RM666.H33 (ebook) | DDC 615.8/8--dc23
LC record available at http://lccn.loc.gov/2016002562

ISBN-13: 978 1 78064 559 9

Commissioning editor: Nicki Dennis/Rachael Russell
Editorial assistant: Emma McCann
Production editor: Lauren Povey

Typeset by SPi, Pondicherry, India.
Printed and bound in the UK by CPI Group (UK) Ltd, Croydon, CR0 4YY.

Contents

About the Editors

Dawn C.P. Ambrose PhD is Principal Scientist at the Central Institute of Agricultural Engineering (Indian Council of Agricultural Research, ICAR), Regional Centre, Coimbatore, India. She has specialized in agricultural process engineering and obtained her PhD in that field at Tamil Nadu Agricultural University, India. Before joining ICAR in 1997, she had experience as an academician and also as a researcher in various other organizations. She has published four books, 13 book chapters, 30 research papers and 47 papers presented at conferences. She is a life member of many professional societies.

Annamalai Manickavasagan PhD, PEng (Canada) obtained a PhD (Biosystems Engineering) from the University of Manitoba, Canada. He is a licensed professional engineer (PEng) in the province of New Brunswick, Canada. After obtaining his PhD, he worked with McCain Foods Limited (Canada) as a Scientist. At present, he is working as an Assistant Professor at the College of Agricultural and Marine Sciences, Sultan Qaboos University, Oman. He has published three books, eight book chapters and more than 60 scientific papers in peer-reviewed journals and presented at international conferences. He has diverse research and management experience with academic institutions and industries in Canada, Malaysia, India and Oman.

Ravindra Naik PhD works as a Principal Scientist at the Central Institute of Agricultural Engineering, (Indian Council of Agricultural Research, ICAR), Regional Centre, Coimbatore, India, and specializes in agricultural structure and process engineering. He obtained his PhD in 1998 from Tamil Nadu Agricultural University, Coimbatore, India and joined the ICAR Directorate of Rice Research, Hyderabad, India, in 1998. His major areas of professional interest are in the postharvest technology of paddy, modified atmosphere storage of fruits and vegetables, nutraceuticals, postharvest machinery for sugarcane and banana, and value addition of horticultural crops, particularly fruits and vegetables. He has been involved in the patenting of six technologies, which are now at various stages of the patenting process. He has operated many externally funded projects and underwent training in the field of nutraceuticals at Virginia Tech, USA. He has also won many awards at national level. He is a member of many national and international scientific associations. He has published many books, book chapters, extension bulletins, and articles in reputed national and international journals.

Contributors

Aregbesola Oladipupo Abiodun, Department of Microbiology, Faculty of Science, Obafemi Awolowo University, Ile-Ife, 22005, Nigeria.

Rashid Abdullah Al-Yahyai, Department of Crop Sciences, Sultan Qaboos University, PO Box 34, Al Khod 123, Muscat, Oman. E-mail: alyahyai@squ.edu.om

Akindele Folarin Alonge, Department of Agricultural and Food Engineering, University of Uyo, PMB 1017, Uyo, Akwa Ibom State, 52003, Nigeria. E-mail: akindelealonge@uniuyo.edu.ng

Dawn C.P. Ambrose, ICAR – Central Institute of Agricultural Engineering, Regional Centre, Coimbatore 641 007, Tamil Nadu, India. E-mail: dawncp@yahoo.com

Gopal Amuthaselvi, Krishi Vigyan Kendra, Sirugamani, Tiruchirappalli 639 115, Tamil Nadu, India. E-mail: g.amuthaselvi@gmail.com

S. Jacob K. Annamalai, ICAR – Central Institute of Agricultural Engineering, Regional Centre, Coimbatore 641 007, Tamil Nadu, India.

Neşet Arslan, Ankara University, Faculty of Agriculture, Department of Field Crops, 06110 Ankara, Turkey.

R. Arul Mari, Department of Food and Agricultural Process Engineering, Agricultural Engineering College and Research Institute, Tamil Nadu Agricultural University, Coimbatore 641003, Tamil Nadu, India.

K. Hüsnü Can Başer, Faculty of Pharmacy, Anadolu University, Eskisehir 26470, Turkey, Near East University, Nicosia, N. Cyprus. E-mail: khcbaser@gmail.com

Hülya Çakmak, Department of Food Engineering, Ege University, Bornova 35100, Izmir, Turkey. E-mail: hulya.cakmak@ege.edu.tr

Salome Amarachi Chime, Department of Pharmaceutical and Industrial Pharmacy, University of Nigeria, Nsukka 410001, Nigeria. E-mail: emmymarachi@yahoo.com; salome.chime@unn.edu.ng

Ghazi Daradkeh, Department of Food Science and Nutrition, College of Agricultural and Marine Sciences (CAMS), Sultan Qaboos University, Muscat, Oman; Hamad Medical Corporation, Doha, Qatar.

Kuntal Das, Department of Pharmacognosy and Regulatory Affairs, Krupanidhi College of Pharmacy # 12/1, ChikkaBellandur, Carmelaram Post., VarthurHobli, Bangalore 560035, India. E-mail: drkkdsd@gmail.com

Alice L. Daunty, Mcrennet Foods, Anna Nagar East, Chennai 102, Tamil Nadu, India.

E-mail addresses are included only for corresponding authors.

Milda E. Embuscado, Senior Principal Scientist, Materials and Process Technology, McCormick & Company, Inc., 204 Wight Avenue, Hunt Valley, MD 21031-1501, USA. E-mail: milda_embuscado@mccormick.com

Musthafa Mohamed Essa, Department of Food Science and Nutrition, College of Agricultural and Marine Sciences (CAMS), Sultan Qaboos University, PO Box 34, Muscat, PIN 123, Oman. E-mail: drmdessa@squ.edu.om

Maria do Carmo Ferreira, Chemical Engineering Department, Federal University of São Carlos, Rodovia Washington Luís, Km 235, Zip Code 13565-905, São Carlos, SP, Brazil. E-mail: mariaf@ufscar.br

Ahmad Ghorbani, Pharmacological Research Center of Medicinal Plants, Medical School, Pardis Campus, Mashhad University of Medical Sciences, Mashad 9177948564, Iran. E-mail: ghorbania@mums.ac.ir

Muhammad Asif Hanif, Department of Chemistry and Biochemistry, University of Agriculture, Faisalabad 38040, Pakistan.

Seher Kumcuoğlu, Department of Food Engineering, Ege University, Bornova 35100, Izmir, Turkey.

Darach Lupton, Oman Botanic Garden, PO Box 808, Al-Khod 122, Muscat, Oman.

Terrence Madhujith, Department of Food Science and Technology, University of Peradeniya, Peradeniya 20400, Sri Lanka. E-mail: madujith@yahoo.com

Muhammad Mumtaz Khan, Department of Crop Sciences, College of Agricultural and Marine Sciences (CAMS), Sultan Qaboos University, PO Box 34, Al Khod 123, Muscat, Oman.

Ravindra Naik, ICAR – Central Institute of Agricultural Engineering, Regional Centre, Coimbatore 641 007, Tamil Nadu, India. E-mail: naikravindra@gmail.com

Anthonia O. Oluduro, Department of Microbiology, Faculty of Science, Obafemi Awolowo University, Ile-Ife 22005, Nigeria. E-mail: aoluduro2003@yahoo.co.uk

Ikechukwu V. Onyishi, Department of Pharmaceutical and Industrial Pharmacy, University of Nigeria, Nsukka 410001, Nigeria.

Ramanathan Parimalavalli, Department of Food Science and Nutrition, Periyar University, Salem 636 011, Tamil Nadu. E-mail: parimala1996@gmail.com

Maripillai Munusamy Pragalyaashree, Department of Food and Agricultural Process Engineering, Agricultural Engineering College and Research Institute (AEC & RI), Tamil Nadu Agricultural University, Coimbatore 641003, India. E-mail: shreepragalyaa@gmail.com

S. Radhai Sri, Department of Nutrition and Dietetics, PSG College of Arts and Science, Coimbatore 641 014, Tamil Nadu, India.

H.G. Ramya, Processing and Food Engineering, Punjab Agricultural University, Ludhiana, Punjab 141004, India. E-mail: ramyarinda@gmail.com

Alagirisamy Reni, Department of Food Processing and Preservation Technology, Faculty of Engineering, Avinashilingam University, Coimbatore 641 043, Tamil Nadu, India.

Aline de Holanda Rosanova, Chemical Engineering Program, Federal University of São Carlos, Rodovia Washington Luís, Km 235, Zip Code 13565-905, São Carlos, SP, Brazil.

J.S. Rutra Priya, ICAR – Central Institute of Agricultural Engineering, Regional Centre, Coimbatore-641 007, Tamil Nadu, India.

A. Sangamithra, Department of Food Technology, Kongu Engineering College, Perundurai, 638052 Erode, Tamil Nadu, India.

Subajiny Sivakanthan, Department of Agricultural Chemistry, Faculty of Agriculture, University of Jaffna, Sri Lanka.

Svein Øivind Solberg, Nordic Genetic Resource Center, Smedjevägen 3, 230 53 Alnarp, Sweden. E-mail: svein.solberg63@gmail.com

S. Reetha Subashini, ICAR – Central Institute of Agricultural Engineering, Regional Centre, Coimbatore 641 007, Tamil Nadu, India.

Paramadhas Sudha, Food Processing Research and Training Centre, Dryland Agricultural Research Station, Tamil Nadu Agricultural University, Chettinad 630 102, Sivagangai District, Tamil Nadu, India. E-mail: sudha.raj2000@gmail.com

Periyasamy Suganya, Department of Food Processing and Preservation Technology, Faculty of Engineering, Avinashilingam Institute for Home Science and Higher Education for Women University, Coimbatore 641 108, Tamil Nadu, India. E-mail: suganya.abe@gmail.com

Şebnem Tavman, Department of Food Engineering, Ege University, Bornova 35100, Izmir, Turkey.

Venkatachalam Thirupathi, Department of Food and Agricultural Process Engineering, Agricultural Engineering College and Research Institute (AEC & RI), Tamil Nadu Agricultural University, Coimbatore 641 003, India.

Preface

Herbs are plants used since time immemorial for flavouring, food, medicine and perfumes. Leafy medicinal herbs are a good source of minerals, vitamins and antioxidants. They are presently considered to be safe alternatives to modern medicines owing to their healing properties and hence are gaining potential in domestic and international markets across the globe. Although medicinal and aromatic crops cover a broad spectrum of types, little attention has been focused on leafy medicinal herbs as such. This book is written in an attempt to unveil and compile information from various sources on leafy medicinal plants. The book discusses 23 different leafy medicinal plants with the purpose of highlighting their importance, and covers aspects ranging from the botany of these plants to their end uses. Each chapter focuses on the botany, chemistry and postharvest technology of an individual leafy medicinal herb, as well as on its general and pharmacological uses. The book will serve as a platform for researchers, students, farmers and industrialists in the area of leafy medicinal herbs.

Dawn C.P. Ambrose
Annamalai Manickavasagan
Ravindra Naik

Introduction

Herbs are any plant with leaves, seeds or flowers used for flavouring, food, medicine or perfume. The term 'medicinal plants' is used to describe plants (or plants from which products are obtained) that are used by human beings in protection against or treatment of illnesses. The attraction of medicinal herbs has grown as a result of the demand created by consumer interest in these plants for culinary, medicinal and other anthropogenic applications. Herbal remedies are considered the oldest forms of healthcare known to humankind on this earth. Several hundred genera of plants are used in herbal remedies and in traditional or folkloric medicines throughout the world. It has been reported that between 40,000 and 50,000 species are known to be used in traditional and modern medicine systems throughout the world. The medicinal properties of many plants are found in the leaves, which are used as alteratives, tonic diuretics and blood purifiers. Hence, leafy medicinal herbs have found an important place in the medicinal and aromatic plant kingdom.

In recent years, there has been a growing focus on the importance of medicinal plants and traditional health systems in solving the healthcare problems of the world. Medicinal herbs are packed with nutrients such as vitamins, minerals, antioxidants, etc., that are essential to humankind for healthy living. Besides their content of important nutrients, plants also synthesize secondary metabolites, for example alkaloids and volatiles, that have disease-preventive properties. Drug discovery from medicinal plants involves botanical, phytochemical, biological and molecular techniques in a multidimensional approach. Furthermore, medicinal plants and their various products can be viewed as important commodity items for the sustainable economic development of a country. There is a need to produce quality raw materials for the promotion of these leafy medicinal herbs – raw materials that have a demand in terms of domestic and international markets worldwide. This could be achieved by various processing techniques, viz. drying, granulating, extraction, etc. Beside their traditional use as medicine, these plants are now industrially processed for various products used for pharmaceutical, food and perfumery and cosmetics applications. In this book, information has been compiled on the botany, chemistry, postharvest technology and uses of 23 leafy medicinal herbs.

1 Aloe Vera

Ravindra Naik[1]*, J.S. Rutra Priya[1] and R. Arul Mari[2]
[1]*ICAR – Central Institute of Agricultural Engineering, Regional Centre, Coimbatore, India;* [2]*Tamil Nadu Agricultural University, Coimbatore, India*

1.1 Botany

1.1.1 Introduction

Aloe vera (*Aloe vera* (L.) Burm.f., syn. *A. barbadensis* Mill.), a traditional medicinal plant, is used in the food, pharmaceutical and cosmetic industries. The word 'aloe' has its roots in the Arabic word 'alloeh', which means 'radiance'. The innermost part of the leaf is a clear, soft, moist and slippery tissue that consists of large thin-walled parenchyma cells in which water is held in the form of viscous mucilage (Newton, 2004; Naik and Annamalai, 2013). Therefore, the thick fleshy leaves of aloe plants contain not only cell wall carbohydrates such as cellulose and hemicellulose, but also storage carbohydrates such as acetylated mannans (Ni *et al.*, 2004).

Aloe vera is an industrial crop and in the food industry it has been utilized for the preparation of health food drinks, beverages such as tea and milk, and ice cream and confectionery. The gel from aloe vera also finds application in the cosmetic and toiletries industry for the preparation of creams, lotions, soaps, shampoos and facial cleansers. Aloe has antitumour and anti-tyrosine properties in addition to efficacy in healing wounds and burns and in the treatment of gastric ulcers. The potential use of aloe vera products often involves some type of processing, such as heating, dehydration and grinding (Chang *et al.*, 2006).

Unfortunately, because of improper processing procedures, aloe products can contain very little or virtually no active ingredients (Ramachandra and Rao, 2006; Naik and Annamalai, 2013). So, it has become very important to evolve a better method of preservation for increasing the shelf life and maintaining the quality of aloe vera gel. This involves proper handling and treatment of the aloe vera leaves before processing, proper processing to produce the gel and proper gel treatment and storage.

1.1.2 History/origin

Aloe belongs to the Liliaceae, a family of perennial tropical plants of African origin. Species of aloe that have been used as folk medicine include Curaçao aloe (*A. vera*), Cape aloe (*A. ferox* Mill.) and Socotra aloe (*A. perryi* Baker). The use of aloe vera as folk medicine dates to antiquity, and there is an early account from around 1500 BC.

*Corresponding author, e-mail: naikravindra@gmail.com

© CAB International 2016. *Leafy Medicinal Herbs: Botany, Chemistry, Postharvest Technology and Uses* (eds D.C.P. Ambrose *et al.*)

The name aloe is from the Greek word and refers to the bitter juice from the leaves of these plants. It is probably derived from the earlier Arabic word alloeh or the Hebrew word allal, both meaning shining bitter substance (Sung, 2006). In biblical times, the Egyptians hailed aloe vera as the plant of immortality. The Chinese called it their elixir of youth. Aloe vera has many common names and is often referred to as the burn plant, first aid plant or medicine plant, as well as being called true aloe or plain aloe vera. It is considered to be a native plant of Somalia with a history dating back to the 4th century BC (Sung, 2006).

Fig. 1.1. An aloe vera plantation.

1.1.3 Location

Although aloe is reported to have originated in tropical Africa, it is now cultivated in the warm climatic areas of Asia, Europe and America (Coats, 1979). The majority of species occur in southern Africa and on the eastern side of that continent. Many other species are found on the Arabian Peninsula and in Madagascar. The Arabian species have strong relationships with the species of north-east Africa. Madagascan species appear not to be closely related to those of mainland Africa and so active speciation seems to have occurred since the separation of these two land masses.

There are over 350 species of aloe grown the world over. However, only two species are grown commercially: *A. vera* (Fig. 1.1) and *A. arborescens* Mill. There are at least two other species that have medicinal properties, namely *A. perryi* and *A. ferox*. Most aloe vera plants are non-toxic, but a few are extremely poisonous and contain a hemlock-like substance (Atherton, 1998). *A. variegata* L. is a dwarf species that is only a few centimetres in diameter and is a popular houseplant.

1.1.4 Morphology

Aloe vera is a spiky cactus-like xerophyte. It is a clump-forming perennial plant with a thick fibrous root and large basal leaves, usually 12–16 per plant. It weighs up to 1.5 kg when mature. The plant matures when it is about 4 years old and has a lifespan of about 12 years. The leaves are up to 0.5 m long and 8–10 cm across at the base, tapering to a point, with saw-like teeth along their margins. In transverse section, the plant shows a slightly concave appearance on the adaxial surface and a distinctly convex appearance on the lower abaxial surface (Grindlay and Reynolds, 1986; Naik and Annamalai, 2013). The leaves are usually covered with thick cuticle which is differentiated into an upper chlorenchyma and a lower parenchyma. As the rosette matures, successive leaves have fewer whitish spots and become grey to greenish in colour (Eshun and He, 2004).

The plant can be harvested every 6–8 weeks by removing 3–4 leaves per plant. Red, yellow, purple or pale striped flowers are present for most of the year growing in a long raceme at the top of the flower stalk which originates from the centre of the basal leaves. The flower stalk grows up to 1.5 m in height. The fruit is a triangular capsule containing numerous seeds (Grindlay and Reynolds, 1986).

Aloe vera leaf structural composition

The aloe vera leaf can be divided into two major parts, namely the outer green rind, including the vascular bundles, and the inner colourless parenchyma containing the aloe gel. Different terms that are used interchangeably include inner pulp, mucilage tissue, mucilaginous gel, mucilaginous jelly, inner gel and leaf parenchyma tissue. Technically, the term 'pulp' or 'parenchyma tissue' refers to the intact fleshy inner part of the leaf including the cell walls and organelles, while 'gel' or 'mucilage' refers to the viscous clear liquid within the parenchyma cells (Ni *et al.*, 2004). The three structural components of the aloe vera pulp are the cell walls, the degenerated organelles and viscous liquid contained within the cells.

The raw pulp of aloe vera is approximately 98.5% water, while the mucilage or gel is about 99.5% water (Eshun and He, 2004). The whole plant contains 99–99.5% water, and the remaining 0.5–1.5% solid material consists of a range of compounds, including water-soluble and fat-soluble vitamins, minerals, enzymes, polysaccharides, phenolic compounds and organic acids (Boudreau and Beland, 2006). It has been hypothesized that this heterogeneous composition of the aloe vera pulp may contribute to the diverse pharmacological and therapeutic activities that have been observed for aloe gel products (Talmadge *et al.*, 2004).

1.2 Chemistry

1.2.1 Chemical composition

Aloe vera's parenchyma tissue or pulp is reported to contain proteins, lipids, amino acids, vitamins, enzymes, inorganic compounds and small organic compounds in addition to the different carbohydrates. Table 1.1 summarizes the various chemical and enzymatic composition of aloe vera leaf pulps and exudates.

1.2.2 Phytochemistry

There are as many as 200 different types of molecules in aloe vera (Davis, 1997). The leaf gel has been reported as containing 98–99.5% water (Eshun and He, 2004; Bozzi *et al.*, 2007). The total solid content of aloe vera gel is 0.66% and soluble solids are 0.56%, with some seasonal fluctuation. Carbohydrates are derived from the mucilage layer of the

Table 1.1. Chemical and enzymatic composition of *Aloe vera* leaf pulp and exudates. From Raksha *et al.* (2014).

Anthraquinone	Aloe emodin, aloectic acid, anthranol, aloin A and B (or collectively known as barbaloin), isobarbaloin, emodin and ester of cinnamic acid
Carbohydrate	Pure mannan, acetylated mannan, acetylated glucomannan, glucogalactomannan, galatan, galactogalacturan, arabinogalactan, galactoglucoarabinomannan, pectic substance, xylan, cellulose, chromones, isoaloeresin- D, isoarabaichromone and neoaloesin A
Enzymes	Phosphatase, amylase, carboxypeptidase, catalase, cyclooxidase, cyclooxygenase, lipase, oxidase, phosphoenolpyruvate carboxylase and superoxide dismutase
Inorganic compounds	Calcium, chlorine, chromium, copper, iron, magnesium, manganese, potassium, phosphorous, sodium and zinc
Non-essential and essential amino acids	Hydroxyproline, isoleucine, leucine, lysine, methionine, phenylalanine, proline, threonine, tyrosine and valine
Organic compounds and lipids	Arachidonic acid, linolenic acid, triglycerides, triterpenoids, gibberellins, lignin, potassium sorbate, salicylic acid and uric acid
Proteins	Lectins and lectin-like substances
Saccharides	Mannose, glucose, rhamnose and aldopentose
Sterols	Campesterol, cholesterol and β-sitosterol
Vitamins	B_1, B_2, B_6 ,C, β-carotene, choline, folic acid, α-tocopherol

plant under the rind surrounding the inner parenchyma or gel. They comprise both monosaccharides and polysaccharides. Bitter aloe (the dried yellow exudates), which consists of free anthraquinones and their derivatives, exerts powerful purgative effects when ingested in large amounts, while at low concentrations, it appears to aid absorption from the gut and is a potent antimicrobial and powerful analgesic agent (Saccu *et al.*, 2001).

About 20 out of 22 amino acids and seven of the eight essential amino acids required by the human body are also present in aloe vera gel. Aloe vera juice has been evaluated for its antioxidant potential and the study showed significant presence of antioxidant in aloe extracts. It is suggested that the growth stage of the aloe plant plays a vital role in the composition and antioxidant activity of its extracts (Hu *et al.*, 2003). Aloe vera juice also has antibacterial properties against Gram-positive bacteria (Alemdar and Agaoglu, 2009). Antiviral and antifungal properties of aloe vera, as well as many other medicinal properties, have also been described (Anonymous, 2008).

The leaves of *A. ferox* contain various phenolic secondary metabolites, including the anthrone-*C*-glycoside homonataloin and three isomers of aloeresin – A, B and C (Speranza *et al.*, 2005). The amounts of homonataloin in exudates from cut leaves of various *Aloe* spp. were reported by Beaumont *et al.* (1984).

1.3 Postharvest Technology

1.3.1 Commercial processing of aloe gel

The commercial processing of aloe vera gel can be broken down into three basic steps, viz. preliminary processing, intermediate processing and final processing (see Fig. 1.2). In preliminary processing, the leaves are harvested and scrubbed, the rind is removed to yield the gel fillet, the fillets are lightly ground and the cellulosic pulp is removed. Next, the crude gel is subjected to intermediate processing to kill bacteria and, if desired, to remove anthraquinones. Intermediate processing yields the materials that, with the addition of preservatives, comprise the basic gel products. Final processing yields preserved liquid products, concentrates and powders. We acknowledge the paper by Ramachandra and Rao (2008) as a major source of the information on these steps that is presented below.

Preliminary processing

Preliminary processing begins with washing the freshly harvested leaves and sanitizing the outer surface and proceeds through removal of the outer rind and expression of the gel fillet, either manually or by machine to produce the gel. Ideally, leaves should be harvested and washed within 2–4 h. In controlled studies it was found that if 24 h elapse between harvesting and washing, some biological activity of the gel is lost. Delaying washing after harvesting while preserving freshness requires meticulous planning and attention. Leaves must not be bruised and refrigeration should be immediate.

FILLETING AND DEPULPING. The next step in processing is production of the gel fillet. Filleting is accomplished by one of two methods: manual removal of the rind with a knife or filleting by machine. In either case, the tip of the leaf is first removed, the butt is trimmed off and the sides of the leaf are trimmed. Tip removal is usually accomplished using a knife at the culling table. The butt can either be trimmed with a knife at the culling table or by wire at the filleting table. Manual filleting generally takes place on a stainless steel table approximately 1 m wide with raised edges approximately 10 cm high. The flow of cleaned leaves is generally at the head of the table where the rind is removed and discarded to the side. Fillets with exuded pseudoplastic gel are removed from the foot of the table. Trimming and filleting tables are often equipped with stainless steel wire, set up on pegs about 1 cm above the surface. These provide a cutting edge for removal of the butt and trimming of the sides. After trimming of the tip, butt and sides, the upper and lower surfaces of the

Aloe vera leaves
↓
Washing
(200 ppm hypochlorite wash and 20 ppm hypochlorite rinse)
↓
Culling and trimming
(Hand filleting or machine filleting)
↓
Gel fillets
↓
Grinding and pulp extraction
↓
Pulp and crude aloe gel
↓
Pasteurization
(Batch pasteurization or HTST/UHT or membrane filtration)
↓
Pasteurized aloe gel
↓
Batch-activated carbon absorption and filtration or activated carbon column
↓
Thin film vacuum evaporation or membrane filtration
↓
Concentrated aloe
↓

Freeze-drying	Addition of 50% maltodextrin
↓	↓
Freeze-dried aloe powder	Spray-drying
	↓
	Spray-dried aloe powder with maltodextrin

Fig. 1.2. Steps in the commercial processing of aloe gel. Key: HTST, high-temperature/short-time; UHT, ultra-high-temperature.

leaf are removed. The rind on the flat side of the leaf is then often removed with a knife, discarding the rind. The gel is then scraped or scooped with the knife away from the rind on the rounder side of the leaf. The rinds are discarded to the side and the gel fillet passed down the table.

The removing of aloe vera gel is done manually by using a sharp knife as in the hand-filleting method. Mechanical methods such as whole leaf processing method (see Section 1.3.2) or passing through the crushing roller method are also adopted for the extraction of aloe vera gel. In the hand-filleting method the worker has to operate in an uncomfortable posture and the time taken for the process is quite long. Also, because aloe vera gel is slippery in nature, there are chances of damage to the operator's hand by the sharp knife that is used. In the mechanical methods, the unwanted constituents also get mixed with the extracted gel, thus making it unsuitable for human consumption. The extracted gel has to be treated with neutralizing chemicals and passed through a series of filters before it becomes fit for human consumption.

A continuous feed aloe vera whole gel extraction unit has been developed (Fig. 1.3), that peels off both top and bottom rinds in a single pass using stainless steel blades. There is a set of stainless steel rods which are provided at the gel outlet which allows the top rind layer of the leaves and the gel to be separated. The whole gel is directly collected in a food grade tray which is partially filled with clean water. The top and the bottom layers of the leaves are collected separately. The equipment consists of a set of two pressure rollers on the top and a set of two

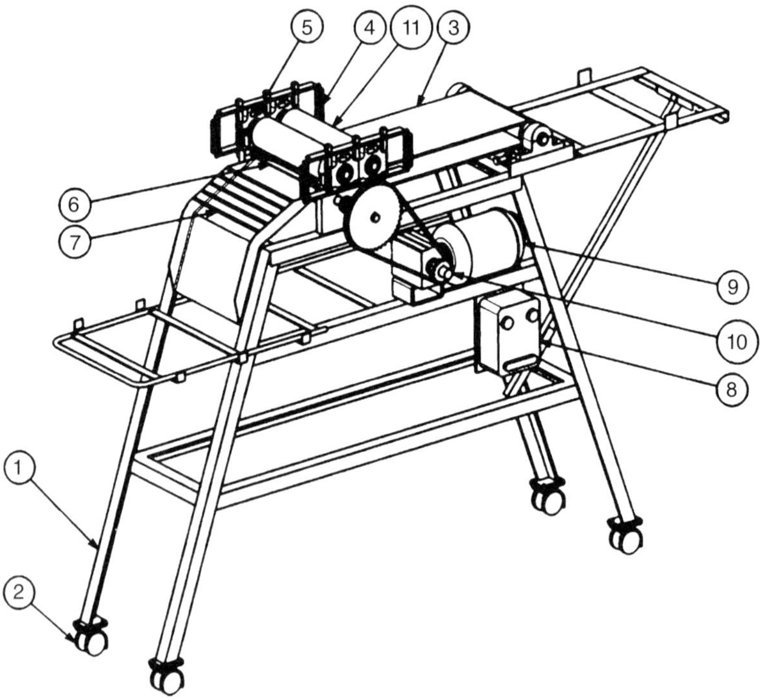

1. Outer frame
2. Castor wheel
3. Continuous feed conveyor
4. Pressure spring
5. Sliding bearing
6. Cutting blade
7. Whole gel diversion rod
8. Starter
9. Motor (1hp, 3 phase)
10. Drive mechanism
11. Top roller assembly

Fig. 1.3. An isometric view of aloe vera gel extraction equipment.

rollers at the bottom. When the aloe vera leaf is fed between rollers, they flatten it and make it devoid of curvature. The bottom set of pressure rollers can be rotated manually by means of a handle or by means of a motor through gear transmission mechanism. Sliding bearing arrangements are provided to adjust the gap between the two sets of rollers based on the average thickness of the aloe vera leaf. Pressure springs are provided between the top and bottom set of rollers for fine adjustment of the gap between the rollers to match the curvature of the leaf. Two blades made of high carbon steel, one each, just above the bottom set of rollers and just below the top set of rollers, are provided. The outer rind at the bottom of the aloe vera leaf is peeled as the leaf moves forward between the rollers. The upper blade is fixed on the spring-loaded top roller, so that the blade is positioned just below the top layer of the leaf. Peeling of the upper rind of the aloe vera leaf takes place simultaneously with that of the bottom rind as the leaf is moved forward by means of a conveyer belt (Anonymous, 2011).

The next and last step in preliminary processing involves removing the cellulosic fibres from the gel fillet. This is accomplished by very coarsely chopping the fillet with an industrial-grade pulverizing unit. The coarsely chopped fillet is then passed through a depulper. Particles that are smaller than 200 μm in diameter will readily pass through the final screen of the depulper. The removal of the fibre yields the crude gel product.

Intermediate processing

DESTRUCTION OF BACTERIA AND FUNGI. Batch pasteurization in unsealed kettles is the traditional method for reducing bacterial numbers. Usually, a solution of aloe vera gel is fed into a steam-jacketed, electrically heated or gas-fired kettle and the temperature raised to 65°C. Attaining this temperature can take from 15 to 60 min depending on the equipment used in the processing facility. In standard pasteurization, the material is then held at 65°C for 15 min. Owing to the limitations of the above methods, larger and more modern processing plants are employing high-temperature/short-time (HTST) pasteurization. This is done in sealed but non-pressurized systems that employ a series of heat exchangers to rapidly (within seconds) raise and then lower the temperature of the liquids. In the regeneration cycle, product entering the HTST unit is initially heated in the first stage of the exchanger by product leaving the HTST. It then passes through a heater section where its temperature is raised to 90–95°C, and afterwards proceeds into a set of holding coils where pasteurization occurs. The pasteurized product then re-enters the regenerating heat exchanger where it is used to raise the temperature of the incoming raw gel to about 50–60°C while its own temperature decreases to about 35–45°C. A final heat exchanger cools the leaving gel to the desired temperature with chilled glycol or water.

The processes described above yield a very slightly yellow to almost water–white (approaching colourless) liquid. This material, usually with added preservatives, is marketed as '1:1 Aloe Vera Gel' (IASC, 2004). Ideally, when mesophyll contamination is low, pasteurization does not affect the colour of aloe gel and pasteurized aloe vera gel is commonly marketed without adsorption on to activated charcoal. This material is called 'non-decolorized' because it has not been treated (decolorized) with activated carbon. The most commonly employed preservatives to prevent bacterial growth are benzoate (up to 0.1%), and sulfite (up to 0.1% for cosmetics). Sorbate in concentrations up to 0.1% is used to retard the growth of fungi. Antioxidants are also added in an attempt to prevent colour change. Chief among these are ascorbate (up to 0.1%). It should be noted that sulfite also has antioxidant properties. Citrate (up to 0.2%), or other food-approved acids, are usually added as a buffer to keep the pH in the proper range of less than 4.5; citrate itself has a mild bacteriostatic effect.

DECOLORIZATION WITH ACTIVATED CHARCOAL. Most activated charcoal adsorption is done in a batch process fashion. Pasteurized aloe vera gel is run into a mixing tank while activated charcoal (0.05–2% w/v) is added with mixing. After 15–60 min, the activated charcoal is removed by filtration. Theoretically, the temperature at which adsorption occurs is critical. However, in most aloe vera industry applications, the temperature of adsorption is determined by convenience. If adsorption is conducted as part of batch pasteurization, then activated charcoal is added immediately after the pasteurization holding period is finished and temperature reduction begins. Filtration is performed when the desired final temperature is attained. Thus, adsorption may begin at a temperature of 65°C and continue for 1 h until a temperature of 45°C is reached. At this point, the product is filtered and finished.

There are four sets of parameters to be optimized for successful adsorption: (i) time, in the range of 15–60 min; (ii) temperature: in the range of 40–60°C; (iii) amount of colour change potential; and (iv) amount (0.05–2% w/v) and type of activated carbon employed. Once adsorption is finished, the activated charcoal must be removed. This is generally done with a filter press using paper with a nominal pore size of 20 μm, rather than by sedimentation or centrifugation. The most active carbon systems generally have a small particle size in order to maximize surface area. When the particle size is significantly below 50 μm, it is advisable to employ 0.05–0.5% of a clarifying agent such as Celite Filteraide (diatomateous earth) to minimize filter clogging.

Final processing – concentration and drying

CONCENTRATION OF ALOE VERA GEL. Pasteurized aloe vera gel (Product M) and pasteurized, decolorized aloe 'gel' are 99% water. An industrial thin film vacuum evaporation system with temperatures maintained at 35–45°C and a throughput usually in the range of several hundred litres per hour results in 'concentrated aloe'. Pressure filtration across membranes can also be used.

FREEZE-DRIED OR LYOPHILIZED ALOE VERA. Freeze-drying is certainly the most elegant method for stabilizing aloe vera materials. Material with a solids content of 5–20 g/dl is poured into stainless steel trays and frozen to low temperature (below −40°C). The frozen material is then put under a high vacuum (50 mTorr = 6.67 Pa). Water gradually sublimes from the frozen material as it is gradually heated. The rate at which the trays are heated controls the rate at which water sublimes from the frozen blocks and thus the degree of vacuum is controlled. After a cycle time of 36–72 h, most of the water has been removed, a high vacuum is sustained (<25 mTorr = <3.34 Pa), and the temperature of product remains at ambient (30°C). Freeze-drying is potentially the method of choice for the production of the finest quality finished product to be used in the manufacture of cosmetics.

SPRAY-DRIED ALOE VERA GEL. Spray-drying is potentially an excellent method of concentrating and preserving aloe vera extracts. Spray-drying can be a two-step process. First, the aloe concentrate is mixed with a matrix and then sprayed into a stream of hot air, which dries the mix of matrix and aloe. The process begins when matrix is added to the aloe liquid concentrate. Matrix, either the disaccharide, lactose, or maltodextrin, a higher molecular weight saccharide, is used to provide a rapidly forming, readily dried nucleus around which the aloe can accrete and dry. The matrix-like maltodextrin is added at a ratio of about 1 g matrix/1 g aloe solids. The fluid is then sprayed in a downward direction out of a series of nozzles as a fine mist. The spray tower, which is an enclosed space, has a positive flow of air heated to 50–90°C. The dried product, which consists of tiny granules, falls into a conical collector from which it is continuously removed.

1.3.2 Processing of aloe vera leaves

The three basic methods of processing aloe vera leaves are the traditional hand-filleted processing, whole leaf processing and total processing.

Traditional hand-filleted aloe vera

In order to avoid contamination of the internal fillet with the yellow sap, the traditional hand-filleting method of processing aloe vera leaves is adopted. In this method, the lower 25 mm of the leaf base (the white part attached to the large rosette stem of the plant), the tapering point (25–50 mm) of the leaf top and the short, sharp spines located along the leaf margins are removed with a sharp knife. The knife is then introduced into the mucilage layer below the green rind, avoiding the vascular bundles, and the top rind is removed. The bottom rind is similarly removed, and the rind parts, to which a significant amount of mucilage remains attached, are discarded.

The mucilage (gel) layer is accumulated on the top of the filleting table, and can later be used in total aloe vera processing (see last part of this section). This layer is of critical concern because the highest concentration of potentially beneficial aloe constituents are found in the mucilage, which represents the constituents synthesized by the vascular bundle cells empowered by energy developed in the green (chlorophyll-containing) rind cells through sun-induced photosynthesis.

Subsequent to their synthesis, the materials of the mucilage layer are distributed to the storage cells (cellulose-reinforced hexagons) of the fillet, a process that is accompanied by dilution owing to the large amount of water (the major fillet constituent), which is stored in the fillet cells (the fillet consists of more than 99% water). The fillet itself is now further washed to ensure that there is no possibility of bacterial contamination, after which it is put into the pulper. The pulper refrigerates the juice for optimum conversion when the holding tank is full; it is then left for 24 h to decant.

The way that the inner gel is extracted from the leaf is very important. The latex portion of the leaf is located between the rind and the inner gel. The gel should be removed from the leaf without disrupting this area so that little or no latex (aloin) gets into the gel. If latex does get into the gel, it makes the gel very bitter. This taste can be distinguished from the vegetable taste of the inner gel with very little experience. If the gel is extracted by mechanical methods, then there is the possibility that the latex can mix with the inner gel, resulting in a loss of purity. By hand-filleting the leaf, it is possible to cleanly separate the gel from the rind. The gel is then ground to a liquid and the pulp is removed. The hand-filleting method is very labour intensive, and because of this, machines have been designed and used that attempt to simulate the hand-filleted technique; however, in general, the product then contains higher amounts of anthraquinone laxatives than are obtained using the traditional hand-filleting approach.

Whole leaf aloe vera processing

This whole leaf process employed in the making of aloe vera juice allows the cellulose (skin) to be dissolved, as well as measurable amounts of aloin to be removed. This total procedure is done entirely by a cold process treatment. Maximum efficiency is thus assured, resulting in a product that is rich in polysaccharides.

In this process, the base and tip of the leaf are removed and then the leaf is cut into sections and ground into a particulate slurry. The method for producing whole leaf aloe vera begins by placing the whole leaf in a grinding unit that pulverizes the entire leaf into a soup-like structure (see Fig. 1.4). The material is then treated with special chemical products that break down the hexagonal structure of the fillet, releasing the constituents. The rind particles are removed using a series of coarse and screening filters, or passage through a juice press, and the expressed juice is then passed through various filtering columns which remove the undesirable laxative agents. This liquid is then pumped into large, stainless steel holding tanks that have been thoroughly cleaned and sanitized.

Once a holding tank is filled, it is sent to a depulping extractor. This machine removes the large pieces of pulp and leaves that were generated by the initial grinding process. The second phase of processing consists of passing the aloe liquid through a series of filters that remove the aloin and aloe emodin (bitter-tasting, harsh laxatives), as well as any microscopic traces of leaves and/or other particles. The filter press is used during this phase. The press is attached to the storage tank containing the pre-filtered aloe liquid, and its carbon-coated plates absorb the aloin and aloe emodin that are by-products of grinding the whole leaf. The aloe liquid is continually passed through the filter press until 99% of the aloin and aloe emodin are removed, and the filtered product is then placed in a second holding tank.

Total aloe vera processing

In total aloe vera processing, the aloe vera leaves are hand-filleted by the traditional, old fashioned, labour-intensive method described above. Then the green rinds and the mucilage layer from the table top are processed by a newly developed proprietary methodology. A combination of the products produced by these two procedures produces an aloe vera product called 'Total Process Aloe', which contains a high concentration of desirable constituents, and is virtually free from undesirable laxative anthraquinones.

Cleaning
↓
(High-pressure sprayers and scrub brushing)
↓
Base and tip removed
↓
Leaf cut into sections
↓
Ground into slurry
↓
Chemical treatment
↓
Juice press (rind particles removed)
↓
Filter press (5 μm filter paper)
(Carbon-coated plates absorb aloin and aloe emodin)
↓
Series of filters
↓
Removal of aloin and aloe emodin, sand and other particles
↓
Final purification
↓
Aloe vera juice
↓
Pasteurization
↓
Flash cooling
↓
Flavours and sugar added
↓
Preservatives added
↓
Packing

Fig. 1.4. Process flow diagram for whole leaf aloe vera processing.

The traditional hand-fillet methodology, coupled with the newly developed proprietary handling of the waste (green rinds and table top mucilage) from the traditional method, and a geographical area where aloe plants thrive, have been combined to achieve the superior quality of Total Process Aloe. Total Process Aloe contains considerably higher concentrations of total solids, calcium, magnesium and malic acid, the major parameters of quality that are utilized and recommended by the International Aloe Science Council (IASE) for certification.

1.3.3 Gel stabilization technique

Aloe vera gel is the mucilaginous jelly obtained from the parenchyma cells of the aloe vera plant. When exposed to air, the gel rapidly oxidizes, decomposes and loses much of its biological activity. Gel stabilization can be carried out either by cold processing or heat treatment. Regardless of the relative quality of the plant, the best results are obtained when leaves are processed immediately after harvesting. This is because derivative decomposition of the gel matrix begins as a result of natural enzymatic reactions, as well as the growth of bacteria, in the presence of oxygen.

In the cold processing technique, the entire processing steps are accomplished without the application of heat. Coats (1979) described the use of enzymes such as glucose oxidases and catalase. Other sterilization steps reported in cold processing include exposing the gel to ultraviolet light, followed by microfiltration.

In heat treatment processing, sterilization is achieved by subjecting the aloe vera liquid obtained from the activated carbon treatment to pasteurization at high temperature. Aloecorp (http://www.aloecorp.com) has reported that the biological activity of aloe vera gel remains essentially intact when the gel is heated at 65°C for periods less than 15 min. Extended periods or higher temperatures have resulted in greatly reduced biological activity. However, it has been suggested that the best method of pasteurization is HTST, followed by flash cooling to 5°C or below.

In both of these processing techniques, stabilization can be achieved by the addition of preservatives and other additives. The use and efficacy of sodium benzoate, potassium sorbate, citric acid and vitamin E have been reported (Chang *et al.*, 2006).

Reception of raw material (aloe vera leaves)
↓
Washing operation
↓
Filleting operation
↓
Grinding/homogenization
↓
Enzyme addition
↓
Filtration
↓
Unpasteurized juice
↓
Addition of vitamin C and citric acid
↓
Deaeration
↓
Pasteurization
↓
Flash cooling
↓
Packaging
↓
Storage

Fig. 1.5. Processing flow diagram for the production of single-strength aloe vera gel juice.

1.3.4 Major unit operations in processing aloe vera leaf gel

Reception of raw materials

After harvesting, aloe vera leaves are preferably transported at reduced temperature from the field to the processing place. The leaves should be sound, undamaged, mould/rot free and mature (3–4 years old) in order to keep all the active ingredients at full concentration (Lawless and Allan, 2000). One important factor that must be considered is the handling/treatment of the leaves after harvesting because the decomposition of the matrix occurs on cutting as a result of natural enzymatic reactions and the activity of bacteria that are normally present on the leaves. This derivative process can adversely affect the quality of the end product. Therefore, there is a need to carefully work towards refrigerating the freshly removed leaves within 4–6 h or getting the raw material directly into the production line (Fig. 1.5).

Filleting operation

The loss of biological activity appears to be the result of enzymatic activity after the aloe vera leaf has been removed from the plant, and it has been demonstrated that once extracted from the leaf, aloe gel has greater stability than the gel left in the leaf. In order to avoid the decomposition of the biological activity in the leaf, the filleting operation must be completed within 36 h of harvesting (Robert, 1997). Another important factor, which leads to non-enzymatic browning of the aloe gel product is the presence of anthraquinone (Zhang *et al.*, 2002).

Grinding/homogenization

The major steps in the production process of aloe vera gel include crushing or grinding. The aloe gel fillets should be crushed and homogenized using a commercial high-speed tissue crusher at room temperature (25°C).

Due to the enzymatic browning reaction, the longer the crushing/grinding time, the greater the browning index of aloe vera gel juice (Liu *et al.*, 2001). Therefore, the gel crushing or grinding should be shortened to within 10–20 min in order to avoid the browning reaction.

Addition of pectolytic enzyme

Treatment of aloe vera gel with pectolytic enzymes for a long time before processing is detrimental to biologically active compounds such as polysaccharides, which are the single most important constituent in aloe (Waller *et al.*, 1978; Gowda *et al.*, 1980; Yagi *et al.*, 1982). It has been reported that enzymatic treatment at 50°C and within 20 min does not induce loss of the biological activity of polysaccharides in aloe vera gel (Maughan, 1984).

Filtration

This is an important process and the operation influences the stability of aloe vera gel juice.

Addition of vitamin C and citric acid

The unpasteurized aloe gel juice is fortified with vitamin C and citric acid to avoid the browning reaction, to improve the flavour of the aloe vera gel juice and to stabilize the juice (Eison-Perchonok and Downes, 1982; Tramell *et al.*, 1986; Kennedy *et al.*, 1992). The pH of aloe gel juice is adjusted to between 3.0 and 3.5 by adding citric acid to improve its flavour.

Deaeration

The aim of the deaeration step is to avoid the oxidation of ascorbic acid (Chan and Cavaletto, 1986), and this eventually improves the shelf life of aloe vera gel juice.

Pasteurization

Like the processing steps for other vegetable juices, the pasteurization step may affect the taste, appearance and content of biological activity of the aloe gel product. HTST treatment (at 85–95°C for 1–2 min) is an effective method to avoid poor flavour and the loss of biological activity (Eshun, 2003).

Flash cooling

After pasteurization, the juice is flash cooled to 5°C or below within 10–15 s. This is a crucial step to preserve the biological activity of aloe vera gel (Eshun, 2003).

Storage

Relative humidity and temperature are two of the most important environmental factors that affect product quality. They can also affect the amount of the volatile substances of the juice that are absorbed by the packaging material (Hernandez and Giacin, 1998) and, consequently, affect the shelf life of the product (Hirose *et al.*, 1988; Sadler and Braddock, 1990).

1.3.5 Time, temperature and sanitation (TTS) process

An innovative process based on time, temperature and sanitation has been developed to eliminate sources of degradation of aloe vera gel during processing. This method is described below.

Timing of the TTS leaf process

Leaves show losses of biological activity beginning at 4–6 h following harvest when the leaves are stored at ambient temperatures. A decrease in activity is also evident when the leaves are stored under refrigerated conditions, even though the rate of activity loss is greatly reduced. The losses of activity appear to be the result of enzymatic activity after the leaf is removed from the plant. In fact, as noted earlier, it has been shown that the gel, once extracted from the leaf, has greater stability than gel which is left in the leaf. The overall timing of TTS production phase is extremely critical: the processing must be completed within 36 h of harvesting the leaves.

Leaf harvesting and handling

Loss of biological activity is also due to the microbial decay of the gel. The first exposure of the inner gel to microbes is when the leaves are harvested from the plant. To prevent contamination of the gel, the leaves are handled carefully and soaked in a food-grade sanitizer which effectively reduces the microbial count on the leaf exterior to acceptable levels.

Flash cooling

As a crucial step in preserving the biological activity of the product, the gel should be cooled to below 50°C in 10–15 s following gel extraction. Rapid cooling leads to enzymatic and microbial deterioration in the gel, and also aids in reducing the microbial counts in the product.

Pasteurization

Biological activity of the product remains active when the gel is heated at 65°C for periods of less than 15 min, but extended periods of heating or higher temperatures will result in greatly reduced activity levels. The best method of pasteurization is HTST, which exposes the gel to elevated temperature for periods of 1–3 min. Once heated, the gel is flash cooled to 5°C or below.

Concentration

The gel obtained after using the pasteurization and flash cooling methods can be concentrated under vacuum without the loss of its biological activity. The concentration operation must be conducted under 125 mm mercury vacuum at a temperature below 50°C and must not exceed 2 min. A higher vacuum and temperature will cause activity loss, as will extended concentration times.

Freeze- or spray-drying

The concentrated product can then be freeze-dried at temperatures between −45 and −30°C, or can be spray-dried at a product temperature below 60°C without loss of its biological activity.

1.3.6 The desiccant dehydration process

This system has been used for many years to dehydrate foods. The pure intact aloe fillets are first washed so that the remaining aloin is removed. Then they are placed in a desiccant dehydration chamber where the desired level of relative humidity and temperature are maintained. The material is then ground to a powder of the desired fineness and packed.

1.3.7 The Qmatrix process (Aloecorp)

Qmatrix drying is a 4th generation dehydration technology that also includes microwave and radio frequency drying. The Qmatrix process is a novel proprietary method of dehydration of aloe gel in that it enables dehydration while maintaining integrity with respect to flavour, colour and nutrients. It is comparable to freeze-drying in quality aspects but without the high operation costs (http:www.aloecorp.com).

Advantages of this process

The advantages include:

- gentle low temperature/short time drying;
- superior sensory attributes retained;
- superior retention of nutrients and bioactivity;
- atmospheric pressure (no vacuum);
- energy and environmental efficiency ('greenness'); and
- superior solubility characteristics.

1.4 Uses

1.4.1 General uses

General uses include the following:

- Traditionally, aloe vera was used topically to heal wounds and for various skin conditions, and orally as a laxative.
- Today, in addition to its traditional uses, aloe vera is also taken orally to treat diabetes, asthma, epilepsy and osteoarthritis. It is also still used topically to treat osteoarthritis, burns, sunburn and psoriasis.

- Aloe vera gel can be found in hundreds of skin products, including lotions and sunblocks.
- The US Food and Drug Administration (FDA) has approved aloe vera as a natural food flavouring.
- The clear gel from aloe leaves is often used as an ointment.
- The green part of the leaf that surrounds the gel can be used to produce a juice or a dried substance (called latex) that is taken orally.

1.4.2 Pharmacological uses

The health benefits of some of the components of aloe vera gel are summarized in Table 1.2. Some of these, and other benefits of aloe vera, are further described below.

Skin penetration enhancement

Aloe vera gel has been shown to increase *in vitro* skin penetration of compounds depending on their molecular weight, with an apparent inverse correlation between enhancement ratio and molecular weight of the compound (Moser *et al.*, 2001; Cole and Heard, 2007).

Aloe vera leaf gel as an excipient in modified-release dosage forms

Gums and mucilages of natural origin that contain complex polysaccharides have found a wide range of pharmaceutical applications, such as functional excipients in various dosage forms, which include binders, disintegrants, emulsifiers, suspending agents, gelling agents and sustaining agents in modified-release tablets (Kulkarni *et al.*, 2005; Jani *et al.*, 2007).

Table 1.2. Novel components of aloe vera along with their health benefits.

Chemical component	Health benefits
Acemannan	Accelerates wound healing, modulates immune system, has antineoplastic and antiviral effects
Alprogen	Anti-allergic
C-glycosyl chromone	Anti-inflammatory
Bradykinase	Anti-inflammatory
Magnesium lactate	Anti-allergic
Salicylic acid	Analgesic, anti-inflammatory

Antidiabetic effects

Several trials have shown a blood glucose lowering effect following the consumption of aloe vera gel. It significantly reduced fasting blood glucose, hepatic transaminases, plasma and tissue cholesterol, triglycerides, free fatty acids and phospholipids, and in addition also significantly increased plasma insulin levels. It was reported by Rajasekaran *et al.* (2006) that the mechanism of action of aloe vera extracts in reducing blood glucose levels is by enhancing glucose metabolism due to its antioxidant mechanism.

Immunomodulatory effects

Immunomodulation activities of the polysaccharides in aloe vera gel may be via activation of macrophage cells to generate nitric oxide, secrete cytokines (e.g. tumour necrosis factor-alpha or TNF-α, interleukin-1 or IL-1, interleukin-6 or IL-6, and interferon-γ or INF-γ) and present cell surface markers. It is reported that relatively high concentrations of acemannan are required to achieve modest activation of macrophages by aloe vera.

Anti-inflammatory effects

The anti-inflammatory activity of mannose-6-phosphate is believed to resemble the effects observed for acetylated mannan in aloe gel. Aloe gel is reported to reduce agent-induced inflammation via the promotion of prostaglandin synthesis as well as via increased infiltration of leucocytes. It has also been shown to significantly reduce leucocyte adhesion and levels of TNF-α (Prabjone *et al.*, 2006).

Antioxidant effects

It has been reported that different fractions of aloe vera, as well as unfractionated whole gel, have antioxidant effects. Glutathione peroxidase activity, superoxide dismutase enzymes and a phenolic antioxidant have been found to be present in aloe vera gel, and may be responsible for these antioxidant effects.

Wound-healing effects

Wound healing is a response to injured tissue that results in the restoration of tissue integrity. Healing by application of aloe vera gel is thought to be promoted by keeping the wound moist, increasing epithelial cell migration, more rapid maturation of collagen and reducing inflammation (Reynolds and Dweck, 1999).

Anticancer effects

The two fractions from aloes that are claimed to have anticancer effects include glycoproteins (lectins) and polysaccharides. Various studies have indicated antitumour activity for aloe vera gel in terms of reduced tumour burden, tumour shrinkage, tumour necrosis and prolonged survival rates. In addition to these effects, aloe vera gel has also shown chemopreventive and anti-genotoxic effects on benzo[α]pyrene-DNA adducts (Steenkamp and Stewart, 2007).

Effect on gastric acid secretion and ulcers

Aloe vera gel has the ability to cure gastric ulcers or protect against their formation. The anti-ulcer activity of the gel has been attributed to several possible mechanisms, including its anti-inflammatory properties, healing effects, mucus stimulatory effects and regulatory effects on gastric secretions (Suvitayavat *et al.*, 2004).

Skin hydration effects

It has been found that aloe vera gel-containing products improve skin hydration possibly by means of a humectant mechanism and increasing the water content of the stratum corneum (Dal'Belo *et al.*, 2006).

Hepatoprotective activities

An aqueous extract of dried aerial parts of aloe vera significantly reduced hepatic damage induced by carbon tetrachloride and reversed certain biochemical parameters. Histopathological studies confirmed the curative efficacy of the water extract of aloe vera against carbon tetrachloride-induced liver damage as indicated by the reversal of centrilobular necrosis, macro-vascular fatty changes and scattered lymphomononuclear cell infiltration into the hepatic parenchyma (Chandan *et al.*, 2007).

Antimicrobial activities

Aloe vera inner gel is effective against both Gram-positive and Gram-negative bacteria. Anthraquinones isolated from the exudate of aloe vera have shown a wide spectrum of antimicrobial activity. The antibacterial activity of emodin against *Escherichia coli* is reported to be through the inhibition of solute transport in membranes. Many anthraquinones have shown antiviral and/or virucidal effects on enveloped viruses (Alves *et al.*, 2004).

1.5 Summary

Aloe vera is becoming popular in developing and developed countries owing to its natural origin and low side effects. It is an industrial crop, and in the food industry it has been utilized for the preparation of health food drinks. Aloe vera gel also finds application in the cosmetic and toiletries industry for the preparation of creams, lotions, soaps, shampoos and facial cleansers. It has been used extensively by many people because of its effectiveness in treating burns, healing wounds, relieving aches and pains, and in a whole range of internal and external disorders. It contains a plethora of antioxidant components, including flavonoids, vitamin C, β-carotene and vitamin E. It also acts as a beauty enhancer and helps in treating skin allergies. Aloe vera has a long history as a medicinal plant with diverse therapeutic applications. It has anti-inflammatory, antioxidant and antibacterial properties. The presence of all of the essential elements in aloe vera may account for most of its therapeutic efficiencies, including gastrointestinal problems, arthritis, stress, diabetes, cancer, ulcers, piles, liver disorders, asthma, etc. The medicinal value of aloe vera is found in the gel obtained by peeling its leaves and is a storehouse of nutrients and phytochemicals. The presence of phytochemicals such as tannins, saponins, flavonoids, alkaloids and anthraquinones in high concentrations is a strong indication of the medicinal value of this plant. The processing of aloe vera includes harvesting, scrubbing and removal of the rind to yield the gel fillet. The fillets are lightly ground and the cellulosic pulp is removed. The crude gel is then subjected to intermediate processing to kill bacteria and, if desired, to remove anthraquinones. Intermediate processing yields materials which, with the addition of preservatives, comprise the basic gel products. Final processing yields preserved liquid products, concentrates and powders. Today, aloe vera is available in many forms, such as juices, candies, jellies, powder, gel, pills and sprays. There are also many equipment and processing protocols available for the primary, secondary and tertiary processing of aloe vera at small scales.

References

Alemdar, S. and Agaoglu, S. (2009) Investigations of *in-vitro* antimicrobial activity of aloe vera juice. *Journal of Animal and Veterinary Advances* 8, 99–102.

Alves, D.S., Perez Fons, L., Estepa, A. and Micol, V. (2004) Membrane related effects underlying the biological activity of the anthrax quinines emodin and barbaloin. *Biochemical Pharmacology* 68, 549–561.

Anonymous (2008) Aloe Vera: History, Science and Medicinal Uses. Available at: http://www.healingaloe.com (accessed 9 July 2015).

Anonymous (2011) Continuous-feed aloe vera gel extraction machine. *CIAE News* 20(1), 2.

Atherton, P. (1998) Aloe vera revisited: review of aloe gel. *The British Journal of Phytotherapy* 4, 176–183.

Beaumont, J., Reynolds, T. and Vaughan, J.G. (1984) Homonataloin in *Aloe* species. *Planta Medica* 50, 505–508.

Boudreau, M.D. and Beland, F.A. (2006) An evaluation of the biological and toxicological properties of *Aloe barbadensis* (Miller), aloe vera. *Journal of Environmental Science and Health, Part C: Environmental Carcinogenesis and Ecotoxicology* 24, 103–154.

Bozzi, A., Perrin, C., Austin, S. and Arce Vera, F. (2007) Quality and authenticity of commercial aloe vera gel powders. *Food Chemistry* 103, 22–30.

Chan, H.T. Jr and Cavaletto, C.G. (1986) Effects of deaeration and storage temperature on quality of aseptically packaged guava puree. *Journal of Food Science* 51, 165–168.

Chandan, B.K., Saxena, A.K., Shukla, S., Sharma, N., Gupta, D.K., Suri, K.A., Suri, J., Bhadauria, M. and Singh, B. (2007) Hepatoprotective potential of *Aloe barbadensis* Mill. against carbon tetrachloride induced hepatotoxicity. *Journal of Ethnopharmacology* 111, 560–566.

Chang, X.L., Wang, C., Feng, Y. and Liu, Z. (2006) Effects of heat treatments on the stabilities of polysaccharides substances and barbaloin in gel juice from *Aloe vera* Miller. *Journal of Food Engineering* 75, 245–251.

Coats, B.C. (1979) *The Silent Healer. A Modern Study of Aloe Vera.* Bill C. Coats, Garland, Texas.

Cole, L. and Heard, C. (2007) Skin permeation enhancement potential of aloe vera and a proposed mechanism of action based upon size exclusion and pull effect. *International Journal of Pharmaceutics* 333, 10–16.

Dal'Belo, S.E., Gaspar, L.R. and Maia Campos, P.M. (2006) Moisturising effect of cosmetic formulations containing aloe vera extract in different concentrations assessed by skin bioengineering techniques. *Skin Research and Technology* 12, 241–246.

Davis, R.H. (1997) *Aloe Vera: A Scientific Approach.* Vantage Press, New York.

Eison-Perchonok, M.H. and Downes, T.W. (1982) Kinetics of ascorbic acid oxidation as a function of dissolved oxygen concentration and temperature. *Journal of Food Science* 47, 765–767, 773.
Eshun, K. (2003) Studies on aloe vera gel: its application in beverage preparation and quality assessment. Thesis submitted in partial fulfilment of the requirements for the Degree of Master of Science to Food Science and Technology School of Southern Yangtze University (Jiangnan University), Wuxi, Jiangsu, China.
Eshun, K. and He, Q. (2004) Aloe vera: a valuable ingredient for the food, pharmaceutical and cosmetic industries: a review. *Critical Reviews in Food Science and Nutrition* 44, 91–96.
Gowda, D., Neelisiddaiah, B. and Anjaneyalo, Y. (1980) Structural studies of polysaccharides from *Aloe saponaria* and *Aloe vanbalenii*. *Carbohydrate Research* 83, 402–405.
Grindlay, D. and Reynolds, T. (1986) The *Aloe vera* phenomenon: a review of the properties and modern uses of the leaf parenchyma gel (review article). *Journal of Ethnopharmacology* 16, 117–151.
Hernandez, R.J. and Giacin, J.R. (1998) Factors affecting permeation, sorption and migration processes in package-product systems. In: Taub, I.A. and Singh, R.P. (eds) *Food Storage Stability*. CRC Press, Boca Raton, Florida, pp. 269–329.
Hirose, K., Harte, B.R., Giacin, J.R., Miltz, J. and Stine, C. (1988) Sorption of D-limonene by sealant films and effects on mechanical properties. In: Hotchkiss, J.H. (ed.) *Food and Packaging Interactions*. ACS Symposium Series Volume 365, ACS Publications, Washington, DC, pp. 28–41.
Hu, Y., Xu, J. and Hu, Q. (2003) Evaluation of antioxidant potential of aloe vera (*Aloe barbadensis* Miller) extracts. *Journal of Agricultural and Food Chemistry* 51, 7788–7791.
IASC (2004) *How Large is the Aloe Market*? International Aloe Science Council News, October 2004, Silverspring, Maryland. Available at: http://www.iasc.org/aloemarket.html (accessed 18 August 2009).
Jani, G.K., Shah, D.P., Jain, V.C., Patel, M.J. and Vithalan, D.A. (2007) Evaluating mucilage from *Aloe barbadensis* Miller as a pharmaceutical excipient for sustained-release matrix tablets. *Pharma Technology* 31, 90–98.
Kennedy, F.C., Rivera, Z.S., Lloyd, L.L., Warner, F.P. and Jumel, K. (1992) L-ascorbic acid stability in aseptically processed orange juice in tetra brick cartons and the effect of oxygen. *Journal of Food Chemistry* 45, 327–331.
Kulkarni, G.T., Gowthamarajan, K., Dhobe, R.R., Yohanan, F. and Suresh, B. (2005) Development of controlled release spheroids using natural polysaccharide as release modifier. *Drug Delivery* 12, 201–206.
Lawless, J. and Allan, J. (2000) *Aloe Vera – Natural Wonder Cure*. HarperCollins, London, pp. 5–12.
Liu, C., Qian, H. and Liu, J. (2001) Study on preservatives in the aloe gel juice system. *Journal of Wuxi University Light Industry [now Journal of Food Science and Biotechnology]* 2001(5), 480–484.
Maughan, R.G. (1984) Method to increase color fastness of stabilized aloe vera. US Patent 4,465,629 A. Available at: http://www.google.com/patents/US4465629?printsec=description#v=onepage\&q\&f=false (accessed 10 July 2015).
Moser, K., Kriwet, K., Naik, A., Kalia, Y.N. and Guy, R.H. (2001) Passive skin penetration enhancement and its quantification *in vitro*. *European Journal of Pharmaceutics Biopharmaceutics* 52, 103–112.
Naik, R. and Annamalai, S.J.K. (2013) Gel textural properties of aloe vera. *The Madras Agricultural Journal* 100, 232–235.
Newton, L.E. (2004) Aloes in habitat. In: Reynolds, T. (ed.) *Aloes: The Genus Aloe*. Medicinal and Aromatic Plants – Industrial Profiles, Volume 38. CRC Press, Boca Raton, Florida, pp. 3–15.
Ni, Y., Turner, D., Yates, K.M. and Tizard, I. (2004) Isolation and characterization of structural components of *Aloe vera* L. leaf pulp. *International Immunopharmacology* 4, 1745–1755.
Prabjone, R., Thong-Ngam, D. and Wisedopas, N. (2006) Antiinflammatory effects of *Aloe vera* on leukocyte-endothelium interaction in the gastric microcirculation of *Helicobacter pyroli*-infected rats. *Clinical Hemorheology and Microcirculation* 35, 359–366.
Rajasekaran, S., Ravi, K., Sivagnanam, K. and Subramanian, S. (2006) Beneficial effects of *Aloe vera* leaf gel extract on lipid profile status in rats with streptozotocin diabetes. *Clinical and Experimental Pharmacology and Physiology* 33, 232–237.
Raksha, B., Pooja, S. and Babu, S. (2014) Bioactive compounds and medicinal properties of *Aloe vera* L.: an update. *Journal of Plant Sciences* (Science Publishing Group, New York) 2, 102–107.
Ramachandra, C.T. and Rao, P.S. (2006) Processing of aloe vera leaf gel: a focus on the present and innovative process technologies. In: *Proceedings of the International Conference on Innovations in Food and Bioprocess Technologies, 12–14 December 2006*. Asian Institute of Technology, Pathumthani, Thailand, pp. 358–377.

Ramachandra, C.T. and Rao, P.S. (2008) Processing of aloe vera leaf gel: a review. *American Journal of Agricultural and Biological Sciences* 3, 502–510.
Reynolds, T. and Dweck, A.C. (1999) Aloe vera leaf gel: a review update. *Journal of Ethnopharmacology* 68, 3–37.
Robert, H.D. (1997) *Aloe Vera: A Scientific Approach*. Vantage Press, New York.
Saccu, D., Bogoni, P. and Procida, G. (2001) Aloe exudate: characterization by reversed phase HPLC and head-space GC-MS. *Journal of Agricultural and Food Chemistry* 49, 4526–4530.
Sadler, G.D. and Braddock, R.J. (1990) Oxygen permeability of low density polyethylene as a function of limonene absorption. An approach to modeling flavour (scalping). *Journal of Food Science* 55, 587–590.
Speranza, G., Morelli, C.F., Tubaro, A., Altinier, G., Duri, L. and Manitto, P. (2005) Aloeresin I, an anti-inflammatory 5-methylchromone from Cape aloe. *Planta Medica* 71, 79–81.
Steenkamp, V. and Stewart, M.J. (2007) Medicinal applications and toxicological activities of aloe products. *Pharmaceutical Biology* 45, 411–420.
Sung, C.K. (2006) The history of Aloe. In: Park, Y.I. and Lee, S.K. (eds) *New Perspectives of Aloe*. Springer, New York, pp. 7–18.
Suvitayavat, W., Sumrongkit, C., Thirawarapan, S.S. and Bunyapraphatsara, N. (2004) Effects of aloe preparation on the histamine-induced gastric secretion in rats. *Journal of Ethnopharmacology* 90, 239–247.
Talmadge, J., Chavez, J., Jacobs, L., Munger, C., Chinnah, T., Chow, J.T., Williamson, D. and Yates, K. (2004) Fractionation of *Aloe vera* L. inner gel, purification and molecular profiling of activity. *International Immunopharmacology* 4, 1757–1773.
Tramell, D.J., Dalsis, D.E. and Malone, C.T. (1986) Effect of oxygen on taste, ascorbic acid loss and browning for HTST-pasteurized, single-strength orange juice. *Journal of Food Science* 51, 1021–1023.
Waller, G.R., Mangiafico, S. and Ritchey, C.R. (1978) A chemical investigation of *Aloe barbadensis* Miller. *Proceedings of the Oklahoma Academy of Science* 58, 69–76.
Yagi, A., Shibata, S., Nishioka, I., Iwadre, S. and Ishida, Y. (1982) Cardiac stimulant action of constituents of *Aloe saponaria*. *Journal of Pharmaceutical Science* 71, 739–741.
Zhang, T., Qian, H. and Liu, C. (2002) Study on non-enzymatic browning of aloe products and its inhibition methods. *Journal of Wuxi University of Light Industry* [now *Journal of Food Science and Biotechnology*] 2002(5), 496–498, 502. [Often incorrectly cited as He, Q.C., Liu, C. and Zhang, T. (2002) Study on non-enzymatic browning of aloe products and its inhibition methods. *Food Science (Chenses)* 23(10), 53–56.]

2 Ashwagandha

Paramadhas Sudha[1]* and Alagirisamy Reni[2]
[1]*Tamil Nadu Agricultural University, Chettinad, India;*
[2]*Avinashilingam University, Coimbatore, India*

2.1 Botany

2.1.1 Introduction

Withania somnifera (L.) Dunal, or ashwagandha, is an erect, evergreen, perennial shrub and member of the Solanaceae family. In Ayurvedic and indigenous medicine, it has been considered to be a medicinal plant for over 3000 years. The genera *Withania* and *Physalis* have played an important role in the traditional Unani medicine system of South-east Asia. Ashwagandha is believed to be an aphrodisiac and to have rejuvenating properties that are useful for the treatment of inflammatory conditions and as an antitumour agent.

Apart from ashwagandha, common names of the plant include withania, winter cherry, Indian winter cherry and Indian ginseng. It also has many vernacular names in different languages (see Table 2.1). The roots and leaves are the parts of ashwagandha that are used for medicinal purposes. The plant is cultivated mainly in the drier parts of India.

Ashwagandha, or winter cherry, is considered by herbalists as the Ayurvedic answer to ginseng because of its rejuvenating properties, and it is often referred to as Indian ginseng by herbalists from Western countries. In Ayurvedic medicine, the herb is used in a way that is similar to how Asian ginseng is used in traditional Chinese medicine. In Sanskrit, ashwagandha means 'the smell of a horse', which is taken to mean that the herb may impart the vigour and strength of a stallion. Some (men and women) even use it to support sexual function. Behind its clumsy and hard-to-pronounce name, ashwagandha holds a few secrets that need to be uncovered. It is known as an adaptogen, a class of medicinal herbs that work to normalize physiological function in various, sometimes unknown, ways.

In India, the estimated annual production of ashwagandha roots is more than 1500 t, while the annual requirement is about 7000 t, so that increased cultivation and higher production have become a necessity (Umadevi *et al.*, 2012). In the Ayurvedic system of medicine there are several products in which ashwagandha is used as a single plant-based formulation, but there is also a huge demand for root raw material for industrial use, thus necessitating the large-scale cultivation of the plant (Fig. 2.1).

*Corresponding author, e-mail: sudha.raj2000@gmail.com

© CAB International 2016. *Leafy Medicinal Herbs: Botany, Chemistry, Postharvest Technology and Uses* (eds D.C.P. Ambrose *et al.*)

Table 2.1. Vernacular names of ashwagandha in different languages.

Arabian	Bahman
Danish	Blærebæger
English	Winter cherry, Indian ginseng
Hindi	Asgandh
Japanese	Ashwangandha
Nepalese	Aasoganda
Sanskrit	Ashvagandha ('horse smelling')
Sinhalese	Amukkara
Tibetan	Ba-dzi-gandha
Unani	Asgandh volaite

Fig. 2.1. A *Withania somnifera* plant showing the leaves and fruits.

2.1.2 History/origin

Ashwagandha is native to the dry regions of India, and the species is also a native of Australia, East Asia and Africa. This herb has been used for over 4000 years in India, and is very important in Ayurveda, a traditional Indian system of medicine. In Sanskrit, ashwagandha means 'horse's smell', and the name probably originates from the odour of its root which resembles that of a sweaty horse. The species epithet of *somnifera* means 'sleep-bearing' in Latin. Traditional uses of ashwagandha among tribal peoples in Africa include the treatment of fevers and inflammatory conditions. The use of ashwagandha in Ayurvedic medicine extends back over 3000–4000 years to the teachings of an esteemed rishi (sage), Punarvasu Atriya. It has been described in the sacred texts of Ayurveda, including the Charaka and Sushruta Samhitas. Robin Lane Fox, in his biography of Alexander the Great, claims that the herb was used in wine in ancient times (http://www.neurosoup.com/ashwagandha-withania-somnifera). According to Anne Van Arsdall, *W. somnifera* was called apollinaris and also glofwyrt in *The Old English Herbarium*, and had a legend that Apollo found it first and gave it to the healer Aesculapiu (http://www.neurosoup.com/ashwagandha-withania-somnifera).

2.1.3 Location

Ashwagandha is a xerophytic plant that is found mostly in the drier parts of India, Sri Lanka, Afghanistan, Baluchistan and Sind, but it is also distributed in the Mediterranean regions, the Canary Islands of Spain and the Cape of Good Hope in South Africa (Uddin *et al.*, 2012). The plant itself is beautiful in appearance, with deep green leaves and branched limbs topped by seeded yellow-coloured flowers; its berries are red. The 23 known *Withania* species are broadly distributed in the drier parts of tropical and subtropical zones, ranging from the Canary Islands, the Mediterranean region and northern Africa to south-west Asia (Bhattacharya *et al.*, 1997a; Dhuley, 1997; Christina *et al.*, 2004; Girdhari and Rana, 2007). Ashwagandha has been in use for thousands of years in Ayurvedic medicine as a stress reliever and also for strengthening the immune system. It is grown as a late rainy-season crop in all the parts of India, but the major ashwagandha-producing states are Rajasthan, Punjab, Haryana, Uttar Pradesh, Gujarat, Maharashtra and Madhya Pradesh (Bhattacharya, 1998).

2.1.4 Morphology

The plant is an erect branching shrub reaching a height of about 30–150 cm. The leaves grow up to 10 cm long and are simple, ovate and glabrous (Fig. 2.1). The flowers are greenish or a lurid yellow, about 1 cm long), few in number (usually about five) and are borne together in axillary, umbellate cymes. The fruits are globose berries, 6 mm in diameter, turn orange red when mature (see Fig. 2.1) and are enclosed in an inflated and membranous calyx. The seeds are yellow, reniform and 2.5 mm in diameter. The crop is harvested 180–210 days after planting (Bhatia *et al.*, 1987; Andallu and Radhika, 2000).

2.2 Chemistry

2.2.1 Chemistry and biochemical composition of active constituents

The major constituents of ashwagandha are alkaloids, such as withanine alkaloids, and steroidal lactones (withanolides). Other constituents include amino acids, choline, β-sitosterol, chlorogenic acid and scopoletin. The leaves of Indian species of the plant are reported to contain withanolides, alkaloids, chlorogenic acid, glycosides, tannins, flavonoids and other compounds (Khare, 2007).

Leaves of the plant from different regions differ in their withanolide content. The leaf extract and dried roots of ashwagandha contain withaferin A, which is resistant to heat but is insoluble in water. The compound is extracted from the leaves with cold alcohol; the extract is purified and dried, and finally crystallized from aqueous alcohol (yield 0.18% on an air-dry basis). Ashwagandha from South Africa is reported to give a high yield of withaferin A, and it is to this compound that the curative properties of the herb are attributed (Chadha, 1976).

The major biochemical constituents of ashwagandha root are the steroidal alkaloids, but much of the plant's pharmacological activity is ascribed to its two main withanolides, withaferin A and withanolide D. Withaferin A is the therapeutically active withanolide present in the leaves. The pharmacological activity of the root is attributed to both its alkaloids and its steroidal lactones (withanolides). The total alkaloid content of the roots of Indian types has been reported to vary between 0.13 and 0.3%. Other biochemically heterogeneous alkaloids found include choline, tropane, pseudotropane alkaloids and cuscohygrene, along with several steroidal lactones. In addition to alkaloids and steroidal lactones, the roots are reported to contain starch, reducing sugars, glycosides and various other compounds (Kulkarni *et al.*, 1991).

2.3 Postharvest Technology

2.3.1 Processing and value addition

For the preparation of leaf extract, the leaves are pulverized into a coarse powder, and the ground powder is soaked in 60 ml 95% ethanol for a day, with occasional shaking, after which the mixture is filtered. The filtrate is evaporated using a rotary evaporator to yield dried extract, which is stored under refrigerated conditions in an airtight screw-capped tube. The yield of withaferin A is 0.2–0.3% of the dry weight of the leaves extracted. Gupta *et al.* (1996) performed a quantitative analysis of an Indian chemotype of *W. somnifera* from Uttar Pradesh by thin-layer chromatography (TLC) densitometry and (in contrast to the results reported above in Section 2.2.1) observed that withaferin A was totally absent from the roots, stems, seeds and persistent calyx of the fruits of intact plants, but was present in the leaves. The leaf extracts exhibited high inhibitory effects on the lipid peroxidation of linoleic acid due to the abundant presence of active antioxidant compounds, and are, therefore, effective agents in retarding Fe^{2+} catalysed lipid oxidation. In contrast to the distribution of withaferin A, the same authors found alkaloids and other withanolides in the roots, fruits and leaves of *Withania* spp. Leaf extraction with 45% alcohol yielded the highest percentage of alkaloids.

Ashwagandha leaf extract has also been found to kill cancer cells owing to its selective inhibitory effect on lipid peroxidation. Other components of the leaf extract, withaferin A and withanone, have different gene targets and mechanisms of action. Whereas withaferin A was toxic to normal cells, withanone was safe and hence can be recruited for safe cancer therapeutics. Furthermore, a combinational approach using withaferin A and withanone against the anticancer and anti-ageing effects of cancer and normal cells, respectively, is warranted.

After washing and peeling, ground roots of ashwagandha were dried under different conditions (in sun, shade or a cabinet drier) to determine the effect of temperature on the withanolide content. Maximum withanolide content was observed after shade drying and the minimum after cabinet drying at an air velocity of 1.8–2.0 m/s (Agrawal *et al.*, 2014).

According to Asthana and Raina (1989), it is important to investigate the side effects of the bioactive compounds of ashwagandha and their possible interactions before experimental clinical research is conducted. Also, although withanolide production by *in vitro* cultures is still far from the levels required for economic exploitation, the study of such cultures is a useful tool for obtaining greater understanding of the withanolide metabolic pathway and allowing the application of plant metabolic engineering techniques to improve the biotechnological production of bioactive compound(s) of *Withania*.

Isolation of flavonoids from leaves

According to Bashir *et al.* (2013), air-dried leaves of *W. somnifera* are ground to a fine powder and extracted with chloroform to remove chlorophyll and resinous and waxy material. The ground material is then extracted with 85% aqueous methanol and the slurry stood at room temperature for 24 h with occasional stirring. The solvent containing the extract is then decanted and filtered by vacuum filtration. The extraction is repeated twice with the same solvent and twice with 50% aqueous methanol. The four filtrates from each extraction are combined and excess solvent evaporated under reduced pressure at 40°C to give crude extracts as a solid mass, which is kept in a glass container. The dry extract is loaded on to an 80 × 3 cm column of polyamide 6S for the chromatographic isolation of flavonoids.

Extraction of bioactive components

Ashwagandha has been used as an antioxidant, adaptogen, antitumour agent, aphrodisiac, liver tonic, anti-inflammatory agent and astringent, and for its immunomodulatory activity (Tripathi, 2003). The major constituents of the plant are withanolides and alkaloids. These bioactive components of are prepared by extraction. During this process, the desirable soluble constituents are separated from those that are not required by using different solvents. The various methods of extraction include maceration, hot continuous extraction, percolation, decoction, ultrasound extraction, supercritical fluid extraction, microwave-assisted extraction, etc. Two of these are outlined below.

EXTRACTION OF BIOACTIVE COMPONENTS USING THE SOXHLET EXTRACTION METHOD. The Soxhlet method is one of the most used conventional methods for the extraction of withanolides from ashwagandha. Cell permeation followed by solubilization of the active constituents by the extracting solvent is the underlying principle. The powdered plant material can be effectively extracted in a Soxhlet extractor at elevated temperature (40–60°C) using 200 ml of distilled petroleum ether followed by *n*-hexane, ethanol, chloroform and methanol. The crude extracts are individually filtered, evaporated in a rotor evaporator, dried and stored in stock vials that are kept refrigerated. However, this type of conventional method is time- and solvent-consuming, thermally unsafe and not economically viable (Jyothi *et al.*, 2010).

EXTRACTION OF BIOACTIVE COMPONENTS USING A MICROWAVE-ASSISTED EXTRACTION METHOD. With increasing demand for more environmental friendly methods, microwave-assisted extraction (MAE) has been developed and optimized for the fast extraction of withanolides, and seems to be a good option. Solvent consumption is less and there is also time saving compared with traditional methods. In this method, the solvent, while in contact with sample, is heated using microwave energy. During this process, the disruption of hydrogen bonds results from microwave-induced dipole rotation of molecules and the migration of ions that enhance the penetration of the solvent into the matrix, and allows the dissolution of the components to be extracted. The extraction process involves extracting finely powdered samples in methanol in an extraction vessel at the requisite irradiation power level and temperature and for the requisite time (Jyothi *et al.*, 2010). After extraction, the extracts are filtered and evaporated under vacuum in a rotary evaporator.

2.3.2 Value-added products

In addition to the traditional drugs made from ashwagandha, value-added products from the plant include root powder, capsules, root extract, herbal beer, etc. Health drinks, herbal tea, functional foods, nutraceuticals and cosmeceuticals are some of the other value-added products based on which enterprises can be set up. Ashwagandha sweet and salty biscuits, ashwagandha churan balls and ashwagandha beverages are prepared by incorporating ashwagandha root powder, ashwagandha leaf powder and ashwagandha root and leaf powder, respectively, into the products.

Powder preparation

The leaves are cleaned and dried under shade or using a tray dryer, The dried leaves are then ground to a coarse powder using a high-capacity grinding machine. The resulting powder is stored in an air-tight container and kept in a cool, dark and dry place.

Ashwagandha tea

Ashwaganda tea is very beneficial for promoting a relaxed state of mind. The herb is sometimes referred to as 'India's ginseng', because it is used by practitioners of Ayurveda for medicinal reasons. It is an excellent remedy for low levels of energy and does not have the adverse side effects of caffeine. Ashwagandha tea may be drunk as a beverage every day by most people, although practitioners of alternative medicine caution their patients against drinking the tea if they have a history of high blood pressure, because it could elevate this even more. Long-term use of the tea is quite beneficial for keeping people who are stressed in a more calm and relaxed state. The tea also helps to rejuvenate the system after exertion but is still calming.

Bakery products

Bakery products are widely consumed processed food products, with 80% of products (such as bread and biscuits) consumed on a regular basis. The common ingredients in biscuits are refined wheat flour, vegetable shortening, butter, sugar, baking powder and flavouring agents. Biscuits are consumed by all age groups of consumers because they release instant energy and offer various other nutritive benefits, particularly to people with a health disorder. Making biscuits with ashwagandha herbs is an innovative steps in upgrading the nutritive value of the biscuits, along with their medicinal value. The herb is considered to be of GRAS (generally recognized as safe) status and is used as an ingredient of cereals, candies, chewing gum and cookies.

Preparation of ashwagandha-based extruded products

Extrusion cooking is a method used for converting starchy and proteinaceous material into fabricated products during which the material is forced through a die at high temperature in a short time to generate friction by the rotation of a single or twin screw(s) to produce extruded snacks (Alavi *et al.*, 1999; Singh *et al.*, 2007; Gamlath, 2008). Snacks can be produced from ashwagandha powder in combination with other ingredients as an extruded product. The acceptability of extrudate snack is greater than that of the usual snack products available on the market (Moraru and Kokini, 2003).

2.4 Uses

2.4.1 General uses

The leaves of ashwagandha are bitter and are recommended to treat fever and painful swellings. The crude preparation of the plant has also been found to be very active against a number of pathogenic bacteria. Sore eyes, ulcers and swellings can be cured using a fomentation of the leaves. The leaves are also used as a hypnotic and an anthelmintic, can be crushed and applied to tumours and ulcers, and are consumed as a vegetable and used as livestock fodder (Kirtikar and Basu, 1991). Other uses of the leaves are to heal open as well as septic and inflamed wounds and abscesses, and to treat inflammation, haemorrhoids, rheumatism and syphilis.

Ashwagandha as a medicinal herb

Ashwagandha is considered to be one of the greatest rejuvenating agents in Ayurvedic medicine. The leaves are applied externally as a paste or after crushing for carbuncles on to inflamed areas and swellings. Herbal tea, powders, tablets and syrups can be prepared from the leaf extract. The root is claimed to have sex-enhancing properties, and the herb is used to promote a peaceful state of mind. The plant is also used as a liver tonic and anti-inflammatory agent and, more recently, to treat asthma, ulcers and insomnia. The incorporation of ashwagandha into the diet can prevent or decrease the growth of tumours in humans. It is an excellent nerve tonic and is used to nourish the nerves and improve nerve function in order to maintain calm during stressful conditions.

2.4.2 Pharmacological uses

The pharmacological activity of ashwagandha is attributed to the presence of several alkaloids and steroid lactones (Chadha, 1976).

Anti-inflammatory activity

The anti-inflammatory activity of ashwagandha is attributed to the biologically active steroid withaferin A (Khare, 2007), which is used to suppress arthritic syndrome effectively and without any toxic effects. Studies of ashwagandha in animal models have proven that it has anti-inflammatory properties (Kulkarni *et al.*, 1993).

Antibiotic activity

Withaferin A in leaf extracts of ashwagandha inhibits the growth of various Gram-positive bacteria, acid-fast and aerobic bacilli and pathogenic fungi, and also inhibited Ranikhet virus. The antibiotic activity of withaferin A is due to the presence of the unsaturated lactone ring. The extract of the shrub is active against Vaccinia virus and *Entamoeba histolytica* (Chadha, 1976; Rastogi and Mehrotra, 1998; Khare, 2007), and Dhuley (1997) reported that ashwagandha showed a protective action against systemic *Aspergillus* infection.

Immunomodulatory activity

The consumption of ashwagandha significantly increased haemoglobin concentration, red blood cell count, platelet count and body weight in mice (Ziauddin *et al.*, 1996). Withaferin A exhibited specific immunosuppressive effects on both human B and T lymphocytes, while withanolide E had a specific effect on T lymphocytes (Rastogi and Mehrotra, 1998; Aggarwal *et al.*, 1999; Davis and Kuttan, 2000; Gautam *et al.*, 2004; Rasool and Varalakshmi, 2006).

Antioxidant activity

In an investigation of the antioxidant activity of the active principles of ashwagandha in rats, Bhattacharya *et al.* (1997a) found the antioxidant effects were mainly due to the presence of glycowithanolides.

Anti-hyperglycaemic effect

Bhattacharya *et al.* (1997b) reported that a preparation containing ashwagandha decreased streptozocin (STZ)-induced hyperglycaemia and pancreatic islet superoxide dismutase (SOD) activity in type 1 diabetic rats.

Hepatoprotective activity

Withaferin A has also been shown by various researchers to have significant hepatoprotective activity against carbon tetrachloride (CCl_4)-induced hepatotoxicity (Aphale *et al.*, 1998; Rastogi and Mehrotra, 1998; Khare, 2007).

2.5 Summary

Withania somnifera, known as ashwagandha or winter cherry, is a solanaceous herb that is known to soothe and calm the nervous system. The multiple health benefits of this herb make it a perfect rejuvenator of physical and psychological health. Its powerful antioxidant compounds scavenge free radicals reducing the impact of ageing. Ashwagandha is a medicinal plant that can potentially generate global business for India, and scientific studies on the crop need to be increased manyfold in order to put its many medicinal qualities to maximum use.

References

Aggarwal, R., Diwanay, S., Patki, P. and Patwardhan, B. (1999) Studies on immunomodulatory activity of *Withania somnifera* (ashwagandha) extracts in experimental immune inflammation. *Journal of Ethnopharmacology* 97, 27–35.

Agrawal, R., Upadhyay, A. and Nayak, P.S. (2014) Influence of drying on the quality of ashwagandha (*Withania somnifera*). *Journal of Food and Pharmaceutical Sciences* 2, 63–67.

Alavi, S.H, Gogoi, B.K, Khan, M, Bowman, B.J. and Rizvi, S.S.H. (1999) Structural properties of protein-stabilized starch-based supercritical fluid extrudates. *Food Research International* 32, 107–118.

Andallu, B. and Radhika, B (2000) Hypoglycemic, diuretic and hypocholesterolemic effect of winter cherry (*Withania somnifera* Dunal) root. *Indian Journal of Experimental Biology* 38, 607–609.

Aphale, A.A., Chhibba, A.D., Kumbhakarna, N.R., Mateenuddin, M. and Dahat, S.H. (1998) Subacute toxicity study of the combination of ginseng (*Panax ginseng*) and ashwagandha (*Withania somnifera*) in rats: a safety assessment. *Indian Journal of Physiology and Pharmacology* 42, 299–302.

Asthana, R. and Raina, M.K. (1989) Pharmacology of *Withania somnifera* (L.) Dunal – a review. *Indian Drugs* 26, 199–205.

Bashir, H.S., Mohammed, A.M., Magsoud, A.S. and Shaoub, A.M. (2013) Isolation of three flavonoids from *Withania somnifera* leaves (Solanaceae) and their antimicrobial activities. *Journal of Forest Products and Industries* 2(5), 39–45.

Bhatia, P., Rattan, S.I.S., Cavallius, J. and Clark, B.F.C (1987) *Withania somnifera* (ashwagandha) a so-called rejuvenator inhibits growth and macromolecular synthesis of human cells. *Medical Science Research* 15, 515–516.

Bhattacharya, S.K. (1998) Adaptogenic activity of siotone, a herbal formulation against an unpredictable chronic stress induced physiological and behavioral perturbation in rats. In: *National Conference on Recent Trends in Spice and Medicinal Plant Research, Calcutta, 2–4 April 1998.*

Bhattacharya, S.K., Satyan, K.S. and Ghosal, S. (1997a) Antioxidant activity of glycowithanolides from *Withania somnifera*. *Indian Journal of Experimental Biology* 35, 236–239.

Bhattacharya, S.K., Satyan, K.S. and Chakarbarti, A. (1997b) Effect of Tarsina, an Ayurvedic herbal formulation, on pancreatic islet superoxide dismutase activity in hyperglycemic rats. *Indian Journal of Experimental Biology* 35, 297–299.

Chadha, Y.R. (ed.) (1976) *The Wealth of India, Vol. X: Sp–W.* Raw Materials: Original Series, Publications and Information Directorate, Council of Scientific and Industrial Research (CSIR), New Delhi, India, pp. 580–585.

Christina, A.J.M., Joseph, D.G., Packialakshmi, M., Kothai, R., Robert, S.J.H. and Chidambaranathan, N. (2004) Anticarcinogenic activity of *Withania somnifera* Dunal against Dalton's ascitic lymphoma. *Journal of Ethnopharmacology* 93, 359–361.

Davis, L. and Kuttan, G. (2000) Immunomodulatory activity of *Withania somnifera*. *Journal of Ethnopharmacology* 71, 193–200.

Dhuley, J.N. (1997) Effect of some Indian herbs on macrophage functions in ochratoxin A treated mice. *Journal of Ethnopharmacology* 58, 15–20.

Gamlath, S. (2008) Impact of ripening stages of banana flour on the quality of extruded products. *International Journal of Food Science and Technology* 43, 1541–1548.

Gautam, M., Diwanay, S.S., Gairola, S., Shinde, Y.S., Jadhav, S.S. and Patwardhan B.K. (2004) Immune response modulation to DPT vaccine by aqueous extract of *Withania somnifera* in experimental system. *International Immunopharmacology* 4, 841–849.

Girdhari, L.G. and Rana, C.A. (2007) Plant review on *Withania somnifera* (ashwagandha). *Pharmacognosy Magazine* 1, 129–136.

Gupta, A.P., Verma, R.K., Misra, H.O. and Gupta, M.M. (1996) Quantitative determination of withaferin A in different plant parts of *Withania somnifera* by TLC densitometry. *Journal of Medicinal and Aromatic Plants* 18, 788–790.

Jyothi, D., Khanam, S. and Sultana, R. (2010) Optimization of microwave assisted extraction of withanolides from roots of ashwagandha and its comparison with conventional extraction method. *International Journal of Pharmacy and Pharmaceutical Sciences* 2(4), 46–50.

Khare, C.P. (2007) *Withania ashwagandha* Kaul (cultivated var.). In: Khare, C.P. *Indian Medicinal Plants: An Illustrated Dictionary*. Springer, Berlin/Heidelberg, Germany, pp. 717–718.

Kirtikar, K.R. and Basu, B.D. (1991) *Indian Medicinal Plants*, Vol. 3. Shiva Publishers, Dehradun, India, p. 1783.

Kulkarni, R.R., Patki, P.S., Jog, V.P., Gandage, S.G. and Patwardhan, B. (1991) Treatment of osteoarthritis with a herbomineral formulation: a double-blind, placebo-controlled, cross-over study. *Journal of Ethnopharmacology* 33, 91–95.

Kulkarni, S.K., Sharma, A., Verma, A. and Ticku, M.K. (1993) GABA receptor mediated anticonvulsant action of *Withania somnifera* root extract. *Indian Drugs* 30, 305–312.

Moraru, C.I. and Kokini, J.L. (2003) Nucleation and expansion during extrusion and microwave heating of cereal foods. *Comprehensive Reviews in Food Science and Food Safety* 2, 120–138.

Rasool, M. and Varalakshmi, P. (2006) Immunomodulatory role of *Withania somnifera* root powder on experimental induced inflammation: an *in vivo* and *in vitro* study. *Vascular Pharmacology* 44, 406–410.

Rastogi, R.P. and Mehrotra, B.N. (eds) (1998) *Compendium of Indian Medicinal Plants*, 2nd reprint. Central Drug Research Institute, Lucknow and National Institute of Science Communication, Council of Scientific and Industrial Research, New Delhi, India. (see Vol. 1, pp. 434–436; Vol. 2, pp. 708–710; Vol. 3, pp. 682–684; Vol. 4, pp. 765–766; Vol. 5, pp. 889–891; Vol. 6. p. 148).

Singh, S., Gamlath, S. and Wakeling, L. (2007) Nutritional aspects of food extrusion: a review. *International Journal of Food Science and Technology* 42, 916–929.

Tripathi, K.D. (2003) *Essentials of Medicinal Pharmacology*, 5th edn. Jaypee Brothers Medical Publishers, New Delhi, p. 405.

Uddin, Q., Samiulla, L., Singh, V.K. and Jamil, S.S. (2012) Phytochemical and pharmacological profile of *Withania somnifera* Dunal: a review. *Journal of Applied Pharmaceutical Science* 2(1), 170–175.

Umadevi, M., Rajeswari, R., Sharmila Rahale, C., Selvavenkadesh, S., Pushpa, R., Sampath Kumar, K.P. and Bhowmik, D. (2012) Traditional and medicinal uses of *Withania somnifera*. *The Pharma Innovation* 1(9), 102–110.

Ziauddin, M., Phansalkar, N., Patki, P., Diwanay, S. and Patwardhan, B. (1996) Studies on the immunomodulatory effect of ashwagandha. *Journal of Ethnopharmacology* 50, 69–76.

3 Basil

Darach Lupton,[1] Muhammad Mumtaz Khan,[2] Rashid Abdullah Al-Yahyai[2]* and Muhammad Asif Hanif[3]

[1]*Oman Botanic Garden, Muscat, Oman;* [2]*Sultan Qaboos University, Muscat, Oman;* [3]*University of Agriculture, Faisalabad, Pakistan*

3.1 Botany

3.1.1 Introduction

Basil (*Ocimum basilicum* L.) is an annual herb belonging to the mint family (Lamiaceae). It has been utilized for millennia and is an essential ingredient in many cooking traditions and practices (Agarwal *et al.*, 2013). The genus *Ocimum* contains a range of some 50 to 150 species and varieties that are native to the tropical regions of Asia and Central and South Africa (Ghosh, 1995). The uncertainty in the exact number of species within the genus is largely attributed to the enormous variation that is found among the constituent species. The variability is prevalent in the morphology, growth habit, flower colour, leaves, stems and chemical composition (Svecova and Neugebauerova, 2010). Basil cross-pollinates readily, and the resulting diversity and variation has led some authors to reclassify sections of the genus (Paton, 1992). There are a number of plants outside the genus *Ocimum* with the common name basil, including 'basil thyme' (*Acinos arvensis* (Lam.) Dandy) and 'wild basil' (*Clinopodium vulgare* L.), which can sometimes lead to confusion and misunderstanding.

O. basilicum is known by different names depending on the location. In the English language, it is typically called basil, common basil or sweet basil. In India, specifically in Hindi and Bengali, it called *babui tulsi*. Other common names of basil are *basilica* (in French), *basilikum* or *basilienkraut* (in German), *basilico* (in Italian), *rehan* (in Arabic) and *albahaca* (in Spanish). In Arabic, it is known as *hebak* as well as *rihan* (Kirtikar and Basu, 2003, cited by Bilal *et al.*, 2012). Probably the most familiar basil is sweet basil (*O. basilicum*); however, this has a large number of cultivars, varying in flavour, scent and uses. There are more than 160 named cultivars in existence today. Popular examples include, *O. basilicum* 'Cinnamon', *O. basilicum* 'Dark Opal' and holy basil (the species *O. tenuiflorum* L., previously known as *O. sanctum* L.) (Fig. 3.1). Scents and flavours can range from cinnamon, liquorice and lemon to anise. The plants can be shrubby or herbaceous, and vary in size from 20 cm to 3 m tall, depending on the species (and the literature source used). The leaves can be smooth, shiny, hairy or curly, and they can be green to blue/purple. The flower colour ranges from white to purple to lavender (Meyers, 2003). Most of the regular varieties of basil are considered annuals;

*Corresponding author, e-mail: alyahyai@squ.edu.om

© CAB International 2016. *Leafy Medicinal Herbs: Botany, Chemistry, Postharvest Technology and Uses* (eds D.C.P. Ambrose *et al.*)

Fig. 3.1. A basil plant in flower. Basil is usually cultivated as an ornamental in Middle Eastern countries due to its attractive flowers and aroma.

however, in warm tropical regions many perennial varieties exist, e.g. *O. tenuiflorum* (Simon *et al.* 1999; Tilebeni, 2011) (Fig. 3.1). Their wide range of forms, colours and sizes has elevated the ornamental importance of basil in recent years (Svecova and Neugebauerova, 2010) and has increased the plants' economic value globally. It is not uncommon to see basil grown as an ornamental plant in public parks and home gardens (Fig. 3.1).

The essential oil content of basil shows a variability between species and cultivars and is thought to be the result of varying ecological factors, geographic origins, genetic patterns, different chemotypes and differences in the nutritional status of plants. In Finland, 17 different collections, all called sweet basil, were analysed for their morphological traits and chemical make-up. A large amount of variation was recorded for both characteristics. A similar variability was found when ten Italian commercially available basil cultivars were studied. The chemical analyses of the varieties showed correlations with their morphological characters. Two varieties with violet leaves were linalool chemotypes and three of the large-leaved varieties were linalool and methyl chavicol (estragole or *p*-methoxyallyl benzene) chemotypes (Galambosi, 1995, cited by Putievsky and Galambosi, 1999). The bulk of the essential oil of basil plants is concentrated in the leaves and flowers; there are trace quantities of essential oils in the branches and stems, but the amounts are not commercially important (Svecova and Neugebauerova, 2010). Basil is highly variable both morphologically and chemically, and the variations appear to be strongly influenced by ecological factors. The origin, source and growing conditions of basil therefore have an impact on plant uses, and in particular upon its flavours, aromas and medical uses. This variability of basil is reflected in its broad array of uses, which will be discussed in more detail later in the chapter.

3.1.2 History/origin

O. basilicum is indigenous to India and other areas in tropical Asia, where it has been grown for 5000 years. The generic name, *Ocimum*, originates from the ancient Greek word *okimon*, which means smell (Tucker and DeBaggio, 2000). There are numerous suggestions for the origins of the word basil. One is that it stems from the Greek word *basileus*, meaning king, as it is believed to have grown close to the area where St Constantine and his mother St Helen discovered the Holy Cross (Jacqueline, 2001). According to Parkinson, basil's scent was 'fit for a king's house' (Grieve, 1931). Other less fanciful speculation suggests that the origin of the name basil stems from the similarity of the species name, *basilicum*, to the name of the basilisk, the fabled serpent with the deadly gaze (Meyers, 2003).

The history of basil is steeped in legend and mystery. Many believe it was Alexander the Great (356–323 BCE) who brought it to Greece. Basil is thought to have been brought to England from India in the 1500s, eventually arriving in the USA in the early 1600s (Darrah, 1980). Culpeper, Gerard and Dioscordes mention basil in their respective herbals (Meyers, 2003).

Gerard praised basil as a remedy for melancholy but also repeated Dioscorides' warning that too much basil 'dulleth the sight … and is of a hard digestion' (Gerard, 1975, cited by Meyers, 2003). Basil was also alleged to cause the spontaneous generation of scorpions and to cause scorpions to develop in the brain. The link with scorpions is evident today in basil's depiction with the astrological sign of Scorpio (Reppert, 1984, cited by Meyers, 2003).

3.1.3 Location

Although basil is grown in a variety of climatic and environmental conditions, the optimum conditions are found in countries with a warm climate. Warmth, light and moisture are the key ecological requirements for basil cultivation. The herb is susceptible to frost so outdoor cultivation is restricted to frost-free regions of the world. Basil is grown widely in the following countries: India, Pakistan, Comores Islands, Madagascar, Haiti, Guatemala, Réunion, Thailand, Indonesia, Russia (Georgia, East Caucasus) and South Africa, Egypt, Morocco, France, Israel, Bulgaria, the USA (Arizona, California, New Mexico), Italy, Hungary, Poland, Germany, Greece, Turkey, other Balkan countries and Slovakia (Putievsky and Galambosi, 1999).

Absolute figures for basil oil production are difficult to acquire. There are numerous small-scale growers working in local operations whose production figures are not integrated into national statistics. However, there are some gross figures available from the 1990s, when gross world production of basil oils was approximately 93–95 tons/year of which 55 tons was from *O. gratissimum* L. and 43 tons from *O. basilicum*. About 100 kg of oils were produced from *O. canum* (Simms) (preferred name *O. americanum* L.). Basil oils were then produced in the following countries (the quantities that follow in parentheses are tons): India (15), Bulgaria (7), Egypt (5), Pakistan (4.5), the Comoros (4.5), Israel (2), the former Yugoslavia (1), the USA (1), Madagascar (1), Réunion and Albania (each 0.5), Hungary (0.3) and Argentina (0.2) (Lawrence, 1993, cited by Hiltunen and Holm, 1999). The USA is probably the largest market for basil oil, followed by the European countries of Germany, France, the UK and the Netherlands (Robbins and Greenhalg, 1979, cited by Hiltunen and Holm, 1999).

Global statistics for the production of dried basil are also hard to obtain. A large portion of the world production, chiefly in the Mediterranean region, and in India and California, is not sold internationally; most of the basil in these areas is consumed locally. Import statistics also show that the USA is one of the world's biggest users of dried basil (Putievsky and Galambosi, 1999).

Other important areas for basil importation are the European countries. In the 1990s, the total amount of basil herb imported to Europe was about 830–880 t/year. France is the largest importer at 300–350 t/year, followed by the UK (250 t/year), Germany (200 t/year) and the Netherlands (80 t/year). The largest supplier of the Western European countries was Egypt (Putievsky and Galambosi, 1999).

3.1.4 Morphology

O. basilicum is an upright, branching herb, 0.6–0.9 m high with square, glabrous stems and branches, usually green but sometimes purple in colour. The leaves are simple and oppositely arranged on the stem. They are 2.5–5 cm or more long and are ovate with an acute tip; the margins are entire, more or less toothed or lobed (Jayaweera, 1981, cited by Bilal *et al.*, 2012). The petiole is 1.3–2.5 cm long. The leaves have numerous oil glands which exude strongly scented volatile oil. The inflorescence is usually racemose, and the terminal raceme is usually much longer than the lateral ones. The bracts are stalked, shorter than the calyx, ovate and acute. The calyx is 5 mm long, enlarging on the fruit. The fruit has a short pedicel. The calyx lower lip has two central teeth and is longer than the rounded upper lip. The corolla is 8–13 mm long, white, pink or purplish in colour, and glabrous or slightly pubescent. The nutlets (seeds) are about 2 mm long, ellipsoid, black and pitted. There are five flower sepals that remain fused into a two-lipped calyx. The ovary is superior and the fruit consists of four achenes (Jayaweera, 1981, cited by Bilal *et al.*, 2012).

Basil requires warm temperate or Mediterranean conditions. The optimum temperature for germination is 20°C, with growing temperatures of 7 to 27°C (Simon, 1995). The plant develops best in long-day, full-sun conditions. It cannot tolerate drought as the plant tissue is very tender. Basil requires well-drained, fertile soils with a high organic matter content. It grows well in soils with a pH ranging from 4.3 to

8.2 and has an optimum pH of 6.4. Basil has medium, deep roots and a high water requirement (Simon, 1995).

3.2 Chemistry

Basil is an impressively aromatic plant and is used as a sweet-smelling herb. Different phenotypic characters, including taste, aroma and many others, are used to describe a variety of basil ecotypes. The height of plants varies from 30 to 300 cm and leaf colour from green to blue/purple; this depends on the type of species (Hiltunen and Holm, 1999). The name of each basil type often represents its particular flavour, with the exception of the sweet basil, whose taste is bright and pungent; anise basil, lemon basil and cinnamon basil offer unique flavours as indicated by their names. The essential oil present in the leaves and other parts of a number of basil species/cultivars is responsible for its distinctive fragrance and aroma. In most species of basil, methyl chavicol, eugenol and linalool are major components. Different species or cultivars have different amounts of each of these chemical constituents, which hence are responsible for the different taste and aroma of each basil cultivar. As an example, the sweet aroma of methyl chavicol has been compared with that of French tarragon and anise, while a floral scent is produced by linalool and eugenol is reminiscent of cloves. The major component present in sweet basils is methyl chavicol while eugenol is present in large amounts in spicy basils. Other chemical components responsible for flavour include geranial (a rose flavour), thymol (a thyme flavour), camphor, *trans*-methyl cinnamate (a cinnamon flavour) and citral (lemon) (DeBaggio and Belsinger, 1996; Al-Maskri *et al.*, 2011).

3.2.1 Chemical composition

In sweet basil, the fat content and calorific value is low while high amount of minerals and vitamin A are present. In 2.5 g of basil leaves (five fresh leaves), there are 96.6 IU vitamin A, 3.85 mg calcium, less than 1 calorie, 11.55 mg potassium, and smaller proportions of vitamin C and other vitamins, protein, fibre and minerals. The GRAS (Generally Recognized As Safe) list of the US Department of Agriculture includes sweet basil leaf to be used in the range of 2–680 ppm and 0.01–50 ppm for the essential oil. The use of exceedingly large quantities of oil is suggested to have a health risk due to the occurrence of carcinogenic compounds. The GRAS-suggested amount of basil essential oil is very minute, and internal use of a large amount of this oil should be avoided (Hanif *et al.*, 2011; Hosseini-Parvar *et al.*, 2015).

3.2.2 Phytochemistry

O. tenuiflorum has essential oils mostly confined to the green leaves and thus has a particular aroma. This leaf scented volatile oil chiefly comprises phenols, terpenes and aldehydes. Besides its essential or fixed oils, the plant also includes alkaloids, glycosides, saponins and tannins. The leaves also contain particular amounts of carotene and ascorbic acid. The reported chemical properties of basil leaves are based on numerous worldwide studies, and hence edaphic and geographic factors are expected to influence different chemical ingredients. The difference in aroma between different varieties of *O. basilicum* is due to the various compositions of their essential oils. In various parts of the world, basil cultivars are present in large diversity, indicating a diverse range of chemical composition. The essential oil of basil usually contains α-terpineol, eucalyptol, eugenol, methyl eugenol, linalool, β-elemene, germacrene D, α-bergamotene, α-guaiene, cubenol, τ-cadinol, camphor, bornylacetate, α-caryophyllene, β-caryophyllene, elixen, β-cadinene, α-copaene, α-bisabolol, β-farnesene, epibiciclosesquiphelandrene, τ-muralol, δ-gurjunene and δ-cadinene (Hanif *et al.*, 2011). The various types of fatty acids present in three *Ocimum* species are presented in Table 3.1. The presence of cardiac glycosides, saponins and tannins in

Table 3.1. Fatty acid composition in basil species (*O. album, O. basilicum* and *O. tenuiflorum*). From Malik *et al.*, 1987, 1979.

Fatty acid	*O. album*[a]	*O. basilicum*[a]	*O. tenuiflorum*[b]
Arachidonic acid (C20:4)	2.73	–	–
Capric acid (C10:0)	1.30	–	–
Lauric acid (C12:0)	0.78	0.85	2.84
Linoleic acid (C18:2)	36.36	21.18	59.10
α-Linolenic acid (C18:3)	–	48.50	21.27
Myristic acid (C14:0)	0.68	0.36	1.90
Oleic acid (C18:1)	44.16	13.33	6.00
Palmitic acid (C16:0)	11.68	9.70	5.54
Stearic acid (C18:0)	2.33	5.45	3.12

[a]Malik *et al.*, 1987; [b]Malik *et al.*, 1989.

the aqueous extract of *O. basilicum* plants has been shown by phytochemical analysis.

3.3 Postharvest Technology

Conventionally, the best harvesting time for basil is early in the morning just after the evaporation of the dew and before the day temperature starts increasing. The strongest activity of basil essential oil has been observed in the morning. No difference in flavour contents has been reported in some findings between fresh and dried basil, but the flavour complexity and intensity that has been observed in fresh leaves is lost to a large extent in the dried leaves. Fresh basil, when placed in an airtight bag after it is wrapped in numerous paper towels, can be stored for up to a week in a refrigerator.

This herb cannot be stored easily for a longer time unless it is dried, so appropriate drying of leaves is recommended for long-term storage. The leaves should not be shredded or broken during drying because the essential oil content will be lost and the aroma will be reduced in such leaves. For basil leaves, shade drying is more appropriate than sun drying in order to avoid the loss of fragrance due to volatility of the essential oils. If dried basil is kept in closed jars and away from heat and light, it can be stored for a year. The leaves of basil can also be maintained for some time by salt layering. Another type of preferred long-term handling is the freeze storage. If the leaves are chopped and tightly wrapped in plastic sheets for freezing, then they will remain green and blackening of the leaves during freezing can be avoided. The leaves can also be frozen in ice cube trays after mixing with olive oil in a food processor. For domestic use, basil leaves can also be preserved by adding olive oil and salt to the storage jar and keeping in a refrigerator. Bacterial growth during such storage may be a possible problem, and even under refrigerated storage, infection with *Clostridium botulinum* may cause botulism. To avoid food-borne botulism, it is important to strictly follow food safety/sanitation instructions for product receipt, handling, processing and storage. For culinary purposes, single or multiple fresh leaves can be removed and used (Pushpangadan and George, 2012).

3.3.1 Processing

Basil, like other herbal plants, is consumed in a variety of ways and for various purposes. In addition to the use of fresh leaves, other common processed forms of basil include whole dry leaves, frozen or powdered leaves, and extracted essential oils. Whole plants or chopped leaves can be stored frozen, with and without oils, to be used for extended periods of time beyond the fresh shelf life. Alternative traditional methods for preserving basil leaves include storage in salt and in the form of oil concentrates (Meyers, 2003).

The herb is traditionally dried by hanging washed bundles inverted in a dry and shaded place or placing whole spread leaves between two sheets of paper to prevent oxidation and discoloration. Forced warm air drying is used for industrial production. Basil leaves should be dried immediately after harvest because they darken if exposed to the open air for an extended period of time. Drying should be done at a temperature not exceeding 40°C to minimize the evaporation of volatile compounds (Putievsky and Galambosi, 1999). Dried basil can be preserved for a year when it is protected from heat, light and moisture (Meyers, 2003).

Essential oil can be extracted from basil in two forms, as herbal oil that originates from the leaves (0.1–0.25%) or as a superior-quality floral oil that is collected from the flowers (0.4%) (Srivastava, 1980; Putievsky and Galambosi, 1999). In India, flowers are harvested four times during the season and produce 12–13 kg/ha oil yield compared to 18–22 kg/ha from one harvest from the much higher fresh yield of whole plants (Srivastava, 1980). In Israel, plants are harvested when half of them have flowers and the fresh annual yield is 75 tons/ha, which produces an essential oil yield 120–140 kg/ha (Putievsky and Galambosi, 1999). A similar distillation process is used for basil oil extraction to the one that is commercially used for other herbs; this takes about an hour using freshly harvested leaves (Wijesekera, 1986; Denny, 1995).

3.3.2 Value addition

Basil leaves can be mixed with a variety of other herbs, including juniper, garlic, marjoram, oregano, paprika, mustard, parsley, pepper, sage, rosemary and thyme, and can be used in stuffings, soups, stews and rice, and also with fish, vegetables, chicken and meats. They can be a key ingredient in vinegars, jams, teas, cheeses, drinks, oils and liqueurs too. Purple basil vinegars can be produced with a good colour, and according to personal taste, cinnamon and lemon basils are used to make delicious desserts and may increase their taste. Larger leaves can be minced, torn or chopped and consumed. Small leaves are good to add to vegetarian dishes, salads, rice and pasta. For maximum flavour, basil is added at the end of cooking. It is used fresh as well as dried, but the drying reduces the predominant flavours. The uses of basil are diverse and plenteous; it is used with meat, vegetables, fish, dressings, sauces, stews, herbal teas, liqueurs and mixed drinks. It is universally used by both the domestic and the industrialized producer in the preparation of pesto, a varying combination of basil, cheese, garlic, oil and nuts. Basil is often used as an additional flavour with tomatoes. Garden-fresh basil is preserved in vinegar or oil, or frozen. Chilling of basil preserves the flavour of the herb more effectively than does drying. The length of storage of dried basil is far more than that of fresh basil, which lasts for only a short time in the refrigerator.

3.4 Uses

Despite being consumed at relatively low amounts, the high levels of antioxidants and minerals in herbs means that many of them have significant health benefits. It is not fully understood what quantities of basil should be ingested to achieve its health benefits and there are no standards or recommendations as to the precise amounts to use. Nevertheless, basil is almost completely calorie free and contains high quantities of dietary fibre and minerals. Even though there appears to be no logical evidence for its usefulness to human health, basil tea and oil are readily available in many health food stores. Having said that, basil is a popular food additive and provides a distinctive flavour and aroma. Basil is a great addition to any kitchen, it adds both flavour and personality to many dishes (Hosseini-Parvar *et al.*, 2015).

3.4.1 General uses

Ritualistic uses

Basil has many uses ranging from culinary to religious, and these are often steeped in ritual. There are a number of interesting beliefs linked with the historical use of basil. In Europe, it was associated with death and it was considered to be unlucky to dream of it. In contrast, in Italy, women wore it in their hair and young men wore it behind their ears when they went courting (Dymock *et al.*, 2005). Hindus in India believe that if you are buried with basil, it is a guaranteed ticket to heaven. The English used it in food and to repel pests, e.g. flies, and evil spirits. Basil is often called *l'herbe royale* (the royal herb) by the French. Jewish folklore implies that while fasting, basil gives you strength (Miele *et al.*, 2001). In Portugal, potted basil is presented to a loved one on the religious holidays of St John and St Antony. Holy basil has religious worth across a range of belief systems; the Greek, Bulgarian, Romanian and Serbian Orthodox churches utilize basil in the preparation of holy water, in some instances a pot of basil is placed under the church altar (Tilebeni, 2011).

Culinary uses

Basil has been incorporated into culinary preparations for thousands of years. It is a very useful gastronomic herb found in a wealth of dishes, sauces and condiments, soups, stews and stuffing, and also in fish, meats and vegetables. It is easily blended with other herbs, including, garlic, oregano, mustard, parsley, pepper, rosemary and thyme (Hemphill, 2000, cited by Meyers, 2003). It is also an important constituent in teas, oils, cheeses and liqueurs (Darrah, 1980; Simon, 1995). Basil is an important component of many alcoholic beverages, including bitters, liquors and spirits. By adding a blend of mixed essential oils of fennel, basil and coriander to a salt solution of whey, Russian researchers found a method to enhance the storage of a carbonated fermented milk beverage (Askerova *et al.*, 1993). Fresh, frozen or dried basil (1–40 g/1) is also used in spirits, garlic or lemon alcoholic beverages, which may be sweet or dry, according to a German patent (Meier, 1990, cited by Hiltunen and Holm, 1999). Basil essential oil has significant commercial value. It is utilized in a range of industrial products, including beverages, prepared foods, dental products, fragrances and soaps (Darrah, 1980). *O. gratissimum* and *O. basilicum* essential oils are considered economic materials in their own right (Lawless, 1992).

Insecticidal properties

Many synthetic insecticides have significant negative side effects and are expensive to produce. Efforts to develop alternative more environmentally friendly insect repellents are high on the environmental agenda. Some evidence suggests that basil has powerful insecticidal properties. A number of studies have been carried out in this respect on *Ocimum* spp. One hundred per cent repellence of *O. gratissimum* essential oil (2% in acetone) has been observed against *Musca domestica* (the housefly) (Singh *et al.*, 1985). Another study demonstrated that *O. basilicum* essential oil repelled the red flour beetle, *Tribolium castaneum* (Mohiuddin *et al.*, 1987, cited by Nahak *et al.*, 2011).

Traditional medical uses

O. basilicum has more than 50 medicinal uses, from analgesic to anthelmintic, and is supposed to treat fungal infections, acne, headaches and over 100 such conditions (Duke, 2002, cited by Meyers, 2003). The following are just a small sample of the traditional medicinal uses. The traditional Chinese medicine system involves the use of *O. basilicum* for treatment of gum ulcers, kidney problems and as a haemostyptic in childbirth. In India, it is used for problems as diverse as earache, menstrual irregularities, arthritis, anorexia and malaria (Medical Economics Company, 2000, cited by Meyers, 2003). Rihan (*O. basilicum* in Arabic) is used in treatment of colds, cataract and diarrhoea in northern and central Oman (Ghazanfar and Al Sabahi, 1993). Reyhan (the Persian name) is used to treat urinary tract infection, chest and lung problems, ulcers and influenza, and in Iran, basil is employed as a tonic, appetizer and expectorant (Naghibi *et al.*, 2005, cited in Rivera Núñez *et al.*, 2012). In Jordan, an infusion of basil is considered to be anthelmintic, anti-emetic and antidiarrhoeal (Kirtikar and Basu, 2003, cited by Bilal *et al.*, 2012). In Guinea, the leaves and stems are used to treat fever, neuralgia, catarrh and renal troubles (Kirtikar and Basu, 2003, cited in Bilal *et al.*, 2012). In Ethiopia, the leaves are used against malaria, headache and diarrhoea. In homeopathy, the fresh mature leaves are used to treat blood dysentery, inflammation and congestion of the kidney. The roots and the leaves are used to treat bowel complaints in children (Jayaweera, 1981, cited by Bilal *et al.*, 2012; Kirtikar and Basu, 2003, cited by Bilal *et al.*, 2012). There is a long list of traditional basil remedies in the literature and according to folklore. As is seen from this account, the uses of basil are medically and geographically diverse.

There is considerable interest in studying the properties and benefits of basil. The following paragraphs and Section 3.4.2 (Pharmacological uses) examine a number of the pharmacological characteristics of basil and explore some of the current research.

Basil has been found to show effectiveness against many fungal, viral, bacterial and protozoal infections. Current studies suggest that basil is helpful in inhibiting the growth of carcinogenic cells and in HIV. Basil leaves are used specifically to treat many fevers and coughs, flu, asthma, influenza, bronchitis, colds, chicken pox and diarrhoea, and they can lower the cholesterol level in blood and act as anti-stress agents. Basil juice is an effective medicine for inflamed eyes and night-blindness, which is often caused by vitamin A deficiency (Grieve and Marshall, 1982; Boggia *et al.*, 2015; Hosseini-Parvar*et al.*, 2015).

There are frequent studies on the antifungal activity of *Ocimum* leaves, essential oils and their components and extracts. Fresh ripe tomato fruits were treated before and after inoculation with *Aspergillus niger* in the presence of *Drosophila busckii* by an ethanolic extract of *O. tenuiflorum*. The fruits did not show signs of rotting for 5 to 7 days after this treatment (Sinha and Saxena, 1989, cited by Nahak *et al.*, 2011). The essential oil of *O. canum* was successful against the fungi causing damping-off disease, *Pythium aphanidermatum*, *P. debaryanum* and *Rhizoctonia solani*. *O. canum* gave a 50% reduction in damping-off disease of tomato plants in *P. aphanidermatum*-infected soil and up to 43% reduction in *P. debaryanum*-infected soil. Phytotoxicity of this essential oil was not observed and it was superior to common synthetic fungicides such as captan (Pandey and Dubey, 1992, 1994). *O. basilicum* essential oil displayed antifungal properties against an *Aspergillus flavus* strain producing aflatoxin and against *A. parasiticus*. The fungistatic properties of the oil were observed at a dose of 1.5 ml/1 and the fungicidal properties at 6.0 ml/1. These doses are much lower than those of industrial synthetic fungicides and fumigants, and effect of the oil treatment is not altered by storage, temperature or increased inocula (Nahak *et al.*, 2011). Antimicrobial activity of sweet basil has been found against such organisms as *Lactobacillus acidophilus*, *Saccharomyces cerevisiae*, *Mycoderma* sp., *A. niger* and *Bacillus cereus* (Meena and Vijay, 1994, cited by Hiltunen and Holm, 1999).

3.4.2 Pharmacological uses

Basil oil is known to have strong antioxidant properties. Research has shown the oil contains potent anticancer, antiviral and antimicrobial properties (Tilebeni, 2011). Antioxidants are an important part of maintaining a healthy and balanced lifestyle, and basil may be a very important source of these essential compounds (Baritaux *et al.*, 1992, cited by Tilebeni, 2011). However, despite these reputed properties, it is important to be aware that basil contains estragole, which may be carcinogenic. In Germany, for example, basil is not considered safe for pregnant women or children (Aruna and Sivaramakrishnan, 1992, cited by Meyers, 2003).

There is extensive diversity in the phytochemical constituents of basil; these constituents vary significantly with time, cultivation processes and storage. The nutritional and pharmacological properties of the whole herb in natural form, as it has been traditionally used, results from the interaction of many different active phytochemicals, and consequently, the overall benefits of basil cannot be completely duplicated using single isolated constituents (Tewari *et al.*, 2012). There is very little data relating to a standardized dosage available from traditional practitioners, which is problematic for chemists and pharmacists. This raises the issue that there needs to be a greater communication between traditional and orthodox medicine in order to improve our understanding of the interactions and properties of basil (Tewari *et al.*, 2012).

Research into the medicinal properties and effects of basil has been conducted at various levels. A methanolic extract of *O. basilicum* was assessed for its analgesic activity in mice. Choudary *et al.* (2010, cited by Bilal *et al.*, 2012) demonstrated that analgesic activity at a concentration 200 mg/kg was similar to that of the drug aspirin. Benedec *et al.* (2007) examined the effects of an *O. basilicum* tincture in acute inflammation in male rats and showed a small but considerably important inflammatory effect. *O. basilicum* oil was shown to contain significant anti-ulcer activity against aspirin, alcohol, histamine, serotonin and stress-induced gastric ulceration (Singh *et al.*, 1999, cited by Nahak *et al.*, 2011). Zeggwagh *et al.* (2007, cited by Bilal *et al.*, 2012) studied the hypoglycaemic effect of aqueous extract of *O. basilicum* in normal and diabetic rats. Their results showed the extracts had high anti-hypoglycaemic effects in diabetic rats.

In the last decade or two, an increased methodical interest in the health benefits of plant phytochemicals (in herbs and spices, vegetables and fruits) has gained prominence in the wider study of plant-based nutritional research. Although the study of plant compounds is by no means a new area of research, scientists have only recently started to characterize bioactive compounds in order to explore their effects on human health and disease. In animal and cell culture studies, basil has displayed anti-inflammatory, antidiabetic, antimicrobial, antioxidant and anticancer activity (Arfat *et al.*, 2014).

Use as a prophylactic agent

A decoction of basil leaves is used against hepatic and gastritis disorders. Basil leaf juice is used to treat dysentery, night blindness and conjunctivitis. The essential oils of basil have 100% larvicidal properties. Basil has excellent antimalarial properties and eugenol is the main constituent responsible for its mosquito-repellent properties. Basil leaf paste is effective against ringworm infection and to clear marks on the face. The occurrence of urosolic acid in the leaves helps to remove wrinkles and returns skin elasticity. Basil is highly beneficial in healing wounds, cuts and ulcers, and in removing parasites and worms (Bansod and Rai, 2008). It supplies numerous antioxidants and offers generous reinforcement against free radical-induced damage. Oxygen free radicals are naturally occurring physiological products containing one or more unpaired electrons, and along with reactive oxygen species (ROS), are considered to be harmful to important membrane lipids, proteins, carbohydrates and DNA. This damage has been related to several diseases, for example atherosclerosis, liver cirrhosis, cancer and diabetes, etc. (Chiang *et al.*, 2005; Bansod and Rai, 2008). It has been well accepted that dietary antioxidants have great prospects for curing these disease processes. Antioxidants also enhance the activity of superoxide dismutase (SOD) and reduce lipid peroxidation (Rai *et al.*, 1997; Hannan *et al.*, 2006). Basil antioxidants help in maintaining good health and in preventing the chance occurrence of heart diseases, as well as most of the other degenerative diseases, because oxidative stress is the hallmark of such diseases (Hannan *et al.*, 2006).

Anticancer activity

The anticancer activity of basil has been long established and is mentioned by several investigators (e.g. Karthikeyan *et al.*, 1999; Somkuwar, 2003). Protection against cancer at the cellular level is provided by the unique array of flavonoids that are found in basil. Water-soluble flavonoids of basil, including vicenin and orientin, have been shown to defend cell structures and chromosomes against radiation and oxygen-based damage in human white blood cell studies (Madhuri, 2001). Basil leaf alcoholic extracts have a modulatory impact on carcinogen-metabolizing enzymes such as aryl hydrocarbon hydroxylase and glutathione-*S*-transferase, (GST) and the cytochromes P450 and b5. They are important detoxicants of mutagens and carcinogens. Basil anticancer activity has also been established against human fibrosarcoma cell cultures, in which alcohol extracts induced cytotoxicity at 50 mg/ml and above. Morphologically, the cancer cells showed condensed nuclei and shrunken cytoplasm and the DNA was found to be fragmented on agarose gel electrophoresis. Basil considerably decreased the occurrence of 3′-methyl-4-dimethylaminoazobenzene-induced hepatomas in rats and benzo(α)pyrene-induced neoplasia of the forestomach of mice. An alcohol extract of basil leaves was shown to have an inhibitory effect on chemically induced skin papillomas in mice (Devi, 2001). A leaf extract of basil applied to 7,12-dimethylbenz(a) anthracene (DMBA)-induced papillomas in mice considerably reduced tumour incidence, the average number of papillomas per mouse and the cumulative number of papillomas. Eugenol, a flavonoid present in basil and other plants showed similar activity. Oral treatment with basil fresh leaf paste inhibits the early events of DMBA-induced buccal pouch carcinogenesis. Basil leaf extract suppresses or blocks the events related to chemical carcinogenesis by hindering the metabolic activation of the carcinogen (Aggarwal and Shishodia, 2006).

Radioprotective activity

The flavonoids vicenin and orientin from basil leaves exhibited a greater radioprotective effect than synthetic radioprotectors by protecting human lymphocytes from the clastogenic effects of radiation at low, non-toxic dilutions (Devi *et al.*, 2000). Among three plant extracts, viz. *Withania somnifera* (L.) Dunal, *O. tenuiflorum* and *Plumbago rosea* (preferred name *P. indica* L.), tested on experimental mice for bone marrow survival following 2 Gy γ-radiation, *O. tenuiflorum* water extract exhibited maximum radioprotection as measured by an exogenous spleen colony forming unit (CFU-S) assay (Devi *et al.*, 1998).

Antimicrobial activity

It is the volatile/essential oils of a hydrophobic nature that account for the biochemical actions of spices and herbs (Pandey and Madhuri, 2010). Basil contains many aromatic essential oil compounds that fluctuate in proportion and quality depending on the cultivar. The important aromatic compounds present include linalool, eugenol, citral, methyl chavicol/estragole, limonene and methyl cinnamate. These aromatic compounds defend the herb from insects, bacteria and fungi. In similar fashion, they can help in protecting against diseases caused by fungi, bacteria and insects.

In studies involving cell culture, the essential oils of basil have demonstrated antimicrobial activity by damaging bacterial cell walls and triggering cell lysis. Linalool, methyl chavicol and methyl cinnamate are also very efficient in hindering the development of pathogenic bacteria such as *Escherichia coli*, *Staphylococcus aureus*, *Streptococcus faecalis*, *Shigella* spp., *Mycobacterium* spp., *Salmonella* spp. and *Pseudomonas aeruginosa*. Pathogenic bacteria can cause illnesses such as pneumonia, food poisoning, urinary tract infections and dysentery.

Basil is also a well-recognized insecticidal, antiviral and antifungal agent. Although it has long been used to treat microbial infections, there is not sufficient data to fully support its efficacy and safety in humans. Basil has potent antimicrobial activity against *P. aeruginosa*, *Bacillus pumilus* and *B. megaterium*, *S. aureus* and *S. albus*, *M. tuberculosis*, *Micrococcus pyogenes* var. *aureus*, *Helminthosporium* spp., *H. oryzae*, *Alternaria tenuis*, *A. solani*, *Curvularia* spp. and *C. penniseli*, *Candida guillermondii*, *Pseudomonas* spp., *S. aureus*, *Fusarium solani*, *Colletotricum capsici*, *Arthrobacter globiformis*, *E. coli* and *Vibrio cholerae*. The high concentrations of linolenic acid in basil oil are considered to be largely responsible for its antimicrobial activity (Phadke and Kulkarni, 1989; Mondal *et al.*, 2009).

Anti-inflammatory effects

The active role of acute inflammation is a normal and protective process in helping the body to deal with infections, tissue injury and immune reactions. It is not unexpected that basil has been used for centuries as a traditional method of curing inflammatory disorders. The anti-inflammatory activity of basil is largely credited to the presence of eugenol, which can block the activity of the enzyme cyclooxygenase (COX). Basil extracts diminish inflammation by stopping the release of pro-inflammatory cytokines and mediators (most notably nitric oxide). Cytokines are proteins that are passed from one cell to another that sanction direct cell-to-cell communication (Singh and Majumdar, 1997; Singh, 1998).

Immunomodulatory activity

Steam-distilled basil essential oil altered the humoral immune response in albino rats. This response could be attributed to the discharge of mediators of hypersensitivity reactions, antibody production and tissue responses to these mediators in the target organs. Basil bolsters the immune reaction by improving both cellular and humoral immunity (Mukherjee *et al.*, 2005).

Adaptogenic/anti-stress activity

The immunostimulant capacity of basil accounts for its adaptogenic action. Basil ale enhanced the survival time of swimming mice and prevented milk-induced leucocytosis in mice and stress-induced ulcers in rats. Stress is 'non-specific result of any demand upon the body' and is experienced by every individual. It can be either psychological or physical. Extreme stress is harmful for the body and its immediate treatment is required. Stress is also involved in the pathogenesis of a variety of diseases, including psychiatric disorders such as immunosuppression, depression and anxiety, endocrine disorders such as diabetes mellitus, cognitive dysfunction, male impotence, hypertension, peptic ulcers and ulcerative colitis. Basil has good rejuvenating activity and helps to reduce stress, assist the body by improving memory and relax the mind. The anti-hypoxic effect of basil increases the survival time in anoxic stress. Basil reduced oxidative stress in a study conducted with rabbits (Chattopadhyay *et al.*, 1992).

Antidiabetic activity

One of the most important capabilities of basil found in recent times is its antidiabetic activity. The anti-glycaemic properties of basil have been reported by various researchers but the mechanism of this action has not yet been explained. A study with neem (*Azadirachta indica*) and basil leaves blended together showed that this blend significantly lowered the sugar level in diabetic patients. Basil extract also caused a noticeable drop of blood sugar in normal, streptozotocin (STZ)-induced and glucose-fed hyperglycaemic and diabetic rats. A completely randomized, placebo-controlled, cross-over single blind trial of holy basil leaves in humans showed a noticeable drop in postprandial and fasting blood glucose levels of 7.3 and 17.6%, respectively. A similar trend was noted in urine glucose levels. The aldose reductase activity of basil assists in reducing the complications of diabetes, such as retinopathy, cataract, etc. (Mandal *et al.*, 1993; Halder *et al.*, 2003; Kochhar *et al.*, 2009; Nair *et al.*, 2009).

Antipyretic activity

The antipyretic action of basil fixed oil extracted from the seeds was examined in rats against typhoid-paratyphoid A/B vaccine-induced pyrexia. The intraperitoneal (ip) administration of basil fixed oil significantly decreased the febrile response, thereby demonstrating its antipyretic activity. The oil showed comparable antipyretic activity to aspirin at a dose of 3 ml/kg. Additionally, it has prostaglandin inhibitory activity which can be explained by its antipyretic activity (Singh *et al.*, 2007).

Anti-arthritic activity

Formaldehyde-induced arthritis in rats was studied to evaluate the anti-arthritic activity of basil fixed oil. The fixed oil significantly reduced inflamed paw diameter. There was conspicuous progress in the improvement in the arthritic conditions of rats on ip dosage of the oil for 10 days. The anti-arthritic effect of the basil fixed oil at 3 ml/kg dose was analogous to that of aspirin at 100 mg/kg. The oil inhibited inflammatory mediators (e.g. histamine, serotonin, prostaglandin-2 (PGE-2) and bradykinin) and carrageenan-induced inflammation. The end result suggests possibly useful anti-arthritic properties of basil oil in these inflammation models (Singh and Majumdar, 1996; Singh *et al.*, 2007).

Use in cardiovascular disease

The cholesterol and triglyceride-lowering properties of basil offer potential for inhibiting cardiovascular disease. The combination of LDL (low-density lipoprotein) cholesterol (which clogs blood vessels) and high levels of circulating triglycerides (a fat form in the blood) are risk factors for heart attack, stroke and atherosclerosis. Basil extracts slowed down platelet aggregation and thrombosis, suggesting their potential for stroke prevention and heart attack (Sharma *et al.*, 2001).

Antioxidant activity

The unique health benefits of basil are primarily due to its very high antioxidant content, and the antioxidants (e.g. phytochemicals such as phenolics and vitamins) that it contains contribute to disease prevention. The principal subtype of basil phenolics is its flavonoids, which include orientin and vicenin; and the plant also contains eugenol and anthocyanins. The presence of anthocyanins in purple basils is responsible for their deep red–violet pigmentation. All of the cultivars of purple basils contain very high antioxidant activity due to their anthocyanin content (Khan *et al.*, 2008).

3.5 Summary

Basil (*Ocimum basilicum* L.) is an annual to perennial herb that is grown around the world for use as a food flavouring, in essential oil applications and in traditional medicinal practices. Basil contains mostly methyl chavicol (estragole), eugenol and linalool. The amount of each of these chemical constituents differs depending on the type of species or cultivar and the cultivation, such as soil type, weather, irrigation, pruning and other horticultural practices. Basil is a vital component of several industrial applications, ranging from food to cosmetics to pharmaceuticals. More uses and applications of basil by-products are continuously being added.

Basil is a key ingredient in vinegars, oils, cheeses, jams, teas, drinks and liqueurs. It has an extensive list of traditional medicinal uses. The unique health benefits of basil are primarily due to its very high antioxidant content. *O. basilicum* has been utilized to treat kidney problems, gum ulcers, as a haemostyptic in childbirth and for problems as diverse as malaria, arthritis, anorexia, menstrual irregularities and earache.

Further research on maximizing the yield per hectare and on the optimum preservation and oil extraction methods are needed, particularly in the developing world, where basil leaf and flower harvesting and postharvest processing methods are very traditional.

Acknowledgements

The authors acknowledge Sultan Qaboos University and The Research Council of Oman for partially funding this work.

References

Agarwal, C., Sharma, N.L. and Gaurav, S.S. (2013) An analysis of basil (*Ocimum* sp.) to study the morphological variability. *Indian Journal of Fundamental and Applied Life Sciences* 3, 521–525.

Aggarwal, B.B. and Shishodia, S. (2006) Molecular targets of dietary agents for prevention and therapy of cancer. *Biochemical Pharmacology* 71, 1397–1421.

Al-Maskri, A.Y., Hanif, M.A., Al-Maskari, M.Y., Abraham, A.S., Al-Sabahi, J.N. and Al-Mantheri, O. (2011) Essential oil from *Ocimum basilicum* (Omani Basil): a desert crop. *Natural Product Communications* 6, 1487–1490.

Arfat, Y.A., Benjakul, S., Prodpran, T., Sumpavapol, P. and Songtipya, P. (2014) Properties and antimicrobial activity of fish protein isolate/fish skin gelatin film containing basil leaf essential oil and zinc oxide nanoparticles. *Food Hydrocolloids* 41, 265–273.

Askerova, A., Guseinov, I., Azimov, A., Dmitrieva, N. and Shamsizade, R. (1993) Manufacture of the carbonated fermented milk beverage, Airan. USSR Patent, SU 1796122.

Bansod, S. and Rai, M. (2008) Antifungal activity of essential oils from Indian medicinal plants against human pathogenic *Aspergillus fumigatus* and *A. niger*. *World Journal of Medical Sciences* 3, 81–88.

Benedec, D., Pârvu, A.E., Oniga, I., Toiu, A. and Tiperciuc, B. (2007) Effects of *Ocimum basilicum* L. extract on experimental acute inflammation. *Revista Medico-chirurgicala a Societatii de Medici si Naturalisti din Iasi* 111, 1065–1069.

Bilal, A., Jahan, N., Ahmed, A., Bilal, S.N., Habib, S. and Hajra, S. (2012) Phytochemical and pharmacological studies on *Ocimum basilicum* L. – a review. *International Journal of Current Research and Review* 4(23), 73–83.

Boggia, R., Zunin, P., Hysenaj, V., Bottino, A. and Comite, A. (2015) Dehydration of basil leaves and impact of processing composition. In: Preedy, V. (ed.) *Processing and Impact on Active Components in Food*. Academic Press, San Diego, California, pp. 645–653.

Chattopadhyay, R.R., Sarkar, S.K., Ganguly, S., Medda, C. and Basu, T.K. (1992) Hepatoprotective activity of *O. sanctum* leaf extract against paracetamol induced hepatic damage in rats. *Indian Journal of Pharmacology* 24, 163.

Chiang, L.C., Ng, L.T., Cheng, P.W., Chiang, W. and Lin, C.C. (2005) Antiviral activities of extracts and selected pure constituents of *Ocimum basilicum*. *Clinical and Experimental Pharmacology and Physiology* 32, 811–816.

Darrah, H.H. (1980) *The Cultivated Basils*. T.E. Thomas, Independence, Missouri.

DeBaggio, T. and Belsinger, S. (1996) *Basil: An Herb Lover's Guide*. Interweave Press, Loveland, Colorado.

Denny, E.F.K. (1995) *Field Distillation for Herbaceous Oils*. Denny Mckenzie Associates, Lilydale, Tasmania, Australia.

Devi, P.U. (2001) Radioprotective, anticarcinogenic and antioxidant properties of the Indian holy basil, *Ocimum sanctum* (tulasi). *Indian Journal of Experimental Biology* 39, 185–190.

Devi, P.U., Bisht, K. and Vinitha, M. (1998) A comparative study of radioprotection by *Ocimum* flavonoids and synthetic aminothiol protectors in the mouse. *The British Journal of Radiology* 71, 782–784.

Devi, P.U., Ganasoundari, A., Vrinda, B., Srinivasan, K. and Unnikrishnan, M. (2000) Radiation protection by the *Ocimum* flavonoids orientin and vicenin: mechanisms of action. *Radiation Research* 154, 455–460.

Dymock, W., Warden, C.J.H. and Hooper, D. (2005) *A History of the Principal Drugs of Vegetable Origin. Pharmacographica Indica, Vol. III*. Shrishti Book Distributors, New Delhi, pp. 82–85.

Ghazanfar, S.A. and Al Sabahi, A.M.A. (1993) Medicinal plants of northern and central Oman. *Economic Botany* 47, 89–98.

Ghosh, G.R. (1995) Tulasi (genus *Ocimum*). *New Approaches to Medicine and Health (NAMAH)* 3, 23–29.

Grieve, M.A. (1931) *A Modern Herbal, Vol.* 2. Dover, New York.

Grieve, M. and Marshall, M. (1982) *A Modern Herbal: The Medicinal, Culinary, Cosmetic and Economic Properties, Cultivation and Folk-lore of Herbs, Grasses, Fungi, Shrubs, and Trees with All Their Modern Scientific Uses*. Dover, New York.

Halder, N., Joshi, N. and Gupta, S.K. (2003) Lens aldose reductase inhibiting potential of some indigenous plants. *Journal of Ethnopharmacology* 86, 113–116.

Hanif, M.A., Al-Maskari, M.Y., Al-Maskari, A., Al-Shukaili, A., Al-Maskari, A.Y. and Al-Sabahi, J.N. (2011) Essential oil composition, antimicrobial and antioxidant activities of unexplored Omani basil. *Journal of Medicinal Plants Research* 5, 751–757.

Hannan, J., Marenah, L., Ali, L., Rokeya, B., Flatt, P. and Abdel-Wahab, Y. (2006) *Ocimum sanctum* leaf extracts stimulate insulin secretion from perfused pancreas, isolated islets and clonal pancreatic β-cells. *Journal of Endocrinology* 189, 127–136.

Hiltunen, R. and Holm, Y. (eds) (1999) *Basil: The Genus Ocimum*. Medical and Aromatic Plants – Industrial Profiles Volume 10. Harwood Academic Publishers, Amsterdam.152. Also reprinted in 2003 by Taylor & Francis Group, CRC Press, Abingdon, UK.

Hosseini-Parvar, S.H., Matia-Merino, L. and Golding, M. (2015) Effect of basil seed gum (BSG) on textural, rheological and microstructural properties of model processed cheese. *Food Hydrocolloids* 44, 557–567.

Jacqueline, A.S. (2001) Father Kino's herbs: growing and using them today. Tierra Del Sol Institute Press, Tucson, Arizona.
Karthikeyan, K., Ravichandran, P. and Govindasamy, S. (1999) Chemopreventive effect of *Ocimum sanctum* on DMBA-induced hamster buccal pouch carcinogenesis. *Oral Oncology* 35, 112–119.
Khan, N., Afaq, F. and Mukhtar, H. (2008) Cancer chemoprevention through dietary antioxidants: progress and promise. *Antioxidants and Redox Signaling* 10, 475–510.
Kochhar, A., Sharma, N. and Sachdeva, R. (2009) Effect of supplementation of tulsi (*Ocimum sanctum*) and neem (*Azadirachta indica*) leaf powder on diabetic symptoms, anthropometric parameters and blood pressure of non-insulin dependent male diabetics. *Studies on Ethno-Medicine* 3, 5–9.
Lawless, J. (1992) *The Encyclopedia of Essential Oils*. Element Books, Rockport, Massachusetts.
Madhuri, S. (2001) Studies on oestrogen induced uterine and ovarian carcinogenesis and effect of ProImmu in rats. PhD thesis, Rani Durgavati Vishwa Vidyalaya, Jabalpur, India.
Malik, M.S., Rafique, M., Sattar, A. and Khan, S.A. (1987) The fatty acids of indigenous resources for possible industrial applications. Part XII: the fatty acid composition of the fixed oils of *Ocimum sanctum* and *Salvia aegyptica* seeds. *Pakistan Journal of Scientific and Industrial Research* 30, 369–371.
Malik, M.S., Sattar, A. and Khan, S.A. (1989) The fatty acids of indigenous resources from possible industrial applications. Part XVII: the fatty acid composition of the fixed oils of *Ocimum basilicum* and *Ocimum album* seeds. *Pakistan Journal of Scientific and Industrial Research* 32, 207–208.
Mandal, S., Das, D.N., Kamala, D., Ray, K., Roy, G., Chaudhari, S.B. and Sahana, C.C. (1993) *Ocimum sanctum* Linn. – a study on gastric ulceration and gastric secretion in rats. *Indian Journal of Physiology and Pharmacology* 37, 91–92.
Meyers, M. (ed.) (2003) *Basil: An Herb Society of America Guide*. The Herb Society of America, Kirtland, Ohio. Available at: http://www.herbsociety.org/factsheets/Basil%20Guide.pdf (accessed 14 July 2015).
Miele, M., Dondero, R., Ciarallo, G. and Mazzei, M. (2001) Methyleugenol in *Ocimum basilicum* L. Cv. 'Genovese Gigante'. *Journal of Agricultural and Food Chemistry* 49, 517–521.
Mondal, S., Mirdha, B.R. and Mahapatra, S.C. (2009) The science behind sacredness of tulsi (*Ocimum sanctum* Linn.). *Indian Journal of Physiology and Pharmacology* 53, 291–306.
Mukherjee, R., Das, P.K. and Ram, G.C. (2005) Immunotherapeutic potential of *Ocimum sanctum* Linn. bovine subclinical mastitis. *Research in Veterinary Science* 79, 37–43.
Nahak, G., Mishra, R.C. and Sahu, R.K. (2011) Taxonomic distribution, medicinal properties and drug development potentiality of *Ocimum* (tulsi). *Drug Invention Today* 3(6), 95–113.
Nair, V.D., Jaleel, C.A., Gopi, R., Gomathinayagam, M. and Panneerselvam, R. (2009) Antioxidant potential of *Ocimum sanctum* under growth regulator treatments. *EurAsian Journal of BioSciences* 3, 1–9.
Pandey, V.N. and Dubey, N.K. (1992) Effect of essential oils from some higher plants against fungi causing damping-off disease. *Biologia Plantarum* 34, 143–147.
Pandey, V.N. and Dubey, N.K. (1994) Antifungal potential of leaves and essential oils from higher plants against soil phytopathogens. *Soil Biology and Biochemistry* 26, 1417–1421.
Pandey, G. and Madhuri, S. (2010) Pharmacological activities of *Ocimum sanctum* (tulsi): a review. *International Journal of Pharmaceutical Science: Reviews and Research* 5, 61–66.
Paton, A. (1992) A synopsis of *Ocimum* L. (Labiatae) in Africa. *Kew Bulletin* 47, 403–435.
Phadke, S. and Kulkarni, S. (1989) Screening of *in vitro* antibacterial activity of *Terminalia chebula, Eclapta alba* and *Ocimum sanctum*. *Indian Journal of Medical Sciences* 43, 113–117.
Pushpangadan, P. and George, V. (2012) Basil. In: Peter, K.V. (ed.) *Handbook of Herbs and Spices*, 2nd edn. Woodhead Publishing, Cambridge, UK, pp. 55–72.
Putievsky, E. and Galambosi, B. (1999) Production systems of sweet basil. In: Hiltunen, R. and Holm, Y. (eds) *Basil: The Genus* Ocimum, Medical and Aromatic Plants – Industrial Profiles, Volume 10. Harwood Academic Publishers, Amsterdam, pp. 39–65.
Rai, V., Iyer, U. and Mani, U. (1997) Effect of tulasi (*Ocimum sanctum*) leaf powder supplementation on blood sugar levels, serum lipids and tissues lipids in diabetic rats. *Plant Foods for Human Nutrition* 50, 9–16.
Rivera Núñez, D., Matilla Séiquer, G., Obón, C. and Alcaraz Arixa, F. (2012) *Plants and Humans in the Near East and the Caucasus. Ancient and Traditional Uses of Plants as Foods and Medicine. An Ethnobotanical Diachronic Review. Vol. 1: The Landscapes, The Plants: Ferns and Gymnosperms*. Ediciones de la Universidad de Murcia, Murcia, Spain.

Sharma, M., Kishore, K., Gupta, S.K., Joshi, S. and Arya, D.S. (2001) Cardioprotective potential of *Ocimum sanctum* in isoproterenol induced myocardial infarction in rats. *Molecular and Cellular Biochemistry* 225, 75–83.

Simon, J.E. (1995) Basil. Newcrop factsheet. Purdue University, Centre for New Crops and Plant Products, West Lafayette, Indiana. Available at: http://www.hort.purdue.edu/newcrop/CropFactSheets/basil.html (accessed 29 May 2003).

Simon, J.E., Morales, M.R., Phippen, W.B., Vieira, R.F. and Hao, Z. (1999) Basil: a source of aroma compounds and a popular culinary and ornamental herb. In: Janick, J. (ed.) *Perspectives on New Crops and New Uses.* ASHS Press, Alexandria, Virginia, pp. 499–505. Available at: https://hort.purdue.edu/newcrop/proceedings1999/v4-499.html (accessed 5 February 2016).

Singh, S. (1998) Comparative evaluation of antiinflammatory potential of fixed oil of different species of *Ocimum* and its possible mechanism of action. *Indian Journal of Experimental Biology* 36, 1028–1031.

Singh, S. (1999) Evaluation of gastric anti-ulcer activity of fixed oil of *Ocimum basilicum* Linn. and its possible mechanism of action. *Indian Journal of Experimental Biology* 37, 253–257.

Singh, S. and Majumdar, D.K. (1996) Effect of fixed oil of *Ocimum sanctum* against experimentally induced arthritis and joint edema in laboratory animals. *International Journal of Pharmaceutics* 34, 218–222.

Singh, S. and Majumdar, D.K. (1997) Evaluation of antiinflammatory activity of fatty acids of *Ocimum sanctum* fixed oil. *Indian Journal of Experimental Biology* 35, 380–383.

Singh, S., Taneja, M. and Majumdar, D.K. (2007) Biological activities of *Ocimum sanctum* L. fixed oil – an overview. *Indian Journal of Experimental Biology* 45, 403–412.

Somkuwar, A.P. (2003) Studies on anticancer effects of *Ocimum sanctum* and *Withania somnifera* on experimentally induced cancer in mice. PhD thesis, Jawaharlal Nehru Krishi Viswavidyalaya, Jabalpur, India.

Srivastava, A.K. (1980) *French Basil and Its Cultivation in India.* Farm Bulletin No. 16. Central Institute Medicinal and Aromatic Plants, Lucknow, India.

Svecova, E. and Neugebauerova, J. (2010) A study of 34 cultivars of basil (*Ocimum* L.) and their morphological, economic and biochemical characteristics, using standardized descriptors. *Acta Universitatis Sapientiae, Alimentaria* 3, 118–135.

Tewari, D., Pandey, H.K., Sah, A.N., Meena, H.S., Manchanda, A. and Patni, P. (2012) Pharmacognostical, biochemical and elemental investigation of *Ocimum basilicum* plants available in western Himalayas. *International Journal of Research in Pharmaceutical and Biomedical Sciences* 3, 840–845.

Tilebeni, H.G. (2011) Review to basil medicinal plant. *International Journal of Agronomy and Plant Production* 2, 5–9.

Tucker, A.O. and DeBaggio, T. (2000) *The Big Book of Herbs: A Comprehensive Illustrated Reference to Herbs of Flavor and Fragrance.* Interweave Press, Loveland, Colorado.

Wijesekera, R.O.B. (1986) *Practical Manual on the Essential Oils Industry.* United Nations Industrial Development Organization, Vienna.

4 Bay

Hülya Çakmak,* Seher Kumcuoğlu and Şebnem Tavman
Ege University, Bornova, Izmir, Turkey

4.1 Botany

4.1.1 Introduction

The bay, *Laurus nobilis* L., is indigenous to Asia Minor and the Mediterranean basin, and its dried aromatic leaves are used as a culinary herb. The plant is also used medicinally, especially in traditional folk medicine. Cultivated bay plants are pruned to different topiary shapes and used as ornamental plants in temperate zones. The species is categorized as a non-timber (secondary) forest product.

L. nobilis is a perennial, evergreen tree or shrub and is known as sweet bay, true bay, bay laurel, Turkish laurel, Roman laurel, sweet laurel or true laurel (Kumar *et al.*, 2001; Charles, 2013; FAO, 2014). Common names of bay in other languages are: *ghar* (Arabic), *yueh kuei* (Chinese), *feuille de laurier* (French), *Lorbeerblatt* (German), *daphni* (Greek), *foglia di alloro* (Italian), *gekkeiju* (Japanese), *louro* (Portuguese), *lawr* (Russian), *hojas de laurel* (Spanish) and *defne yaprağı* (Turkish) (Seidemann, 2005; Raghavan, 2007). Indian bay (*Cinnamomum tamala* (Buch.-Ham.) Th. G.G. Nees), Californian bay (*Umbellularia californica* (Hook. & Arn.) Nutt.), West Indian bay (*Pimenta racemosa* (Mill.) J.W. Moore), Indonesian bay (*Eugenia polyantha*, preferred name *Syzygium polyanthum* (Wight) Walp.) and cherry laurel (*Prunus laurocerasus* L.) are other commercially available herbal 'bay' plants and can be confused with true *L. nobilis* (Buttery *et al.*, 1974; Kumar *et al.*, 2001; Weiss, 2002; Raghavan, 2007).

Dried bay leaves are a popular herb that is widely used in pickles, soup, meat and fish dishes, casseroles and sauces; it is also used in spice mixes such as bouquet garni (Akgul, 1993; Raghavan, 2007; Charles, 2013). Apart from the food industry, the essential and fixed oils of the leaves and fruits are utilized in soap, perfumery, cream, cosmetics, pharmaceutical formulations and candle making (Acar, 1988; Akgul, 1993; Elzebroek and Wind, 2008).

Turkey is the third largest herb and spice producer after India and Bangladesh (FAO STAT, 2014), and bay leaf provides one of highest sources of total export revenues from herbs and spices after thyme, cumin, sage and nigella (black cumin, *Nigella sativa* L.) (BAKA, 2012). Some 90% of high-quality dried bay leaf sales throughout the world are supplied by Turkey (BAKA, 2012), although the domestic consumption is quite high as well. According to the export statistics of the Aegean

*Corresponding author, e-mail: hulya.cakmak@ege.edu.tr

© CAB International 2016. *Leafy Medicinal Herbs: Botany, Chemistry, Postharvest Technology and Uses* (eds D.C.P. Ambrose *et al.*)

Exporters Association (AEA), the total dried bay leaf export of Turkey is gradually increasing by year, as shown in Fig. 4.1 (AEA, 2014). The biggest importers of bay leaf are Vietnam, the USA, Russia, Poland and Germany. Besides export of the bay leaf as a culinary herb, the essential oil of the leaves and fruits is also exported, although the statistics for this are not included in Fig. 4.1.

4.1.2 History/origin

In the plant kingdom, Mediterranean bay (*L. nobilis*) is in the division Magnoliophyta, class Magnoliopsida, order Laurales, family Lauraceae, subfamily Lauroideae, tribe Litseae, subtribe Laurineae and genus *Laurus* (Acar, 1987; Ceylan, 1997; Weiss, 2002; Seidemann, 2005; FAO, 2014). The Lauraceae family includes 50 genera and 2000 species of trees and shrubs (Wiart, 2006). The genus *Laurus* includes two varieties of *L. nobilis*, *L. nobilis* itself and *Laurus nobilis* var. *angustifolia* (Acar, 1987). Bay belongs to the same family as avocado, sassafras and cinnamon.

In Greek mythology, Daphne, who is escaping from the love of the Greek God Apollo, prayed to the river God that she should disappear and she was turned into a sacred laurel tree. Bay leaves are therefore symbols of immortality, grace, victory, peace, prosperity, merit, nobility and glory in Ancient Greek and Roman mythology, and even today. Because of this, heroes and wise men were rewarded with wreaths made from bay leaves and branches in ancient times (Elzebroek and Wind, 2008; Charles, 2013). The winners of the Pythian Games (the forerunner of the Olympic Games) were honoured with a laurel wreath in the memory of Apollo (Raghavan, 2007). The meaning of 'Laurus' is 'crown of laurel' and 'triumph'. Bay leaves are also believed to protect from lightning strikes (Weiss, 2002). Government bodies in Turkey, such as the Turkish Armed Forces, the Gendarmerie General Command, Ministry of Education, Turkish Police Department, High Council of Judges and Prosecutors, Ministry of Youth and Sports and some of the municipalities use a bay leaf or laurel wreath as an emblem to represent the superior attributes of those bodies.

4.1.3 Location

The motherland of bay is Asia Minor and the Balkan Peninsula (Acar, 1987). In ancient

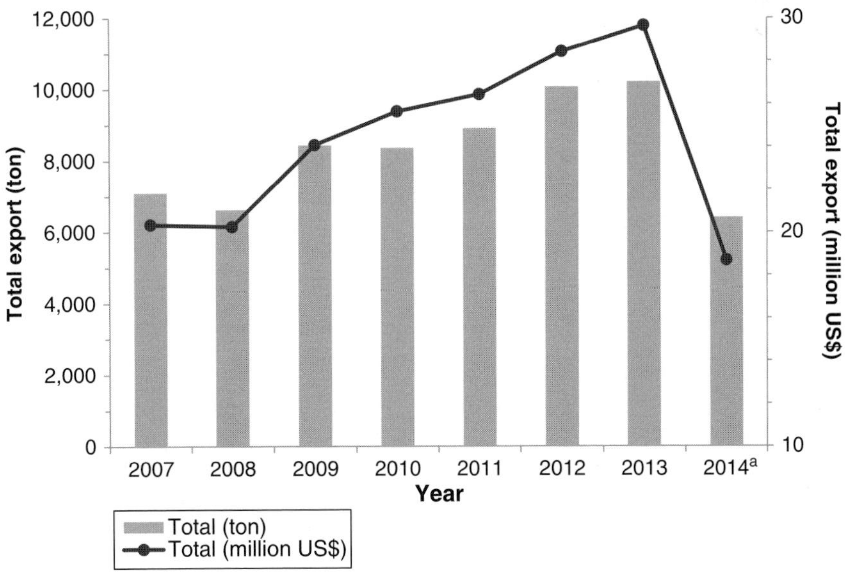

Fig. 4.1. Turkey's total dried bay leaf export; [a]2014 figures for January–July only. From AEA, 2014.

times, the plant spread to the whole of the Mediterranean coast. Today, wild plants grow in the Mediterranean basin, Asia Minor (Anatolia), the Balkan Peninsula and western, middle and southern parts of Europe, and the coast of the Black Sea in Russia where the climate is quite favourable for maquis vegetation (Acar, 1987). Wild plants have adapted to the climate of the whole coastline of Turkey, namely the Black Sea, Aegean Sea and the Mediterranean Sea. Plants are also cultivated in Turkey, Greece, France, Spain, Italy, Portugal, China, Israel, Asia Minor, Central America and Arabic countries from Libya to Morocco (Kumar *et al.*, 2001; Board, 2010; Charles, 2013). According to the classification of European Spice Association (ESA), the bay tree grows in temperate climate zones. The climate biome classification for bay is tropical wet, tropical wet and dry, steppe or semiarid, subtropical dry summer, temperate oceanic, temperate continental and temperate with dry winters (UNIDO and FAO, 2005). A short summer drought period with sufficient temperature and rainfall promotes vegetative production and the leaves become bigger, thinner and tender, while in regions that have a longer summer drought, plant growth reaches a standstill and the leaves become smaller, thicker and tougher (Acar, 1988).

The optimum growing conditions for the west Anatolian bay plant are given as: from sea level to 600 m altitude, sometimes up to 1000 m in the south; located on a plain, at the bottom of a hill, by the mouth of a stream or by a hill stream; shade exposure to the east, north, north-west or north-east; 600 mm to 2000 mm annual rainfall; warm temperatures (~15°C), but able to withstand several degrees of frost; sites where the soils were formed in alluvium and colluvium derived from limestone and mica schist; and soil texture moderate (loam) with the pH neutral or slightly alkaline (Erden, 2005; Elzebroek and Wind, 2008; EFRI, 2014). Bay plants can be propagated from seed, by cuttings or by layering (Elzebroek and Wind, 2008). However, propagation from seed is utilized as it provides greater harvest efficiency (Acar, 1987). The plant shows periodicity in seeding, and the seeds exhibit double dormancy because of the characteristics of both the seed coat and embryo (EFRI, 2014). The pericarp is scarified and removed from the seed to aid further germination.

Wild or cultivated bay can sometimes be affected by damage caused by the following insects: *Trioza alacris*, *Aonidia lauri*, *Dynaspidiotus britannicus*, *Ceroplastes japonicus*, *C. rubens*, *Pulvinaria regali*, *Cacoecimorpha pronubana*, *Archips rosanus* and *Otiorhynchus ovalipennis* (Weiss, 2002; EFRI, 2014). Chemical pesticides may reduce the harm of these insects, although pesticide residues can remain and result in a decrease in the yield of essential oil.

4.1.4 Morphology

The bay is a dioecious leafy evergreen tree or shrub up to 10 m high for wild plants and up to 3 m for cultivated plants (Spratt, 1830; Acar, 1987; Akgul, 1993; Weiss, 2002; Elzebroek and Wind, 2008). Cultivated plants, owing to their shorter height, can be ranked as shrubs, but wild plants can be ranked as trees (Spratt, 1830). The stem is branched, covered with shiny, smooth, olive green or dark brown bark (Spratt, 1830; Elzebroek and Wind, 2008; Parthasarathy *et al.*, 2008). The plant has a great ability to generate shoots from the roots and stools (Acar, 1987). The flowers are yellow and white and appear in clusters; most of the male flowers have 10–12 stamens, while the female flowers have four non-functional staminoids (Acar, 1987). The small, oval-shaped greenish fruits turn deep purple or black after ripening (Spratt, 1830; Acar, 1987; Akgul, 1993; Kumar *et al.*, 2001). The fruits are 1–1.5 cm in diameter; 28% of the fruit weight is flesh and 72% seed (Baytop, 1984; Acar, 1987; Erden, 2005).The leaves are thick, firm and leathery with short stalks, lanceolate or lanceolate oblong in shape with an acuminate leaf tip, rounded leaf base; they are about 5–12 cm long and 2–5 cm wide, with entire wavy margins. The upper surface of the leaf is glossy, yellowish/olive green to brown and the lower surface is pale olive to brown with pinnate venation, as can be seen in Fig. 4.2 (Baytop, 1984; Akgul, 1993; Kumar *et al.*, 2001; Raghavan, 2007; Elzebroek and Wind, 2008).

Fig. 4.2. A cultivated bay leaf plant in the yard of Ege University campus in Turkey.

4.2 Chemistry

4.2.1 Chemical composition

The major and minor nutrients of dried bay leaf are listed in Table 4.1. Apart from the listed nutrients, dried leaves also include essential oils, tannins, resin and mucilage (Baytop, 1984; Akgul, 1993). The fresh leaves have about 50% moisture content (Çakmak *et al.*, 2013) and the dried leaves have a moisture content of 5–10% (Elzebroek and Wind, 2008). Wild and cultivated bay leaves show a different nutrient composition. For example, the total sugar content of wild samples was significantly higher ($p < 0.05$) than that of cultivated samples (Dias *et al.*, 2014). However, as far as the total carbohydrate level is concerned, the wild and cultivated samples were statistically in the same group. The lipid

Table 4.1. Nutrient contents of bay leaf. From USDA, 2014.

Nutrients	Unit	Value/100 g portion
Proximate analysis		
Energy	kcal	313
Fibre, total dietary	g	26.30
Protein	g	7.61
Total lipid (fat)	g	8.36
Water	g	5.44
Carbohydrate (by difference)	g	74.97
Lipids		
Cholesterol	mg	0
Fatty acids, total saturated	g	2.28
Fatty acids, total monounsaturated	g	1.64
Fatty acids, total polyunsaturated	g	2.29
Minerals		
Calcium	mg	834
Iron	mg	43
Magnesium	mg	120
Phosphorus	mg	113
Potassium	mg	529
Sodium	mg	23
Zinc	mg	3.70
Vitamins		
Folate	μg	180
Niacin	mg	2.005
Riboflavin	mg	0.421
Thiamine	mg	0.009
Vitamin A, IU	IU	6185
Vitamin A, RAE	μg	309
Vitamin B_6	mg	1.740
Vitamin C (total ascorbic acid)	mg	46.5

(also known as fixed oil) content and composition of bay leaves and fruits are important, and the fruit oil has an important economic value, especially in soap making. The fruits of the bay plant contain 25–30% fixed oil (Acar, 1987; Akgul, 1993; Çelik and Yılmaz, 1996; Erden, 2005). The main fatty acids of the fixed oil of commercial fruits (from Turkey) are oleic, linoleic, palmitic and lauric acids (Çelik and Yılmaz, 1996). From a similar place of harvest, Ozcan *et al.* (2010) determined that linoleic, lauric and palmitic acids are the main components of the fruit oil. For plants of Tunisian origin, lauric, oleic, linoleic and palmitic acid had the highest contents in the fruit oil (Marzouki *et al.*, 2008). Hafizoğlu and Reunanen (1993) determined similar dominant components in fruit oil of Turkish origin; analysis by GC-MS (gas chromatography–mass spectrometry) demonstrated 54.2% lauric acid, 17.2% linoleic acid and 15.1% oleic acid. The fixed oil content of leaves is somewhat lower than the fruits (EFRI, 2014). The fixed oil extracted from the leaves from Turkish origin includes linolenic, palmitic, linoleic and oleic acids (Çelik and Yılmaz, 1996). Dias *et al.* (2014) determined 24 different fatty acids in both wild and cultivated plants. Linolenic acid was the major component of wild samples, whereas the palmitic acid content was higher in cultivated plants.

The major active compounds of the plant are present in the essential (volatile) oil, which is extracted from the leaves, and the essential oil content of leaves is 0.8–4.0% (Baytop, 1984; Ceylan, 1997; Raghavan, 2007). The extraction yield and the composition of essential oil can be affected by several factors, such as: the place of harvest (edaphic, climatic and genetic differences), plant type (wild or cultivated), time/season of harvest, the age of shoots, the extraction and drying method and temperature (Acar, 1988; Diaz-Maroto *et al.*, 2002; Kilic *et al.*, 2004; Erden, 2005; Zekovic *et al.*, 2009; Kandi and Sefidkon, 2011; Sellami *et al.*, 2011; Özek, 2012). Acar (1988) determined that the lowest essential oil content was obtained from leaves harvested in June, and the highest in August, for all of the plants located in the different geographical regions of Turkey. Lira *et al.* (2009) did not find any significant difference in the composition of essential oil in leaves from Argentina, although the total amount was the lowest in the December harvest. Drying temperature and type of drier can influence the composition and amount of essential oil and other bioactive compounds in the leaves (Diaz-Maroto *et al.*, 2002; Erden, 2005; Kandi and Sefidkon, 2011; Sellami *et al.*, 2011). Depending on the drying time, significant changes were obtained in the overall essential oils (Diaz-Maroto *et al.*, 2002); the composition of the essential oil extracted from leaf samples dried in ambient air or shade (at 22–25°C) gave the closest results to the fresh samples; the losses from this drying method were minimal compared those from oven, infrared (IR) and microwave drying (Diaz-Maroto *et al.*, 2002; Kandi and Sefidkon, 2011; Sellami *et al.*, 2011). Sellami *et al.* (2011) concluded from their study that drying of bay leaves resulted in the losses of some essential oils together with the appearance of new substances that were not found in fresh samples.

Hydrocarbons, alcohols, esters, phenol and phenol ethers, and terpene oxides are the main chemical constituents identified in bay leaf oil (Acar, 1987). The monoterpenoid 1,8-cineole (25–63%), which is also known as eucalyptol, comprises the major component of the essential oil; however, α-terpinyl acetate, sabinene, α-pinene, linalool, β-pinene, α-terpineol, 4-terpineol, methyl eugenol, eugenol and limonene can also be found at various concentrations depending on the aforementioned factors (Acar, 1987; Kumar *et al.*, 2001; Raghavan, 2007; Board, 2010). The six major essential oil components found in bay leaves according to their geographic origin are listed in Table 4.2, and the chemical structure of the five major essential oil components is shown in Fig. 4.3.

Fruits, flowers, buds and even stems of the bay plant, along with the leaves, have been investigated for their volatile compounds and essential oil composition. Viridiflorene (Fiorini *et al.*, 1997), α-eudesmol (Kilic *et al.*, 2004) and 1,8-cineole (Moghtader and Salari, 2012) are the major volatiles in flowers of bay that have been described in the recent literature. For the fruits, (*E*)-β-ocimene is the major compound (Kilic *et al.*, 2004; Marzouki *et al.*, 2008). For the buds and stem bark,

Table 4.2. Major compounds in essential oil of bay leaves from different plant origins, in decreasing order of content.

Origin	Compound in decreasing order of content from columns 1 to 6						Reference
	1	2	3	4	5	6	
Algeria	1,8-cineole	linalool	camphene	isovaleraldehyde	β-phellandrene	α-pinene	Jemaa *et al.*, 2012
Argentina	1,8-cineole	linalool	sabinene	α-terpinyl acetate	α-pinene	β-pinene	Lira *et al.*, 2009
Croatia	1,8-cineole	methyl eugenol	α-terpinyl acetate	linalool	sabinene	eugenol	Politeo *et al.*, 2007
Croatia	1,8-cineole	α-terpinyl acetate	methyl eugenol	sabinene	α-terpineol	α-pinene	Zekovic *et al.*, 2009
Northern Cyprus	1,8-cineole	α-terpinyl acetate	4-terpineol	α-pinene	sabinene	β-pinene	Yalçın *et al.*, 2007
France	1,8-cineole	α-terpinyl acetate	methyl eugenol	linalool	sabinene	limonene	Fiorini *et al.*, 1998
Iran	1,8-cineole	α-terpinyl acetate	sabinene	β-pinene	α-pinene	α-terpineol	Kandi and Sefidkon, 2011
Iran	1,8-cineole	sabinene	α-pinene	β-pinene	camphene	α-terpineol	Moghtader and Salari, 2012
Italy	1,8-cineole	α-terpinyl acetate	linalool	methyl eugenol	sabinene	α-terpineol	Caredda *et al.*, 2002
Morocco	1,8-cineole	isovaleraldehyde	linalool	α-terpineol	2-carene	α-pinene	Jemaa *et al.*, 2012
Morocco	1,8-cineole	2-carene	*trans*-ocimene	sabinene	*cis*-ocimene	*trans*-caryophyllene	Cherrat *et al.*, 2014
Portugal	1,8-cineole	α-terpinyl acetate	linalool	methyl eugenol	sabinene	carvacrol	Ramos *et al.*, 2012
Spain	1,8-cineole	linalool	α-terpinyl acetate	sabinene	methyl eugenol	α-pinene	Diaz-Maroto *et al.*, 2002
Tunisia	1,8-cineole	α-terpinyl acetate	α-pinene	β-pinene	sabinene	methyl eugenol	Bouzouita *et al.*, 2001
Tunisia	1,8-cineole	methyl eugenol	eugenol	4-terpineol	sabinene	linalool	Sellami *et al.*, 2011
Tunisia	1,8-cineole	linalool	eugenyl methyl ether	isovaleraldehyde	camphene	β-phellandrene	Jemaa *et al.*, 2012
Turkey	1,8-cineole	sabinene	α-terpinyl acetate	α-pinene	β-pinene	α-terpineol	Kilic *et al.*, 2004
Turkey	1,8-cineole	α-terpinyl acetate	sabinene	4-terpineol	β-pinene	α-pinene	Bayramoglu *et al.*, 2009
Turkey	1,8-cineole	α-terpinyl acetate	sabinene	α-pinene	β-pinene	4-terpineol	Özek, 2012
Turkey	1,8-cineole	α-terpinyl acetate	sabinene	α-pinene	4-terpineol	β-pinene	El *et al.*, 2014

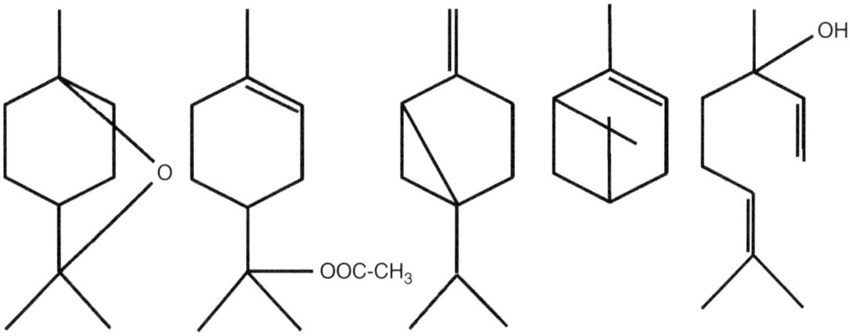

Fig. 4.3. Major components of bay leaf essential oil. Chemical formulae from Akgul, 1993.

1,8-cineole is again the major component of the volatiles (Fiorini *et al.*, 1997; Kilic *et al.*, 2004). The main components in the wood of the stems are α-terpinyl acetate, methyl eugenol and α-copaene (Fiorini *et al.*, 1997).

4.2.2 Phytochemistry

Antioxidant activity

Several studies have been conducted on the bioactive compounds of bay leaves, fruits, roots, branches, bark, leaf extracts and leaf/fruit essential oils. These physiologically active chemicals have antioxidant and antimicrobial activity. Phenolic compounds, especially flavonoids and tocopherols, are natural antioxidants that are found in bay tissue. These antioxidant compounds are extracted by various methods, such as extraction with chemicals (chloroform, ethanol, ether, ethyl acetate, hexane, methanol, *n*-butanol), hydrodistillation (steam distillation), infusion and decoction (Demo *et al.*, 1998; Kang *et al.*, 2002; Conforti *et al.*, 2006; Elmastaş *et al.*, 2006; Politeo *et al.*, 2007; Kaurinovic *et al.*, 2010; Ouchikh *et al.*, 2011; Albayrak *et al.*, 2012; Ramos *et al.*, 2012; Muñiz-Márquez *et al.*, 2013; Cherrat *et al.*, 2014; Dias *et al.*, 2014; El *et al.*, 2014).

Antioxidant activities of the extracts are determined according to the scavenging of DPPH (2,2-diphenyl-1-picrylhydrazyl), ABTS (2,2′-azino-bis(3-ethylbenzothiazoline-6-sulfonic acid) and superoxide anion radicals, and by ferric reducing ability (FRAP). Other methods include the β-carotene bleaching assay, phosphomolybdenum assay and linoleic acid peroxidation (lipid oxidation). The DPPH scavenging activity of leaf extracts can be expressed as the IC_{50} (half maximal inhibitory concentration), EC_{50} (half maximal effective concentration) or inhibition percentage values. Albayrak *et al.* (2012) reported in their study that decoction extracts had the highest scavenging activity (IC_{50} = 25.05 µg/ml), while methanol extracts had moderate activity (IC_{50} = 30.05 µg/ml) and infusion extracts had the lowest activity (IC_{50} = 47.62 µg/ml). Conforti *et al.* (2006) found a little higher radical scavenging activity for ethanolic extracts of wild bay leaves (IC_{50} = 22 ± 0.531 µg/ml), while ethanolic extracts of cultivated samples had an IC_{50} of 29 ± 0.634 µg/ml. Higher contents of eugenol, methyl eugenol and vitamin E contents were associated with the higher antioxidant capacity of wild plants by the authors. In a recent study, methanolic extracts of the cultivated sample (EC_{50} = 150 µg/ml) had higher DPPH scavenging activity than that of the wild sample (EC_{50} = 200 µg/ml) (Dias *et al.*, 2014).

A clear picture of the effect of different solvents used on the DPPH radical scavenging activity of the extracts was presented by Kaurinovic *et al.* (2010). When ether, chloroform, ethyl acetate, *n*-butanol and water were used as solvents, the highest DPPH radical scavenging activities were obtained for the ethyl acetate extracts (IC_{50} = 83.24 µg/ml).

The ethyl acetate extracts had also showed the highest scavenging activity of the superoxide anion and nitric oxide. Depending on the extraction method and time, DPPH inhibition by the essential oil of bay leaf was found to be 83–91.1% (Cherrat *et al.*, 2014; El *et al.*, 2014). Comparably higher DPPH inhibition was obtained from ultrasound-assisted extracts of leaves (94.73 ± 0.49%), while the inhibition of lipid oxidation was 73.55 ± 1.91% (Muñiz-Márquez *et al.*, 2013). A 98.6% inhibition of the lipid peroxidation of linoleic acid emulsion was detected for the maximum concentration (60 μg/ml) of the ethanolic extract of leaves (Elmastaş *et al.*, 2006).

The antioxidant activity of free volatile aglycones derived from bay leaf was evaluated by the FRAP and DPPH assays, but these substances showed lower antioxidant activity than bay essential oil and synthetic antioxidants (Politeo *et al.*, 2007). In the β-carotene bleaching assay, both the methanolic and ethanolic extracts of wild bay leaf samples showed higher antioxidant activity than that of cultivated samples (Conforti *et al.*, 2006; Dias *et al.*, 2014).

The total amounts of phenolic compounds in bay leaves were found to be 12–84.5 mg GAE (gallic acid equivalents)/g (Elmastaş *et al.*, 2006; Muchuweti *et al.*, 2007; Ouchikh *et al.*, 2011; Albayrak *et al.*, 2012). The tannin composition of bay leaves was around 10 mg GAE/g (Muchuweti *et al.*, 2007). As well as the total phenolics, flavonoids and proanthocyanidins are other bioactive compounds that directly affect the total antioxidant capacity of bay leaf (Ouchikh *et al.*, 2011). Interestingly, these compounds were found in higher amounts in the roots than the leaves. In the same study, antioxidant activity was measured by the phosphomolybdenum assay, ferrous ion chelating activity and β-carotene bleaching assay, but the results of these assays did not give the same results; instead, they showed that the extracts of the roots had lower antioxidant activity than the leaves.

Another group of chemicals with antioxidant activity are phenolic acids; so far, vanillic acid, caffeic acid, ferulic acid, *p*-coumaric and 2-hydroxycinnamic acid have been identified in bay leaves (Muchuweti *et al.*, 2007; Muñiz-Márquez *et al.*, 2013). Flavonol derivatives are also present as part of the bay leaf flavonoid compounds; those that have been detected are quercetin, isorhamnetin and kaempferol derivatives (Fiorini *et al.*, 1998; Dias *et al.*, 2014). The total flavonols and total flavones of cultivated bay leaves were found to be significantly higher than those of wild samples in either methanolic extracts or infusions (Dias *et al.*, 2014).

Tocopherols are terpenoids and four different tocopherol isomers (α, β, γ and δ) contribute to vitamin E activity beyond showing antioxidant activity. The α-tocopherol content of bay leaves was found to be 49.69–139.34 mg/100 g fresh leaf (Gómez-Coronado *et al.*, 2004; Ouchikh *et al.*, 2011). The γ-tocopherol content was 0–1.15 mg/100 g fresh leaf (Gómez-Coronado *et al.*, 2004; Ouchikh *et al.*, 2011). When the tocopherol extraction methods used were compared, the amounts of both α- and γ-tocopherol were found to be significantly higher ($p < 0.05$) by the micro-scale saponification method (Ouchikh *et al.*, 2011). Not only the leaves but also the branches and roots of bay are a source of α and γ-tocopherol (Ouchikh *et al.*, 2011). Branches had comparably higher values than roots, and once again the micro-scale saponification method gave better results (Ouchikh *et al.*, 2011). The total tocopherol and α-tocopherol contents were significantly higher ($p < 0.05$) in wild than cultivated bay leaves (Conforti *et al.*, 2006; Dias *et al.*, 2014); the α-tocopherol content was determined to be 370.05 mg/100 g dry weight (dw) for the wild sample, 304.74 mg/100 g dw for the cultivated sample, while the total tocopherol content was 780.12 mg/100 g dw for the wild sample and 655.70 mg/100 g dw for the cultivated sample (Dias *et al.*, 2014). The tocopherol extraction method, plant type (wild or cultivated), plant part and even the solvents used may be responsible for the variation in the results obtained.

Four anthocyanins were identified in the fruits (berries) of the bay leaf plant in a study by Longo and Vasapollo (2005); these were cyanidin-3-*O*-glucoside, cyanidin-3-*O*-rutinoside, peonidin-3-*O*-glucoside and peonidin-3-*O*-rutinoside. Of these four anthocyanins, cyanidin-3-*O*-rutinoside was the major component, at 116 mg/g stoned fresh berry.

Overall, the antioxidant activity and the amount and composition of bioactive substances in bay vary significantly and the factors affecting the composition are similar to those that affect the essential oil composition of the plant.

Antimicrobial activity

The antimicrobial activity of the essential oils and extracts of herbs and spices is a popular subject, as some of them are quite promising. Intensive studies have been done on the antimicrobial activity of extracts made using various solvents (ethanol, ethyl acetate, hexane), and also on hydrodistilled extracts, SFE/SPFE (supercritical fluid extraction) extracts, essential oils and even ground bay leaf (Aktuğ and Karapinar, 1986; Dadalıoğlu and Evrendilek, 2004; Santoyo *et al.*, 2006; Derwich *et al.*, 2009; El Malti and Amarouch, 2009; De Corato *et al.*, 2010; Ceyhan *et al.*, 2012; Ramos *et al.*, 2012; Cherrat *et al.*, 2014; El *et al.*, 2014; Mello da Silveira *et al.*, 2014). In these studies, the effects of the extracts were tested against pathogenic Gram-positive and/or Gram-negative bacteria (Aktuğ and Karapinar, 1986; Dadalıoğlu and Evrendilek, 2004; Ertürk, 2006; Santoyo *et al.*, 2006; Derwich *et al.*, 2009; El Malti and Amarouch, 2009; Ceyhan *et al.*, 2012; Millezi *et al.*, 2012; Cherrat *et al.*, 2014; El *et al.*, 2014; Mello da Silveira *et al.*, 2014), moulds (Simić *et al.*, 2004; Ertürk, 2006; Santoyo *et al.*, 2006; De Corato *et al.*, 2010) and yeasts (Ertürk, 2006; Santoyo *et al.*, 2006; Ozcan *et al.*, 2010; Fukuyama *et al.*, 2011; Ceyhan *et al.*, 2012).

The activity of the essential oil of bay leaf against common food-borne pathogens such as *Staphylococcus aureus* 6538P and *Listeria monocytogenes* Scott-A (Gram-positive), and *Escherichia coli* O157:H7, *Salmonella typhimurium* NRRL E 4463 (Gram-negative) was evaluated, and only in the case of *L. monocytogenes* Scott-A no inhibition was observed (El *et al.*, 2014). Mello da Silveira *et al.* (2014) determined the minimum inhibitory concentration (MIC) of bay leaf essential oil in fresh Tuscan sausages; the lowest MIC was observed against *L. monocytogenes* serotype 2 ATCC 19112 (1.25 g/l) and *Yersinia enterocolitica* ATCC 9610 (1.25 g/l). In the same study, the lowest minimum bactericidal concentrations (MBC) were observed for *E. coli* ATCC 25922 (2.5 g/l) and *Y. enterocolitica* ATCC 9610 (2.5 g/l). The MIC of bay leaf essential oil was 0.5–14 µl/ml for the Gram-positive bacteria tested and 4–14 µl/ml for the Gram-negative bacteria tested (Cherrat *et al.*, 2014). The highest bactericidal effect was determined against *L. monocytogenes* CECT 4031 (8 µl/ml) among the Gram-positive bacteria, and among the Gram-negative bacteria, *Y. enterocolitica* CECT 4315 (14 µl/ml) was the most resistant. There is some disagreement over the resistance of the bacteria tested to the antibacterial activity of bay leaf, because in some of the studies it was claimed that Gram-negative bacteria are more resistant than Gram-positive bacteria (Aktuğand Karapinar, 1986; Derwich *et al.*, 2009; Millezi *et al.*, 2012; Cherrat *et al.*, 2014), whereas in another (El *et al.*, 2014) the opposite result was determined. The inhibitory potential of the bay leaf has been demonstrated both in Gram-negative and Gram-positive bacteria (Dadalıoğlu and Evrendilek, 2004; Ertürk, 2006; Ozcan *et al.*, 2010; Ramos *et al.*, 2012), and as the composition of the essential compounds or active constituents might be different, there is no need to separate the effectiveness in terms of Gram-positive or Gram-negative bacteria.

The antifungal activity of bay leaf essential oil has been studied under both *in vitro* and *in vivo* conditions (Simić *et al.*, 2004; De Corato *et al.*, 2010; Rosello *et al.*, 2015). Essential oil at a concentration of 1000 µg/ml completely inhibited both *Botrytis cinerea* and *Monilinia laxa*, but this concentration partially inhibited *Penicillium digitatum* (De Corato *et al.*, 2010). Simić *et al.* (2004) studied the antifungal activity of bay leaf essential oil on 17 different fungi; among these, *Alternaria alternata* gave the lowest minimum fungicidal concentration (MFC) (10 µl/ml), and *Aspergillus ochraceus* and *A. terrus* showed the highest MFC (50 µl/ml). The antifungal activity of bay leaf oil was also tested on peaches, kiwifruit, citrus fruits (orange and lemon) and rice under *in vivo* conditions (De Corato *et al.*, 2010; Rosello *et al.*, 2015). In the De Corato *et al.* (2010) study, *B. cinerea, M. laxa* and *P. digitatum* were artificially inoculated on to the fruits and the oil was applied at a concentration of 3 mg/ml. In peaches, decay was inhibited by 91% in the preventive treatment (applied before fungal inoculation) and 76% in the curative treatment (applied after inoculation). In various studies, *Candida albicans* was the yeast type that was generally tested for the determination of the antifungal activity of bay leaves (Ertürk, 2006; Santoyo *et al.*, 2006; Ozcan *et al.*, 2010; Fukuyama *et al.*, 2011; Ceyhan *et al.*, 2012). The MIC of leaf essential oil was found to be 0.1 mg/ml (Ozcan *et al.*, 2010), while that for aqueous extracts was 1.56 mg/ml (Ceyhan *et al.*, 2012) and that for ethanolic extracts was 5 mg/ml (Ertürk, 2006).

4.3 Postharvest Technology

4.3.1 Processing

Different parts of plants can be consumed as spices or herbs, including the bark, flower, leaves, rhizomes, roots, stems, seeds, bulbs, tubers, berries or arils (Raghavan, 2007; Schweiggert *et al.*, 2007). The spices can also be consumed in a fresh, dried, whole, ground, crumbled, paste, extract (oil, oleoresin and derivatives) or powder form (Raghavan, 2007; Board, 2010). Harvesting and processing steps are therefore unique for each herb type. Here, the whole dried bay leaf processing steps are summarized, starting from harvesting.

Bay leaves can be harvested all the year round by virtue of the plant being an evergreen. However, for the herbal plants, leaves are generally collected when the plants bear flowers (Baytop, 1984).There are some small differences in the harvesting of bay leaves depending on the type of bay plant – whether they are wild or cultivated. For cultivated bay, one or two harvests a year are recommended for the highest yield and highest quality of dry leaf (Weiss, 2002; Elzebroek and Wind, 2008; Polat *et al.*, 2009). In the Mediterranean region, the optimum harvest time is the autumn season. For example, in Turkey, Greece and the former Yugoslavia, bay leaves are harvested from August to October; for Morocco and Portugal, the recommended harvest time is July and August (Weiss, 2002; Erden, 2005). The harvesting should be under optimum conditions and avoid dew, high humidity and rain (UNIDO and FAO, 2005), as these can accelerate deterioration and discoloration, and thus result in a low-quality product.

Collection of the leaves and flowers of bay is generally carried out by hand or using small farming tools such as rakes (Baytop, 1984; Weiss, 2002). Sometimes, the stems of the plants are cut, and the leaves or fruits removed following the harvesting. Mechanical harvesting is not favoured because small farmers or wild plant collectors are in charge of most of the harvesting, especially in developing countries, but clear-cutting, pollarding or a combination of these two methods can be applied (Polat *et al.*, 2009).

Drying

After the harvesting of herbs and spices, some pre-drying processes are applied, including washing, peeling off the bark (ginger, cinnamon, etc.), puncturing (for red peppers before drying), bleaching (for delaying enzymatic activity), blanching, curing (turmeric, garlic, etc.), threshing, sifting or chemical treatment (Akgul, 1993; UNIDO and FAO, 2005). However, none of these pretreatments are applied to bay leaves. Wild bay leaf collectors or small-scale farmers dry the leaves after harvesting, although large-scale producers who are supplied with dried leaves from the smaller producers carry out cleaning and washing when the leaves are transported to the production facility.

Moisture levels of spices and herbs of up to 80% are decreased to below 10% by using drying to enhance their shelf-life stability (Diaz-Maroto *et al.*, 2002; UNIDO and FAO, 2005). The drying of bay leaves and most of the other leafy herbs and spices is performed by several methods. These methods can be classified as sun drying, shade drying, drying in greenhouse type dryers and drying with hot air (Baytop, 1984; UNIDO and FAO, 2005). Shade drying or artificial drying are especially recommended to meet the highest final quality (Acar, 1987; Grabowski *et al.*, 2003). Sun drying is an easy and comparably cheaper method, but direct sun light or the use of temperatures over 50°C in other drying methods may cause discoloration, loss of volatiles and excessive shrinkage (Baytop, 1984; Acar, 1987; Grabowski *et al.*, 2003; UNIDO and FAO, 2005; Sellami *et al.*, 2011; Çakmak *et al.*, 2013). A bright green colour is appreciated for dried bay leaves that are consumed as a culinary herb, as shown in Fig. 4.4. Hot-air drying is an applicable method when open-air drying cannot be practised owing to the air conditions, and this method also allows considerable reduction of the drying time. In addition to these traditional methods, new drying methods have recently been introduced, such as heat pump drying, microwave oven drying, convection oven drying and freeze-drying; these can both meet food safety and quality regulations and decrease the energy costs of drying (Diaz-Maroto *et al.*, 2002; Kuzgunkaya and Hepbasli, 2007; Sellami *et al.*, 2011; Çakmak *et al.*, 2013). However, freeze-drying is only applied to spices and herbs that have relatively higher economic values (Grabowski *et al.*, 2003).

Fumigation and/or irradiation

Dried herbs and spices are subjected to heat treatment, fumigation and/or irradiation for pest control and to lower the microbiological risks. Heat treatment (pasteurization) by using steam, infrared radiation or a hot air oven can be applied, with different process temperatures and exposure times for different levels of lethality (Akgul, 1993). Fumigation and gamma irradiation and their combination are other alternative processes for implementing microbial safety. Fumigation is an efficient, fast and economic pest control method, but the application should be handled only by authorized personnel (Akgul, 1993; WHO, 1998; UNIDO and FAO, 2005). Chemical pesticides (fumigants) in the gaseous state, such as methyl bromide, ethylene dibromide, ethylene oxide, ethylene chlorohydrin, methyl formate, phosphine, aluminium phosphide or their mixtures are applied to kill the target pests (Akgul, 1993; WHO, 1998; UNIDO and FAO, 2005). The use of some of these fumigants is now limited or totally banned owing to their toxic residues and potential health risks (Akgul, 1993; Raghavan, 2007; US FDA, 2013).

Fig. 4.4. Sun-dried bay leaves.

Irradiation with γ-rays is gaining popularity even though the set-up costs are high and trained personnel are needed for their use (Akgul, 1993). Cobalt 60 is the source for the production of γ-rays, and the radiation dose is expressed by the gray (Gy), where 1 Gy is equal to 1 J of energy absorbed per kilogram of irradiated product (Lacroix *et al.*, 2003). This method is stated to be superior due to its applicability to pre-packed herbs and spices, but there is still some controversy in the literature about the possible health risks and product quality (Akgul, 1993; UNIDO and FAO, 2005). Although it is stated that the loss of volatile or flavour components is either eliminated or minimal in irradiated herbs and spices (Lacroix *et al.*, 2003; UNIDO and FAO, 2005), colour and aroma (volatile) losses are also reported in literature (Polovka and Suhaj, 2010; Kirkin *et al.*, 2014). The maximum average absorbed radiation dose for herbs, spices and seasonings is 10 kGy for the reduction of target pathogenic microorganisms and 1 kGy for insect disinfestation. The foodstuffs that have been treated with ionizing radiation must be labelled as 'irradiated' or 'treated with ionising radiation' (FAO and WHO, 2001; Lacroix *et al.*, 2003; Polovka and Suhaj, 2010; EFSA, 2011; ESA, 2013).

Packaging

Dried herbs or spices are packaged before transportation to the end user. The moisture content of herbs/spices should be decreased to below 10% (UNIDO and FAO, 2005), but the maximum moisture level is declared as 8% (w/w) in the final product (ESA, 2013). Bay leaves are classified according to shape, size, colour and aroma before packaging (Akgul, 1993). The packaging requirements are according to the various quality standards, and need to conform to the international standards for imported goods, as well as consumer preferences. The packaging material should be properly selected to protect the herb/spice from moisture, heat, light and potential pests, and to decrease the gas (oxygen) permeability to preserve the aroma and essential compounds; it should also ensure microbial safety and meet the appropriate quality standards. The packaging material should be food grade and non-polluting (UNIDO and FAO, 2005; ESA, 2013). Paper bags, plastic bags, polyethylene bags, cardboard boxes, tins and jute or sisal sacks are preferred for bay leaf packaging (Baytop, 1984; UNIDO and FAO, 2005). Sometimes, transparent glass jars are preferred to show the premium quality of whole dried bay leaves. The International Trade Centre has produced a manual that covers the packaging requirements; for jute and sisal packaging in particular, these should conform to the standards of CAOBISCO Ref C502- 51-sj of 20-02-95 (Muggeridge and Clay, 2001; Weiss, 2002; UNIDO and FAO, 2005; ESA, 2013). The packaged bay leaves should be kept in a cool and dry place, and contamination and infestation should be prevented; the suggested storage conditions for spices are 10–15°C and 55–65% relative humidity (RH) (Raghavan, 2007). A maximum of 1 year of shelf life is recommended for herbs and spices owing to the loss of bioactive and essential compounds, but they can be stored up to 3 years if the packaging is tightly closed (Baytop, 1984; Raghavan, 2007).

Quality standards

Physical and chemical quality requirements for dried whole bay leaf herb according to the Turkish food codex, notification for herbs and spices are: 'max. 0.1% (w/w) foreign matter, max. 8% (w/w) moisture, max. 7% (w/w) ash, max. 1.5% (w/w) acid-insoluble ash, max. 30% (w/w) cellulose, max. 15% broken and 10% blemished leaf; minimum essential oil should be 1 ml/100 g leaf' (Anonymous, 2013). Moisture, ash and essential oil requirements are similar for the ESA, except that the acid-insoluble ash content is little higher and given as max. 2% (w/w) (ESA, 2013). The cleanliness specifications of the American Spice Trade Association (ASTA) for bay leaves are: '2 (by count) whole dead insects, 1 mg/lb mammalian excreta, 10 mg/lb other excreta, 2% (w/w) mould, 2.5% (w/w) defiled/infested insect, 0.5% (w/w) extraneous/foreign matter and stems are excluded from this specifications' (Muggeridge and Clay, 2001). According to the Turkish food codex, microbiological criteria, spices, herbs and/or their mixtures should meet the following requirements: coagulase-positive staphylococci 10^3–10^4 cfu/g, *Bacillus cereus* 10^3–10^4 cfu/g and *Salmonella* absent in 25 g (Anonymous, 2011). The classification of herbs and spices by microbiological criteria of within Recommendation 2004/24/EC and ESA samples that are listed as 'Satisfactory' should have: *Salmonella* spp. absent in 25 g, *B. cereus* $<10^3$ cfu/g, *Clostridium perfringens* $<10^2$ cfu/g and *E. coli* $<10^1$ cfu/g (Sagoo *et al.*, 2009).

4.3.2 Value addition

Bay is used as a culinary herb in both the whole and powdered form (fresh or dried), and also as a medicinal herb. Dried leaves can be consumed as a herbal tea prepared by either decoction or infusion.

The essential oil of bay leaves and fruits is produced in Russia, Georgia, Albania, the former Yugoslavia, Italy, Portugal, Morocco and Turkey (Ceylan, 1997; Weiss, 2002; Charles, 2013).Various methods are applied to extract essential oil from leaves, such as steam distillation (hydrodistillation), supercritical fluid extraction (SFE/SPFE), solvent extraction or solvent-free microwave assisted steam distillation (SFME) (Demo *et al.*, 1998;

Fiorini *et al.*, 1998; Caredda *et al.*, 2002; Dadalıoğlu and Evrendilek, 2004; Kosar *et al.*, 2005; Parthasarathy *et al.*, 2008; Lira *et al.*, 2009; Zekovic *et al.*, 2009; Sellami *et al.*, 2011; Jemaa *et al.*, 2012; Moghtader and Salari, 2012; Özek, 2012; Cherrat *et al.*, 2014; El *et al.*, 2014). The hydrodistillation method can be applied using a Clevenger-type apparatus (Dadalıoğlu and Evrendilek, 2004; Kosar *et al.*, 2005; Sangun *et al.*, 2007; Bayramoglu *et al.*, 2009; Kandi and Sefidkon, 2011; Özek, 2012; Cherrat *et al.*, 2014), the Likens–Nickerson apparatus (Bouzouita *et al.*, 2001) and the Soxhlet extractor apparatus using water (steam) as a solvent (Zekovic *et al.*, 2009). The Clevenger-type apparatus is also known as the 'European Pharmacopoeian hydrodistillation apparatus', and details are described by Hinneburg *et al.* (2006).

Bay leaf essential oil is sometimes extracted by the SFE/SPFE method with CO_2 as a carrier solvent (Caredda *et al.*, 2002; Zekovic *et al.*, 2009; De Corato *et al.*, 2010). SFME extraction is also a hydrodistillation method in which a Clevenger-type apparatus is placed inside the microwave oven (Kosar *et al.*, 2005; Bayramoglu *et al.*, 2009; El *et al.*, 2014). These solvent-free extraction methods are viewed as green (clean) technology, to which much attention is paid nowadays.

The efficiency of extraction methods is compared according to the quantity of essential oil extracted. The average min./max. yields of the various extraction methods are: 0.23–2.65% for hydrodistillation, 0.60–2.54% for SPFE and 6.5% for the chemical solvent method (Demo *et al.*, 1998; Fiorini *et al.*, 1998; Bouzouita *et al.*, 2001; Kosar *et al.*, 2005; Hinneburg *et al.*, 2006; Lira *et al.*, 2009; Zekovic *et al.*, 2009; De Corato *et al.*, 2010; Kandi and Sefidkon, 2011; Sellami *et al.*, 2011; Jemaa *et al.*, 2012; Moghtader and Salari, 2012; Özek, 2012; Cherrat *et al.*, 2014). The maximum yield of the SFME method at 622 W power was 0.024 ml/g dry leaf, while for the hydrodistillation method it was 0.022 ml/g dry leaf (Bayramoglu *et al.*, 2009; El *et al.*, 2014). However, the difference in essential oil yields was not significant in these studies ($P > 0.05$). The hydrodistillation and SFME extraction methods show remarkable variation in the volatile compounds of the essential oil. In particular, the content of major volatiles such as 1,8-cineole and α-terpinyl acetate was higher in SFME-extracted oil than in hydrodistilled oil (Kosar *et al.*, 2005; Bayramoglu *et al.*, 2009; El *et al.*, 2014).

Volatile compounds can also contribute to the odour and aroma perception of bay leaves. Monoterpene and sesquiterpene hydrocarbons and their derivatives are the main volatiles of Mediterranean bay leaf (Kilic *et al.*, 2004). In a study by Buttery *et al.* (1974), 1,8-cineole, α-pinene, linalool and limonene were the major aroma compounds in the Mediterranean bay leaf oil tested; 1,8-cineole was also dominant in other commercially available so-called bay oils extracted from *U. californica* and *P. racemosa*. The odour compounds are also affected from the age of the shoots that are harvested. Younger shoots can have weaker and non-odour-contributing monoterpene hydrocarbons (Kilic *et al.*, 2004). The odour compounds of fresh bay leaves depend on the age of shoot/part of the plant and were described in a comprehensive study by Kilic *et al.* (2004). In this, odour compounds were ranked using flavour dilution (FD) factors, and eucalyptus (1,8-cineole), fresh green ((*Z*)-3-hexenal), flowery clove (eugenol) and pepper odours were higher in old shoots, while flowery ((*E*)-isoeugenol), eucalyptus (1,8-cineole) and flowery clove (eugenol) odours were higher in the younger shoots. Apart from these major odour-active substances, there are still unidentified compounds in trace amounts in Mediterranean bay leaf and these may act as a signature odour of this species (Buttery *et al.*, 1974; Kilic *et al.*, 2004). Bay leaf aroma is described as a pleasant sweetly aromatic, camphoraceous, cineolic and slightly bitter aftertaste (Charles, 2013).

The aromatic fruit oil is used in pharmacy and veterinary practice (Board, 2010). Bay leaf oil, either alone or mixed with other essential oils, is used as a massage oil in aromatherapy, and in hot baths for mind and body relaxing effects. Bay leaf essential oil has also showed strong pest repellent and fumigant activity, but exposure time, dose tested and oil origin (chemical composition) significantly affected these activities (Jemaa *et al.*, 2012). Turkish producers benefit from the dried leaves as a pest control agent used in raisin and dried fig packages (Acar, 1987; Zeybek and Zeybek, 1994; Polat *et al.*, 2009).

After the leaf harvest, small-scale wild bay leaf collectors or cultivated bay leaf farmers utilize the bark, wood, stem and rest of the plant as firewood, which is then used in the drying of the leaves. The wood of the bay plant, together with the leaves, may be burned for the smoking of meat and fish.

A most interesting study of bay leaves was done recently by Onay (2014). Seeds of the plant were used as a biomass feed stock in catalytic pyrolysis and the catalytic pyrolysis oil, with the addition of catalyst, had comparably lower oxygenated compounds, and therefore could well be an ingredient for transportation fuels (Onay, 2014).

4.4 Uses

4.4.1 General uses

Bay leaf is a popular culinary herb in French, Italian, Greek and Turkish cuisine (Akgul, 1993; Farrell, 1998). Dried leaves are used to garnish fish, meat and chicken dishes, casseroles, soups, sauces, puddings, pickles, canned vegetables, confectionery, baked goods and pastries and impart their fresh, pungent, spicy, cineolic and piney aroma (Ceylan, 1997; Kumar *et al.*, 2001; Raghavan, 2007; Charles, 2013). Dried bay leaves, together with parsley and thyme, are tied with a string to produce a bundle of herbs known as bouquet garni in French cuisine (Raghavan, 2007). Leaves are placed in canned salmon, tuna and sardine to remove or mask the strong fishy odour. Also, in Mediterranean countries, dried leaves are put into olive oil to give it an aromatic flavour and enhance the appetite. Dried bay leaf powder is an ingredient of chicken gravy powder, a mix for beef, North Indian curry blends, fish marinade, and ras-el-hanouth (a Moroccan spice mix) (Raghavan, 2007). Some artisan bread powder mixes include the aromatic powder of bay leaf to improve the flavour.

Oleoresin is a mixture of essential oils, and the resin extracted from herbal plants by solvents, as well as the semi-solid and dark green oleoresin of bay leaves can be used instead of the herb or the essential oil (Akgul, 1993; Farrell, 1998; Weiss, 2002; Charles, 2013). 2.27 kg of bay leaf oleoresin is equal to 45.45 kg fresh ground leaves in terms of odour and flavour (Farrell, 1998).

The ripe fruits of the plant are used in non-alcoholic and alcoholic beverages such as home-made liquors (Akgul, 1993; Ceylan, 1997). The essential and fixed oil of both the leaf and fruit are utilized in soap, candles, cream, skin and haircare products, and in shampoos for preventing hair loss (Akgul, 1993; Charles, 2013). In the Mediterranean Turkish city of Antakya (Hatay) especially, bay is quite popular in the production of soaps that are handmade from bay fruit oil mixed with olive oil; the soap is believed to treat hair loss and hair dandruff. Essential fruit and leaf oil has a soft, floral, spicy and aromatic scent, and the oil is used in perfumes and in cologne as an accent in many masculine and some feminine fragrances. Bay oil is classified as a middle perfumery note. Cineolic and camphoraceous accents are also used in detergents, soaps, home fragrances, and even in incense.

As it is an evergreen plant, bay leaf is grown as ornamental plant and specimen tree in gardens, indoor facilities and landscape designs owing to its bright and aesthetically pleasing appearance. The attractive white or yellow flowers of the plant are in blossom between March and May, and the plant is grown in containers or pruned as hedges along with the wild types.

4.4.2 Pharmacological uses

The extracts or essential oils of bay leaves and bay fruits have an important place in Turkish folk medicine, and they are used in both internal and external treatment of various diseases. A decoction of fresh bay leaves has therapeutic effects on cardiac diseases, joint calcification and rheumatism (Tuzlacı and Sadıkoğlu, 2007). A decoction of the branches and fruits of bay leaves is also used for treating bee and snake bites as well as stomach ulcers (Tuzlacı and Tolon, 2000). The essential oil of bay fruits has gastroprotective and antineuralgic effects, and shows analgesic and wound-healing properties when used as a cream (Zeybek and Zeybek, 1994). In Turkish folk medicine, the fruit essential oil and extract are also used as an antipyretic, digestive, appetite stimulant, as a remedy for toothache, chronic headache and insect bites, as a regulator of the bloodstream and as a diaphoretic. The essential oil of the leaves has emmenagogue, irritant, rubefacient and antirheumatic effects (Ceylan, 1997). Anti-emetic, antiseptic, antipyretic, diuretic, expectorant, appetite-stimulating, digestive, carminative and sedative effects of bay leaves are also reported in the literature (Spratt, 1830; Baytop, 1984; Akgul, 1993; Duke, 2002). In Iranian folk medicine, the leaf essential oils are also a remedy for rheumatism pain and epilepsy (Sayyah *et al.*, 2002, 2003). The highly esteemed Turkish professor and pharmacist Turhan Baytop suggested an infusion formula prepared as a drink for treating acute bronchitis, indigestion and inappetency; this was made by leaving 4 g of dried bay leaf and 8 g of dried orange peel in 200 ml of boiled water for 15 min, filtering and sweetening with honey (if wanted) (Baytop, 1984).

The sesquiterpene lactones isolated from hexane and ethyl acetate extracts of bay leaves are 10-epigazaniolid, gazaniolide, spirafolide, costunolide, reynosin and santamarin, and those substances show cytotoxic activity (Fang *et al.*, 2005). Julianti *et al.* (2012) isolated and determined 22 different sesquiterpenes from bay leaves and determined by spectroscopic analysis that ten of these (new) compounds were eudesmane lactones and their methyl esters. The cytotoxicity of these fractions were tested against K562 leukaemia tumour cells, and most of them showed moderate to significant toxicity; the most active fraction had an LC_{50} of 4.57 ± 2.1 mM. Essential oil from the leaves and seeds also inhibited K562 leukaemia tumour cells with the IC_{50} values of 95 and 75 μg/ml, respectively (Saab *et al.*, 2012). Gazaniolide, costunolide, santamarin, reynosin, 11,13-dehydrosantonin, spirafolide and lauroxepine are sesquiterpene lactones isolated from methanol extracts of the fruits of the plant (Barla *et al.*, 2007). The cytotoxic effect of these substances was tested against A2780 human ovarian cancer cells, and the IC_{50} values were in the range 4–13.5 μg/ml.

Another active fraction of bay leaf is lauroside B, and this metastigmane glycoside suppressed the proliferation of tested human melanoma cell lines (A375, WM115, and SK-Mel-28) and has the potential to reduce apoptosis in aggressive melanoma cancer cells (Panza *et al.*, 2010). The antiproliferative activity of bay leaf and fruit essential oils was also evaluated using breast cancer cell lines (MCF7 and T47D) (Al-Kalaldeh *et al.*, 2010; Abu-Dahab *et al.*, 2014). Ethanolic extracts of bay leaf at a concentration 50 μg/ml gave 52.53% inhibition of MCF7 cell lines; however, the antiproliferative effect was somewhat lower for both the aqueous extract and the essential oil (Al-Kalaldeh *et al.*, 2010). In another study, fruit and leaf extracts obtained using different solvents were compared against MCF7 and T47D breast cancer cell lines, and the lowest IC_{50} value was obtained from ethyl acetate extract of bay fruits with MCF7 cancer cells (20.6 μg/ml), while for the T47D cells the lowest IC_{50} (12.3 μg/ml) was obtained from ethanolic extracts of bay fruits (Abu-Dahab *et al.*, 2014).

The analgesic, anti-inflammatory and anticonvulsant activities of bay leaf essential oil have also been tested in animal studies (Sayyah *et al.*, 2002, 2003). In humans, clinical studies showed that bay leaves reduce plasma glucose levels (20–30%) and total cholesterol levels (20–24%) in people with type 2 diabetes (Khan *et al.*, 2009; Aljamal, 2011). Aqueous extracts of bay seeds and aqueous or methanolic extracts of the fruits showed a nearly 100% gastroprotective effect and thus anti-ulcerogenic activity (Afifi *et al.*, 1997; Gürbüz *et al.*, 2002). In addition to these health-promoting activities, antidiarrheal activity was observed in rats treated with different doses of bay leaf aqueous extracts (Qnais *et al.*, 2012). The neuroprotective effects of bay leaves were studied by Cho *et al.* (2010), and their results (as stated by the authors) showed that the chloroform fraction of methanol extracts of bay leaves had the potential to cure ischaemic neuronal damage and significantly reduced death-associated protein kinase (DAPK) dephosphorylation in human cell lines and organotypic tissues.

There is no specific information regarding the toxicity of bay leaves or oil. According to the *Food Additives Status List* of the US Food and Drug Administration (FDA), bay leaf as a herb and essential oil/oleoresin has GRAS (generally recognized as safe) status, with the GRAS numbers 182.10 and 182.20 (Charles, 2013; US FDA, 2014). In the only study on the toxicity of bay leaf, the toxic dose was determined as 0.3 mg bay leaf extract/g mouse weight, and at this level, inflammation and cell necrosis in the liver and heart, and oxidative stress, were induced (El Malti and Amarouch, 2009). However, there is no predictable risk for public health as the real consumable portion is much lower than the given limits. The only dosage information on bay leaf is given by the American Pharmaceutical Association (APA): 1–2 tablespoons leaf/cup water and 3 times/day; 1–2 drops essential oil added to brandy, honey, or tea (Duke, 2002). Only a few cases have been reported of allergic reactions (contact dermatitis) to bay, and these occurred when the person was exposed to laurel oil (Özden *et al.*, 2001; Bleasel *et al.*, 2002; Adişen and Önder, 2007).

4.5 Summary

The bay plant is an evergreen tree/shrub used as a culinary herb in Mediterranean cuisine. Apart from its aromatic properties, the therapeutic activity of the plant has been extensively studied in both folk and modern medicine. The essential oil of the leaves and the fruits is extracted and the major component compounds and their amounts have been found to change with several factors, such as climate, place of harvest, season of harvest, the age of shoots and the parts of the plant used. Among the constituents that have been evaluated, 1,8-cineole is the main bioactive compound of bay leaf essential oil and it also contributes the characteristic bay leaf odour. The traditional formulas for treating several diseases with bay leaf have been passed through the generations, and some of their health-promoting effects have been proven by clinical studies. The cytotoxic activity of the essential oil and extracts have been determined by several authors against leukaemia, human ovarian cancer, human melanoma cancer and breast cancer cell lines. Besides its health-improving status, the antioxidant and antimicrobial (antibacterial and antifungal) effects of bay and its extracts have been examined. When all these positive effects are taken into account, bay will be acknowledged more than it is today in the future.

References

Abu-Dahab, R., Kasabri, V. and Afifi, F.U. (2014) Evaluation of the volatile oil composition and antiproliferative activity of *Laurus nobilis* L. (Lauraceae) on breast cancer cell line models. *Records of Natural Products* 8, 136–147.

Acar, I. (1987) *Production and Utilization of Bay Laurel (*Laurus nobilis *L.) Leaves and Essential Oil.* Technical Bulletin No. 186, Forest Research Institute, Kamer Matbaacilik, Ankara. [In Turkish.]

Acar, I. (1988) *Studies on Leaf Quality of Bay Laurel (*Laurus nobilis *L.) Distributed in Turkey.* Technical Bulletin No: 202, Forest Research Institute, Kamer Matbaacilik, Ankara. [Iin Turkish.]

Adişen, E. and Önder, M. (2007) Allergic contact dermatitis from *Laurus nobilis* oil induced by massage. *Contact Dermatitis* 56, 360–361.

AEA (2014) Export data of herbs and spices. Aegean Exporters Association, Alsancak, Izmir, Turkey. Obtained from: http://www.egebirlik.org.tr/Asp/Content.asp?MS=1&Id=0 (accessed 21 August 2014 but no longer available via this link).

Afifi, F.U., Khalil, E., Tamimi, S.O. and Disi, A. (1997) Evaluation of the gastroprotective effect of *Laurus nobilis* seeds on ethanol induced gastric ulcer in rats. *Journal of Ethnopharmacology* 58, 9–14.

Akgul, A. (1993) *Spice Science and Technology*. Damla Matbaacilik ve Ticaret, Konya, Turkey. [In Turkish.]
Aktuğ, Ş.E. and Karapinar, M. (1986) Sensitivity of some common food-poisoning bacteria to thyme, mint and bay leaves. *International Journal of Food Microbiology* 3, 349–354.
Albayrak, S., Aksoy, A., Sagdic, O. and Albayrak, S. (2012) Antioxidant and antimicrobial activities of different extracts of some medicinal herbs consumed as tea and spices in Turkey. *Journal of Food Biochemistry* 36, 547–554.
Aljamal, A. (2011) Effects of bay leaves on the patients with diabetes mellitus. *Research Journal of Medicinal Plant* 5, 471–476.
Al-Kalaldeh, J.Z., Abu-Dahab, R. and Afifi, F.U. (2010) Volatile oil composition and antiproliferative activity of *Laurus nobilis*, *Origanum syriacum*, *Origanum vulgare*, and *Salvia triloba* against human breast adenocarcinoma cells. *Nutrition Research* 30, 271–278.
Anonymous (2011) Regulation on Turkish food codex microbiological criteria. Available at: http://www.tarim.gov.tr/Sayfalar/EN/Mevzuat.aspx?OgeId=15 (accessed 7 August 2014).
Anonymous (2013) Turkish food codex notification no. 2013/12 on spices. Available at: http://mevzuat.basbakanlik.gov.tr/Metin.Aspx?MevzuatKod=9.5.17268&sourceXmlSearch=baharat&MevzuatIliski=0 (accessed 7 August 2014).
BAKA (2012) Medicinal and aromatic plants sectorial report. Batı Akdeniz Kalkınma Ajansı, Isparta, Turkey. Available at: http://www.baka.org.tr/dokuman-listesi-tumu.html (accessed 7 August 2014).
Barla, A., Topçu, G., Öksüz, S., Tümen, G. and Kingston, D.G. (2007) Identification of cytotoxic sesquiterpenes from *Laurus nobilis* L. *Food Chemistry* 104, 1478–1484.
Bayramoglu, B., Sahin, S. and Sumnu, S. (2009) Extraction of essential oil from laurel leaves by using microwaves. *Separation Science and Technology* 44, 722–733.
Baytop, T. (1984) *Therapy with Medicinal Plants in Turkey (Past and Present)*. Sanal Matbaacilik, Istanbul, Turkey. [In Turkish.]
Bleasel, N., Tate, B. and Rademaker, M. (2002) Allergic contact dermatitis following exposure to essential oils. *Australasian Journal of Dermatology* 43, 211–213.
Board, N. (2010) Bay or laurel leaves. In: Board, N. (ed.) *Handbook on Spices*. Asia Pacific Business Press, Delhi, pp. 278–280.
Bouzouita, N., Nafti, A., Chaabouni, M.M., Lognay, G.C., Marlier, M., Zghoulli, S. and Thonart, P.H. (2001) Chemical composition of *Laurus nobilis* oil from Tunisia. *Journal of Essential Oil Research* 13, 116–117.
Buttery, R.G., Black, D.R., Guadagni, D.G., Ling, L.C., Connolly, G. and Teranishi, R. (1974) California bay oil I: constituents, odor properties. *Journal of Agricultural and Food Chemistry* 22, 773–777.
Çakmak, H., Kumcuoğlu, S. and Tavman, Ş. (2013) Thin layer drying of bay leaves (*Laurus nobilis* L.) in conventional and microwave oven. *Academic Food Journal/Akademik Gida* 11(1), 20–26.
Caredda, A., Marongiu, B., Porcedda, S. and Soro, C. (2002) Supercritical carbon dioxide extraction and characterization of *Laurus nobilis* essential oil. *Journal of Agricultural and Food Chemistry* 50, 1492–1496.
Ceylan, A. (1997) *Medicinal Plants-II: Essential Oil Plants*. Yayinlari No. 481, Ziraat Fakultesi, Ege Universitesi Basımevi, Izmir, Turkey. [In Turkish.]
Ceyhan, N., Keskin, D. and Uğur, A. (2012) Antimicrobial activities of different extracts of eight plant species from four different family against some pathogenic microorganisms. *Journal of Food, Agriculture and Environment* 10, 193–197.
Charles, D.J. (2013) *Antioxidant Properties of Spices, Herbs and Other Sources*. Springer, New York.
Cherrat, L., Espina, L., Bakkali, M., García-Gonzalo, D., Pagán, R. and Laglaoui, A. (2014) Chemical composition and antioxidant properties of *Laurus nobilis* L. and *Myrtus communis* L. essential oils from Morocco and evaluation of their antimicrobial activity acting alone or in combined processes for food preservation. *Journal of the Science of Food and Agriculture* 94, 1197–1204.
Cho, E.Y., Lee, S.J., Nam, K.W., Shin, J., Oh, K.B., Kim, K.H. and Mar, W. (2010) Amelioration of oxygen and glucose deprivation-induced neuronal death by chloroform fraction of bay leaves (*Laurus nobilis*). *Bioscience, Biotechnology, and Biochemistry* 74, 2029–2035.
Çelik, S. and Yılmaz, Ö. (1996) The fatty acid composition of *Laurus nobilis* L. leaves and fruits. *GIDA* 21(3), 165–167.
Conforti, F., Statti, G., Uzunov, D. and Menichini, F. (2006) Comparative chemical composition and antioxidant activities of wild and cultivated *Laurus nobilis* L. leaves and *Foeniculum vulgare* subsp. *piperitum* (Ucria) Coutinho seeds. *Biological and Pharmaceutical Bulletin* 29, 2056–2064.
Dadalıoğlu, I. and Evrendilek, G.A. (2004) Chemical compositions and antibacterial effects of essential oils of Turkish oregano (*Origanum minutiflorum*), bay laurel (*Laurus nobilis*), Spanish lavender

(*Lavandula stoechas* L.), and fennel (*Foeniculum vulgare*) on common foodborne pathogens. *Journal of Agricultural and Food Chemistry* 52, 8255–8260.
De Corato, U., Maccioni, O., Trupo, M. and Di Sanzo, G. (2010) Use of essential oil of *Laurus nobilis* obtained by means of a supercritical carbon dioxide technique against post harvest spoilage fungi. *Crop Protection* 29, 142–147.
Demo, A., Petrakis, C., Kefalas, P. and Boskou, D. (1998) Nutrient antioxidants in some herbs and Mediterranean plant leaves. *Food Research International* 31, 351–354.
Derwich, E., Benziane, Z. and Boukir, A. (2009) Chemical composition and antibacterial activity of leaves essential oil of *Laurus nobilis* from Morocco. *Australian Journal of Basic and Applied Sciences* 3, 3818–3824.
Dias, M.I., Barros, L., Duenas, M., Alves, R.C., Oliveira, M.B.P.P., Santos-Buelga, C. and Ferreira, I.C.F.R. (2014) Nutritional and antioxidant contributions of *Laurus nobilis* L. leaves: would be more suitable a wild or a cultivated sample? *Food Chemistry* 156, 339–346.
Diaz-Maroto, M.C., Perez-Coello, M.S. and Cabezudo, M.D. (2002) Effect of drying method on the volatiles in bay leaf (*Laurus nobilis* L.). *Journal of Agricultural and Food Chemistry* 50, 4520–4524.
Duke, J.A. (2002) *Handbook of Medicinal Herbs*, 2nd edn. CRC Press, Boca Raton, Florida.
EFRI (2014) *Handbook of Bay Laurel* (Laurus nobilis *L.*). Ege Forestry Research Institute, Izmir, Turkey. [In Turkish.] Available at: http://www.efri.gov.tr/yayinlar/Son_defne_elkitabi.pdf (accessed 1 August 2014).
El, S.N., Karagozlu, N., Karakaya, S. and Sahin, S. (2014) Antioxidant and antimicrobial activities of essential oils extracted from *Laurus nobilis* L. leaves by using solvent-free microwave and hydrodistillation. *Food and Nutrition Sciences* 5, 97–106.
El Malti, J. and Amarouch, H. (2009) Antibacterial effect, histological impact and oxidative stress studies from *Laurus nobilis* extract. *Journal of Food Quality* 32, 190–208.
Elmastaş, M., Gülçin, I., Işildak, Ö., Küfrevioğlu, Ö.İ., İbaoğlu, K. and Aboul-Enein, H.Y. (2006) Radical scavenging activity and antioxidant capacity of bay leaf extracts. *Journal of the Iranian Chemical Society* 3, 258–266.
Elzebroek, A.T.G. and Wind, K. (2008) Spices and flavorings. In: Elzebroek, A.T.G. and Wind, K. (eds) *Guide to Cultivated Plants*. CAB International, Wallingford, UK, pp. 263–266.
Erden, U. (2005) Investigation of seasonal variability and optimum drying methods of bay (*Laurus nobilis* L.). MSc. thesis, University of Cukurova, Adana, Turkey. [In Turkish.]
Ertürk, Ö. (2006) Antibacterial and antifungal activity of ethanolic extracts from eleven spice plants. *Biologia* 61, 275–278.
EFSA (2011) EFSA assesses the safety of food irradiation. European Food Safety Authority, Parma, Italy. Available at: http://www.efsa.europa.eu/en/press/news/cef110406 (accessed 10 September 2015).
ESA (2013) European Spice Association quality minima document. Available at: http://www.esa-spices.org/index-esa.html/publications-esa (accessed 28 October 2014).
Fang, F., Sang, S., Chen, K.Y., Gosslau, A., Ho, C.T. and Rosen, R.T. (2005) Isolation and identification of cytotoxic compounds from bay leaf (*Laurus nobilis*). *Food Chemistry* 93, 497–501.
FAO (2014) *Laurus nobilis* data sheet. Ecocrop, Food and Agriculture Organization of the United Nations, Rome. Available at: http://ecocrop.fao.org/ecocrop/srv/en/cropView?id=1333 (accessed 4 August 2014).
FAO and WHO (2001) *Proposed Draft Revision to the Recommended International Code of Practice for the Operation of Irradiation Facilities Used in the Treatment of Foods*. Food and Agriculture Organization of the United Nations, Rome and World Health Organization, Geneva, Switzerland. Available at: ftp://ftp.fao.org/Codex/Meetings/CCFAC/ccfac33/fa01_12e.pdf (accessed 12 October 2014).
FAOSTAT (2014) Food and agricultural commodities production. Food and Agriculture Organization of the United Nations, Rome. Available at: http://faostat.fao.org/site/339/default.aspx (accessed 4 August 2014).
Farrell, K.T. (1998) Bay (laurel) leaves. In: Farrell, K.T. (ed.) *Spices, Condiments and Seasonings*. Springer, New York, pp. 38–41.
Fiorini, C., Fourasté, I., David, B. and Bessière, J.M. (1997) Composition of the flower, leaf and stem essential oils from *Laurus nobilis* L. *Flavour and Fragrance Journal* 12, 91–93.
Fiorini, C., David, B., Fouraste, I. and Vercauteren, J. (1998) Acylated kaempferol glycosides from *Laurus nobilis* leaves. *Phytochemistry* 47, 821–824.
Fukuyama, N., Ino, C., Suzuki, Y., Kobayashi, N., Hamamoto, H., Sekimizu, K. and Orihara, Y. (2011) Antimicrobial sesquiterpenoids from *Laurus nobilis* L. *Natural Product Research* 25, 1295–1303.

Gómez-Coronado, D.J., Ibanez, E., Rupérez, F.J. and Barbas, C. (2004) Tocopherol measurement in edible products of vegetable origin. *Journal of Chromatography A* 1054, 227–233.
Grabowski, S., Marcotte, M. and Ramaswamy, H.S. (2003) Drying of fruits, vegetables, and spices. In: Chakraverty, A., Mujumdar, A.S., Raghavan, G.S.V. and Ramaswamy, H.S. (eds) *Handbook of Postharvest Technology: Cereals, Fruits, Vegetables, Tea, and Spices.* Marcel Dekker, New York, pp. 653–695.
Gürbüz, I., Üstün, O., Yeşilada, E., Sezik, E. and Akyürek, N. (2002) *In vivo* gastroprotective effects of five Turkish folk remedies against ethanol-induced lesions. *Journal of Ethnopharmacology* 83, 241–244.
Hafizoğlu, H. and Reunanen, M. (1993) Studies on the components of *Laurus nobilis* from Turkey with special reference to laurel berry fat. *Lipid/Fett* [now *European Journal of Lipid Science and Technology*] 95, 304–308.
Hinneburg, I., Dorman, H.J.D. and Hiltunen, R. (2006) Antioxidant activities of extracts from selected culinary herbs and spices. *Food Chemistry* 97, 122–129.
Jemaa, J.M.B., Tersim, N., Toudert, K.T. and Khouja, M.L. (2012) Insecticidal activities of essential oils from leaves of *Laurus nobilis* L. from Tunisia, Algeria and Morocco, and comparative chemical composition. *Journal of Stored Products Research* 48, 97–104.
Julianti, E., Jang, K.H., Lee, S., Lee, D., Mar, W., Oh, K.B. and Shin, J. (2012) Sesquiterpenes from the leaves of *Laurus nobilis* L. *Phytochemistry* 80, 70–76.
Kandi, M.N.H. and Sefidkon, F. (2011) The influense [*sic*] of drying methods on essential oil content and composition of *Laurus nobilis* L. *Journal of Essential Oil Bearing Plants* 14, 302–308.
Kang, H.W., Yu, K.W., Jun, W.J., Chang, I.S., Han, S.B., Kim, H.Y. and Cho, H.Y. (2002) Isolation and characterization of alkyl peroxy radical scavenging compound from leaves of *Laurus nobilis. Biological and Pharmaceutical Bulletin* 25, 102–108.
Kaurinovic, B., Popovic, M. and Vlaisavljevic, S. (2010) *In vitro* and *in vivo* effects of *Laurus nobilis* L. leaf extracts. *Molecules* 15, 3378–3390.
Khan, A., Zaman, G. and Anderson, R.A. (2009) Bay leaves improve glucose and lipid profile of people with type 2 diabetes. *Journal of Clinical Biochemistry and Nutrition* 44, 52–56.
Kilic, A., Hafizoglu, H., Kollmannsberger, H. and Nitz, S. (2004) Volatile constituents and key odorants in leaves, buds, flowers, and fruits of *Laurus nobilis* L. *Journal of Agricultural and Food Chemistry* 52, 1601–1606.
Kirkin, C., Mitrevski, B., Gunes, G. and Marriott, P.J. (2014) Combined effects of gamma-irradiation and modified atmosphere packaging on quality of some spices. *Food Chemistry* 154, 255–261.
Kosar, M., Tunalier, Z., Özek, T., Kürkcüoglu, M. and Baser, K.H.C. (2005) A simple method to obtain essential oils from *Salvia triloba* L. and *Laurus nobilis* L. by using microwave-assisted hydrodistillation. *Zeitschrift für Naturforschung C* 60, 501–504.
Kumar, S., Singh, J. and Sharma, A. (2001) Bay leaves. In: Peter, K.V. (ed.) *Handbook of Herbs and Spices, Vol. 1.* Woodhead Publishing, Cambridge, UK, pp. 52–61.
Kuzgunkaya, E.H. and Hepbasli, A. (2007) Exergetic evaluation of drying of laurel leaves in a vertical ground-source heat pump drying cabinet. *International Journal of Energy Research* 31, 245–258.
Lacroix, M., Marcotte, M. and Ramaswamy, H.S. (2003) Irradiation of fruits, vegetables, nuts and spices. In: Chakraverty, A., Mujumdar, A.S., Raghavan, G.S.V. and Ramaswamy, H.S. (eds) *Handbook of Postharvest Technology: Cereals, Fruits, Vegetables, Tea, and Spices.* Marcel Dekker, New York, pp. 623–653.
Lira, P.D.L., Retta, D., Tkacik, E., Ringuelet, J., Coussio, J.D., van Baren, C. and Bandoni, A.L. (2009) Essential oil and by-products of distillation of bay leaves (*Laurus nobilis* L.) from Argentina. *Industrial Crops and Products* 30, 259–264.
Longo, L. and Vasapollo, G. (2005) Anthocyanins from bay (*Laurus nobilis* L.) berries. *Journal of Agricultural and Food Chemistry* 53, 8063–8067.
Marzouki, H., Piras, A., Marongiu, B., Rosa, A. and Dessi, M.A. (2008) Extraction and separation of volatile and fixed oils from berries of *Laurus nobilis* L. by supercritical CO_2. *Molecules* 13, 1702–1711.
Mello da Silveira, S., Luciano, F.B., Fronza, N., Cunha, A. Jr, Scheuermann, G.N. and Vieira, C.R. (2014) Chemical composition and antibacterial activity of *Laurus nobilis* essential oil towards foodborne pathogens and its application in fresh Tuscan sausage stored at 7°C. *LWT – Food Science and Technology* 59, 86–93.
Millezi, A.F., Caixeta, D.S., Rossoni, D.F., Cardoso, M.D.G. and Piccoli, R.H. (2012) *In vitro* antimicrobial properties of plant essential oils [of] *Thymus vulgaris, Cymbopogon citratus* and *Laurus nobilis*

against five important foodborne pathogens. *Ciência e Tecnologia de Alimentos [Food Science and Technology (Campinas)]* 32(1), 167–172.
Moghtader, M. and Salari, H. (2012) Comparative survey on the essential oil composition from the leaves and flowers of *Laurus nobilis* L. from Kerman Province. *Journal of Ecology and the Natural Environment* 4, 150–153.
Muchuweti, M., Kativu, E., Mupure, C.H., Chidewe, C., Ndhlala, A.R. and Benhura, M.A.N. (2007) Phenolic composition and antioxidant properties of some spices. *American Journal of Food Technology* 2, 414–420.
Muggeridge, M. and Clay, M. (2001) Quality specifications for herbs and spices. In: Peter, K.V. (ed.) *Handbook of Herbs and Spices, Vol. 1.* Woodhead Publishing, Cambridge, UK, pp. 13–21.
Muñiz-Márquez, D.B., Martínez-Ávila, G.C., Wong-Paz, J.E., Belmares-Cerda, R., Rodríguez-Herrera, R. and Aguilar, C.N. (2013) Ultrasound-assisted extraction of phenolic compounds from *Laurus nobilis* L. and their antioxidant activity. *Ultrasonics Sonochemistry* 20, 1149–1154.
Onay, Ö. (2014) Effects of catalyst on pyrolysis of laurel (*Laurus nobilis L.*) seed in a fixed bed tubular reactor. *Chemical Engineering Transactions* 37, 127–132.
Ouchikh, O., Chahed, T., Ksouri, R., Taarit, M.B., Faleh, H., Abdelly, C., Kchouk, M.E. and Marzouk, B. (2011) The effects of extraction method on the measured tocopherol level and antioxidant activity of *L. nobilis* vegetative organs. *Journal of Food Composition and Analysis* 24, 103–110.
Ozcan, B., Esen, M., Sangun, M.K., Coleri, A. and Caliskan, M. (2010) Effective antibacterial and antioxidant properties of methanolic extract of *Laurus nobilis* seed oil. *Journal of Environmental Biology* 31, 637–641.
Özden, M.G., Öztaş, P., Öztaş, M.O. and Önder, M. (2001) Allergic contact dermatitis from *Laurus nobilis* (laurel) oil. *Contact Dermatitis* 45, 178–178.
Özek, T. (2012) Distillation parameters for pilot plant production of *Laurus nobilis* essential oil. *Records of Natural Products* 6, 135–143.
Panza, E., Tersigni, M., Iorizzi, M., Zollo, F., De Marino, S., Festa, C., Napolitano, M., Castello, G., Ialenti, A. and Ianaro, A. (2010) Lauroside B, a megastigmane glycoside from *Laurus nobilis* (bay laurel) leaves, induces apoptosis in human melanoma cell lines by inhibiting NF-κB activation. *Journal of Natural Products* 74, 228–233.
Parthasarathy, V.A., Zachariah, T.J. and Chempakam, B. (2008) Bay leaf. In: Parthasarathy, V.A., Chempakam, B. and Zachariah, T.J. (eds) *Chemistry of Spices.* CAB International, Wallingford, UK, pp. 426–434.
Polat, S., Gulbaba, A.G., Tufekci, S. and Ozkurt, A. (2009) *Determination of the Most Suitable Leaf Harvesting Methods of Bay Laurel (*Laurus nobilis *L.) and its Economy (the Case of Tarsus).* Technical Bulletin No: 34, Eastern Mediterranean Forestry Research Institute, Tarsus, Turkey. [In Turkish.]
Politeo, O., Jukic, M. and Milos, M. (2007) Chemical composition and antioxidant activity of free volatile aglycones from laurel (*Laurus nobilis* L.) compared to its essential oil. *Croatica Chemica Acta* 80, 121–126.
Polovka, M. and Suhaj, M. (2010) The effect of irradiation and heat treatment on composition and antioxidant properties of culinary herbs and spices – a review. *Food Reviews International* 26, 138–161.
Qnais, E.Y., Abdulla, F.A., Kaddumi, E.G. and Abdalla, S.S. (2012) Antidiarrheal activity of *Laurus nobilis* L. leaf extract in rats. *Journal of Medicinal Food* 15, 51–57.
Raghavan, S. (2007) *Handbook of Spices, Seasonings, and Flavorings,* 2nd edn. CRC Press, Boca Raton, Florida.
Ramos, C., Teixeira, B., Batista, I., Matos, O., Serrano, C., Neng, N.R., Nogueira, J.M.F., Nunes, M.L. and Marques, A. (2012) Antioxidant and antibacterial activity of essential oil and extracts of bay laurel *Laurus nobilis* Linnaeus (Lauraceae) from Portugal. *Natural Product Research* 26, 518–529.
Rosello, J., Sempere, F., Sanz-Berzosa, I., Chiralt, A. and Santamarina, M.P. (2015) Antifungal activity and potential use of essential oils against *Fusarium culmorum* and *Fusarium verticillioides. Journal of Essential Oil Bearing Plants* 18, 359–367.
Saab, A.M., Tundis, R., Loizzo, M.R., Lampronti, I., Borgatti, M., Gambari, R., Menichini, F., Esseily, F. and Menichini, F. (2012) Antioxidant and antiproliferative activity of *Laurus nobilis* L. (Lauraceae) leaves and seeds essential oils against K562 human chronic myelogenous leukaemia cells. *Natural Product Research* 26, 1741–1745.
Sagoo, S.K., Little, C.L., Greenwood, M., Mithani, V., Grant, K.A., McLauchlin, J., de Pinna, E. and Threlfall, E.J. (2009) Assessment of the microbiological safety of dried spices and herbs from production and retail premises in the United Kingdom. *Food Microbiology* 26, 39–43.

Sangun, M.K., Aydin, E., Timur, M., Karadeniz, H., Caliskan, M. and Ozkan, A. (2007) Comparison of chemical composition of the essential oil of *Laurus nobilis* L. leaves and fruits from different regions of Hatay, Turkey. *Journal of Environmental Biology* 28, 731–733.
Santoyo, S., Lloria, R., Jaime, L., Ibanez, E., Senorans, F.J. and Reglero, G. (2006) Supercritical fluid extraction of antioxidant and antimicrobial compounds from *Laurus nobilis* L. Chemical and functional characterization. *European Food Research and Technology* 222, 565–571.
Sayyah, M., Valizadeh, J. and Kamalinejad, M. (2002) Anticonvulsant activity of the leaf essential oil of *Laurus nobilis* against pentylenetetrazole- and maximal electroshock-induced seizures. *Phytomedicine* 9, 212–216.
Sayyah, M., Saroukhani, G., Peirovi, A. and Kamalinejad, M. (2003) Analgesic and anti-inflammatory activity of the leaf essential oil of *Laurus nobilis* Linn. *Phytotherapy Research* 17, 733–736.
Schweiggert, U., Carle, R. and Schieber, A. (2007) Conventional and alternative processes for spice production – a review. *Trends in Food Science and Technology* 18, 260–268.
Seidemann, J. (2005) *World Spice Plants: Economic Usage, Botany, Taxonomy.* Springer, Heidelberg, Germany.
Sellami, I.H., Wannes, W.A., Bettaieb, I., Berrima, S., Chahed, T., Marzouk, B. and Limam, F. (2011) Qualitative and quantitative changes in the essential oil of *Laurus nobilis* L. leaves as affected by different drying methods. *Food Chemistry* 126, 691–697.
Simić, A., Soković, M.D., Ristić, M., Grujić-Jovanović, S., Vukojević, J. and Marin, P.D. (2004) The chemical composition of some Lauraceae essential oils and their antifungal activities. *Phytotherapy Research* 18, 713–717.
Spratt, G. (1830) *Flora Medica: Containing Coloured Delineations of the Various Medicinal Plants.* Callow and Wilson, London.
Tuzlacı, E. and Sadıkoğlu, E. (2007) Turkish folk medicinal plants, Part VI: Koçarlı (Aydın). *Journal of Faculty Pharmacy of Istanbul University* 39, 25–37.
Tuzlacı, E. and Tolon, E. (2000) Turkish folk medicinal plants, Part III: Şile (Istanbul). *Fitoterapia* 71, 673–685.
UNIDO and FAO (2005) *Herbs, Spices and Essential Oils: Post-harvest Operations in Developing Countries.* United Nations Industrial Development Organization, Vienna, Austria and Food and Agriculture Organization of the United Nations, Rome.
US FDA (2013) *FDA Draft Risk Profile: Pathogens and Filth in Spices.* US Food and Drug Administration, Silver Spring, Maryland. Available at: http://www.fda.gov/downloads/Food/FoodScienceResearch/RiskSafetyAssessment/UCM367337.pdf (accessed 13 October 2014).
US FDA (2014) Food Additive Status List, Bay, Bay leaves. US Food and Drug Administration, Silver Spring, Maryland. Available at: http://www.fda.gov/food/ingredientspackaginglabeling/foodadditivesingredients/ucm091048.htm#ftnB (accessed 1 August 2014).
USDA (2014) National Nutrient Database for Standard Reference. US Department of Agriculture, Washington, DC. Available at: http://ndb.nal.usda.gov/ndb/search/list (accessed 1 August 2014).
Weiss, E.A. (2002) *Spice Crops.* CAB International, Wallingford, UK.
WHO (1998) *Quality Control Methods for Medicinal Plant Materials.* World Health Organization, Geneva, Switzerland.
Wiart, C. (2006) *Medicinal Plants of Asia and the Pacific.* CRC Press, Boca Raton, Florida.
Yalçın, H., Akın, M., Şanda, M.A. and Çakır, A. (2007) Gas chromatography/mass spectrometry analysis of *Laurus nobilis* essential oil composition of northern Cyprus. *Journal of Medicinal Food* 10, 715–719.
Zekovic, Z.P., Lepojevic, Z.D. and Mujic, I.O. (2009) Laurel extracts obtained by steam distillation, supercritical fluid and solvent extraction. *Journal of Natural Products* 2, 104–109.
Zeybek, N. and Zeybek, U. (1994) *Pharmaceutical Botany: Angiospermae Systematics and Important Items*, 2nd edn. Ege Universitesi Basımevi, Izmir, Turkey. [In Turkish.]

5 Betel Vine

S. Jacob K. Annamalai, S. Reetha Subashini, J.S. Rutra Priya and Ravindra Naik*
ICAR – Central Institute of Agricultural Engineering, Regional Centre, Coimbatore, India

5.1 Botany

5.1.1 Introduction

The betel vine plant is reported to have originated from South and South-east Asia. Its scientific name is *Piper betle* L. It belongs to the family Piperaceae, the black pepper family. The names of betel vine in local (Indian) languages are *vetrilai* in Tamil, *tambula* in Sanskrit, *vettilakkotti* in Malayalam, *villaya* in Kannada, *tamalapaku* in Telugu, *vedechpan* in Marathi, *nagerbel* in Gujrati and *pan* in Hindi and Bangala. In other (foreign) languages, it is called *tanbol* in Arabic and *burg-e-tanbol* in Persian. [There are variations on many of these spellings.] Betel leaves are a special item that is offered to guests in order to show respect and they are traditionally used in Indian society. A well-prepared betel quid (a combination of betel leaf, areca nut and slaked lime, which may also contain tobacco) is regarded as an excellent mouth freshener and mild vitalizer, and is regularly served on social, cultural and religious occasions such as marriage, puja (a religious festival) and the religious function performed after cremation (Mehrotra, 1981; Guha, 1997). The leaves are traditionally used for chewing in their natural raw condition, along with many other ingredients, including sliced areca nut, slaked lime, coriander, aniseed, clove, cardamom, sweetener, coconut scrapings, ashes of diamond, pearl, gold and silver (Ayurvedic preparations), jelly, peppermint, flavouring agent, fruit pulp, etc. (Krishnamurthi, 1969).

The fresh leaves of betel vine (see Fig. 5.1) are popularly known as *paan* in India, and they are consumed by about 20 million people in that country. Betel is cultivated over about 55,000 ha and has an annual production worth about Rs 9000 million (Kaleeswari and Sridhar, 2013). The leaves are nutritive and also contain anticarcinogenic agents, thereby showing much promise for manufacturing of a blood cancer drug. Some disputed reports also claim that the excessive chewing of betel leaves may cause oral cancer. It has been reported that there is a great wastage of the leaves during storage and transportation, and during the glut season. Moreover, the surplus leaves, if not disposed of properly, may cause environmental pollution and health hazards. Such wastage may be minimized in various ways, including the use of surplus betel leaves for the extraction of essential oils. The oil can be used as an industrial raw material for manufacturing

*Corresponding author, e-mail: naikravindra@gmail.com

© CAB International 2016. *Leafy Medicinal Herbs: Botany, Chemistry, Postharvest Technology and Uses* (eds D.C.P. Ambrose *et al.*)

Fig. 5.1. A betel vine plant. From *The Hindu* (http://www.thehindu.com).

medicines, perfumes, mouth fresheners, tonics, food additives, etc. (Guha, 2007) This chapter describes the importance of betel vine leaves and their processing to develop various value-added products.

5.1.2 History

Betel leaves have been used by humans since time immemorial. The use of *tambula* is been mentioned in the *Vedas* and *Ayurveda Sastra*. Betel leaf with a bit of betel nut has been used in Hindu rituals as a pious offering to God on many auspicious occasions and for older people as a mark of respect during ceremonies. The commercial product is the betel leaf, which is mainly used for chewing with areca nut, slaked lime, tobacco and some other ingredients (as the betel quid). The betel chewing habit in Sri Lanka dates back to 340 BC and during that time betel was a prestigious item used by the prestigious society of the country.

The chewing of 'pan' has also been said to be popular among 'aryas' (nobles) and has been credited with many medicinal properties, as indicated in the *Susruta Samhita*, a classical Sanskrit text on medicine. Since then, betel leaves have occupied an important place in daily life of Indian people (Rayaguru *et al.*, 2007). Historically, the word 'pan' in Hindi and other Indian languages is probably a derivative of the Sanskrit word 'pan', meaning leaf. It has been very intimately connected with ancient Indian history, religion and culture, as is evident from many references in the early Sanskrit literature (3000 BC), such as the *Vedas*, *Ramayana*, *Mahabharata*, *Mahavansha*, etc. Marco Polo (1295 AD) took notice of the pan-chewing habit of the people in south India. Over the centuries, pan chewing had become so prevalent that the serving and chewing of pan had been raised to various levels of fine art at the Mughal darbar (the Mughal court), particularly during the Akbar's regime (Chopra *et al.*, 1956).

5.1.3 Location

Betel is believed to have originated from Malaysia or the surrounding East Asian region and other South Asian countries, and have been traded by Chinese and Arab merchants (Rathnasoma and Senavirathna, 2002). Today, betel is grown for local consumption and export, and the major betel-/growing countries are Sri Lanka, India, Thailand and Bangladesh.

Malay culture and tradition hold betel (areca) nut and betel leaves in high esteem, as is evident from their use in many social and religious ceremonies. The exchange of betel leaves and areca nuts, or the presentation of betel leaves or a bunch of areca nuts, bundles of betel leaves and a betel box have earned their own social meaning and importance. *Tambula* is a token of honour, pledge and love, and exchange is a sign of marriage or betrothal. This single tradition is an integral part of the folklore, art, rituals, ceremonies and social intercourse of daily life in Cambodia, India, Malaysia, Myanmar, Thailand and Vietnam (Ahuja and Uma, 2011).

There are more than 100 varieties of betel vine in the world, of which about 40 are found in India and 30 in West Bengal (Maity, 1989; Guha, 1997). In spite of being an alien species, the plant is, and has been since antiquity, much more popular in India than in any other country of the world.

5.1.4 Morphology

The betel leaf is heart shaped and the size of the leaf varies between different cultivars.

The leaves are simple alternate stipulate petiolate with 0.75 to 3.8 cm long, ovate oblong broadly ovate cordate or obliquely elliptic entire glabrous coriaceous 10 to 18 cm long and 5 to 10 cm broad acuminate oblique and rounded base. The general colour of the leaves is yellowish green to dark green in colour with a glossy upper surface and a characteristic and pleasant odour. The leaves are aromatic with a varied taste ranging from sweet to pungent due to the presence of essential oils.

A transverse section of the leaf through the midrib shows a four-layered upper and a two-layered lower epidermis. The cuticle is thick on the upper epidermis and thin on the lower epidermis. The cells of the outer epidermal layers on both sides of the leaf are small and contain tannins and oils. The sub-epidermal cells on the abaxial side are enlarged and store water. Crystal and oil reserves are found in the sub-epidermal cells on both sides of the leaf. The palisade layers are well distinguished, with double-layered short wide compact cells; the mesophyll cells are three–four layered and small lobed. Thick walled irregular secretory cells are seen with dense contents, mostly of essential oil (Pradhan *et al.*, 2013).

5.2 Chemistry

5.2.1 Chemical and nutritional composition

P. betle contains a wide variety of biologically active compounds whose concentration depends on the variety of the plant, season and climate. The aroma of betel leaf is due to the presence of essential oils that consist of phenols and terpenes. The betel leaf is rich in many components that are nutritious and essential for healthy building of the body. The juice of the pan has nutritional elements of various kinds, including vitamins A, B and C, essential oils, alkaloids and arecolin, etc. Tables 5.1 and 5.2 present data on the composition of fresh betel leaves.

Table 5.1. Nutritional composition of fresh betel leaves. From Guha, 2006.

Constituents	Approximate composition
Calcium	0.2–0.5%
Carbohydrate	0.5–6.10%
Chlorophyll	0.01–0.25%
Energy	44 kcal/100 g
Essential oil	0.08–0.2%
Fat	0.4–1.0%
Fibre	2.3%
Iodine	3.4 μg/100 g
Iron	0.005–0.007%
Minerals	2.3–3.3%
Nicotinic acid	0.63–0.89 mg/100 g
Nitrogen	2.0–7.0%
Phosphorus	0.05–0.6%
Potassium	1.1–4.6%
Protein	3–3.5%
Riboflavin	1.9–30 μg/100 g
Tannin	0.1–1.3%
Thiamine	10–70 μg/100 g
Vitamin A	1.9–2.9 mg/100 g
Vitamin C	0.005–0.01%
Water	85–90%

Table 5.2. Percentage chemical composition of fresh betel leaves. From Rekha *et al.*, 2014.

Constituents (in descending order of content)	Approximate composition (%)
Chavibetol	53.1
Chavibetol acetate	15.5
Caryophyllene	3.71
Allylpyrocatechol diacetate	0.71
Campene	0.48
Chavibetol methyl ether	0.48
Eugenol	0.32
Allylpyrocatechol monoacetate	0.23
α-Pinene	0.21
β-Pinene	0.21
D-Limonene	0.14
Safrole	0.11
1,8-Cineol	0.04

5.2.2 Phytochemistry

The specific strong pungent aromatic flavour in betel leaves is due to the presence of phenol and terpene-like compounds. The total phenol content varies with plant gender. The male plant contains threefold higher total phenolic content and twofold higher thiocyanate content compared with female plants. The quality of the leaf depends upon the phenolic content, i.e. the more the phenolic content, the better the leaf quality. Many earlier research works have shown that betel leaves contain starch, diastases, sugars and an essential oil composing of safrole, allyl pyrocatechol monoacetate, eugenol, terpinen-4-ol, eugenyl acetate, etc. as the major components. Phytochemical investigations of leaves also revealed the presence of alkaloids, carbohydrates, amino acids, tannins and steroidal components. The middle part of the main vine contains the largest quantity of tannin. The major terpenoids and phenols found in betel leaf include 1,8-cineole (or 1,8-cineol, also known as eucalyptol), cadinene, camphene, caryophyllene, limonene, pinene, chavicol, allylpyrocatechol, carvacrol, safrole, eugenol and chavibetol. Eugenol has been identified as the antifungal principle in the oil. The fresh new leaves contain a much greater amount of essential oil, the enzyme diastase and sugar than old leaves. Chavicol is four times more potent as an antiseptic agent than carbolic acid (Pradhan *et al.*, 2013).

5.3 Postharvest Technology

5.3.1 Harvesting and processing

As the betel vines reach to a certain height, leaves are harvested from the lower portion of the stem. In India, harvesting is done during March–April in Uttar Pradesh, Madhya Pradesh and Bihar, during May–June in Andhra Pradesh and during January–February or April–May in Tamil Nadu. Mature leaves are plucked by hand along with a portion of petiole. In Karnataka and Tamil Nadu, leaves are plucked from side shoots. In south India, comparatively tender leaves are preferred in the market. After plucking, they are washed thoroughly and made into bundles according to the prevailing custom of the area.

Betel leaves are usually used for chewing fresh and unprocessed, but in certain areas, leaves are subjected to a process known as bleaching or curing. There is good demand for such leaves, which fetch higher prices in the markets. Bleaching is done by successive heat treatments at 60–70°C for 6–8 h (Rayaguru *et al.*, 1999).

Harvesting

Leaves that are sufficiently matured are plucked along with a portion of the petiole by hand without any aid. The level of maturity is decided based on the consumer preference in the local area. The leaves mature in 15–30 days and one or two harvestings can be done every month. In India, the leaves are plucked at an interval of 7–15 days and the yield is about 50–70 leaves per plant each year. About 7–8 million leaves are harvested annually from 1 ha of betel vine garden. Thus, the betel vine provides a continuous source of income, even to small farmers. On average, 60–80 lakh leaves are harvested annually from a 1 ha plantation.

Cleaning, sorting and packaging

The harvested leaves are washed, cleaned and graded according to their size, colour, texture, maturity and quality. Damaged leaves are rejected. The good quality leaves are separated, a portion of the petiole cut off and bundled into groups of 50 or 100, and then packed in bamboo baskets. Packing is mostly into bamboo baskets, and in many places straw, fresh or dried banana leaves, wet cloth or gunny cloth, etc. are used for an inner lining. The packaging is done in a very specific manner. The betel leaf bundles are arranged towards the periphery of the basket in a circular pattern so that a cavity is created at the centre (Fig. 5.2). Then the basket, along with the leaves, is covered with a layer of gunny cloth on top and stitched properly. Freshly plucked leaves are sold in the local pan markets either to local pan vendors or to middlemen and processors.

Conditioning

Conditioning is a process in which the green colour of the leaves will be changed

Fig. 5.2. Betel leaves in baskets for market sale. Photo courtesy of Liz Bordo, from Burma Link (available at: www.burmalink.org/background/burma/ethnic-groups/overview/).

to yellow/white. This is in high demand by a group of consumers. The process not only fetches a high price but also increases the storability of the leaves to a significant extent. In the traditional method of conditioning, a small chamber (called a Bhatti) made of brick and mud is used. The chambers/vats are of various sizes ranging from 1.2 × 1.2 × 2.4 m to 2.7 × 2.7 × 2.4 m depending on the capacity needed. They are usually constructed inside houses. The walls of the vat are constructed of brickwork of mud plaster, sometimes accompanied by a layer of cow dung. The floor is also either cemented or covered by an even layer of clay and dung to make it airtight. An insulated door of 0.75 × 1.8 m is provided at the front that is fabricated of a bamboo mat frame covered with straw and gunny cloth. The door is about 15 cm larger on each side of the dimensions of the door opening. Care is taken to make the chamber airtight as well as insulated by restricting the opening of the door.

Inside the chamber, at a height of 30 cm from ground level, bamboo racks are provided at an interval of 45–60 cm between racks on all sides of the chamber, excluding the entrance side. The bundles are placed in such a manner that the front and backside of the leaves are exposed alternately. Then the baskets are covered with moist gunny cloth and arranged in the racks in order. At a given time, about 10–40 baskets each containing about 2000 leaves can be kept in the chamber. The number varies according to the size of the chamber. In one corner of the chamber, a small chullah is kept, in which wood charcoal is burned during conditioning. About 2–3 kg of charcoal are required for one charging. After firing, the movable chullah is kept in a corner away from the racks to avoid direct heat. The door is also properly covered to avoid any leakage of heat from the chamber. After 10 to 12 h of charging the baskets are taken out for cooling. The cooling period varies from 36 to 72 h depending upon the weather conditions. During the cooling period, the leaves are sorted and reshuffled. Before subjecting the baskets to a second phase of charging, the damaged or rotten leaves are discarded. The process is then repeated until all of the leaves are conditioned, i.e. the green colour of the leaves changes to a yellow colour. After the second phase, the amount of fuel used is gradually reduced. The fuel requirement for obtaining the desired temperature range, as well as to achieve the final conditioned stage of the betel leaves, is mainly decided through personal experience as no rule or scientific formula to calculate it exists. The temperature requirement varies with the variety and quality of the leaves. During the summer season, the leaves are fully conditioned after two or three chargings, but in winter, sometimes as many as five chargings are required. The conditioned leaves are finally packed in bamboo baskets to be transported to other cities (Rayaguru *et al.*, 2007).

5.3.2 Value addition

Several value-added products have been developed from betel leaves, namely betel toothpaste, mouthwash, shampoo, face cream and ointment, instant betel chew, instant betel quid and pellets, etc. The commercial use of betel vine should be cost-effective and safe. The leaves of the plant have a high economic and medicinal value, but since ancient times have been mostly used for chewing purposes and ceremonial events, along with other condiments (Arambewela *et al.*, 2006). The composition of the chewing combination in the form of a betel quid can be varied, and different ingredients are used from country to country.

The oil from betel leaves has multiple potential uses in cottage industry for the manufacturing of numerous commercial products, such as medicine, *gutkha* (chewable mouth freshener), incense sticks, fragrance and flavouring agents, etc. (Fig. 5.3). The establishment of a rural industry for the extraction of essential oil from betel leaves can be achieved at a very reasonable initial investment along with suitable ideas for minimizing the wastage of surplus leaves and increasing the agricultural and industrial employment opportunities in the betel leaf growing regions of India and other countries (Guha, 2007).

Fig. 5.3. Essential oil from betel leaves as a retail product. From SAT Group, Essential Oil.In, New Delhi (available at: http://www.essentialoil.in/betel-leaf-oil.html, where details are also given of the many medicinal uses).

5.4 Uses

5.4.1 General uses

Betel leaves have a strong pungent aromatic flavour and are widely used as a masticatory agent. Generally mature or overmature leaves, which have ceased physiological growth but not yet become brittle, are used for chewing. The basic preparation for chewing purpose consists of betel leaf smeared with hydrated lime and acacia extract to which a scraping of areca nut is added; flavourings such as coconut shavings, clove, cardamom, fennel, powered liquorice, nutmeg and also tobacco are used according to taste. In some places, the prepared pan is covered with silver or gold. A beverage called pan-supari nectar has been developed by the Central Food Technological Research Institute at Mysore, India, and is said to be a good source of calcium (Ramamurthi and Usha Rani, 2012).

General characteristics of betel leaves are as follows:

- The leaves are rich in vitamins B, C and E.
- They have a stimulatory effect on the heart, brain and liver.
- They clean the mouth and throat.
- They help in digestion by increasing salivation and when chewed with lime in neutralizing excess acid.
- They are good for teeth as they contain chlorophyll.
- They are useful in catarrhal, pulmonary infections and night blindness.
- The fresh leaf powder is used as a lotion for patients suffering from smallpox and enlarged glands.
- They are used with honey as a remedy for coughs.
- Betel leaf extract are used as an antioxidant for the storage of oily products such as fish, fish oil, ghee, etc.
- Betel leaves may be used for the manufacture of essential oil, perfume and food additives.

Industrial uses

The essential oil from betel vine is very important in medicines and in the pan masala industry. The oil is used by manufacturers of sweet betel nut, zarda (chewing tobacco) and pan masala sold in the 'Pan shop'.

There is wide scope for using pan oil in the medicine industry and in perfumery, cosmetics, etc., and the prospects are good.

Culinary uses

Betel leaf is a popular spice in South-east Asian cooking, with the leaves being used in their raw and cooked forms. A traditional way of preparing the leaves is as a wrapping for spiced minced meat and other pieces of food. Because the leaves are so attractive, they are often used as a base for decorating platters, with food arranged on top of them. The white flower spikes of the betel plant develop into seeds or fruits that look a little like a green–brown mulberry when ripe and can be eaten; the fruit is a tasty morsel of sweet jelly-like pulp (Palaniappan *et al.*, 2012)

5.4.2 Pharmacological uses

The medicinal properties of pan were recognized as early as 600 AD when the Ayurvedic system of medicine came into practice. Betel leaves are beneficial to the throat and remove viscidity. The leaves also help in digestion and tend to remove bad smells from the mouth. The juice of betel leaves is used as an adjunct to the pills administered in the Ayurvedic medicines. The fresh crushed leaves are used as an antiseptic for cuts and wounds. They are also good for the respiratory system and are used in the treatment of bronchitis, coughs and colds (Chopra *et al.*, 1958). Pan chewing is considered to be a good and cheap source of dietary calcium up to a certain limit. It also increases the digestive capacity when used with lime, neutralizes the acidity and acts as a blood purifier.

As a folk medicine, betel leaf is traditionally known to be useful for the treatment of various problems and diseases, including bad breath, boils and abscesses, conjunctivitis, constipation, headache, hysteria, itches, mastitis, mastoiditis, leucorrhoea, otorrhoea, ringworm, swelling of the gums, rheumatism, abrasion, cuts and injuries, etc., while the root is known for its female contraceptive effects (Chopra *et al.*, 1956; Khanra, 1997). Furthermore, the essential oil contained in the leaves has antibacterial, antiprotozoal and antifungal properties. Therefore, betel leaf oil kills or inhibits growth of bacteria such as those causing typhoid, cholera and tuberculosis, a property that needs proper evaluation and exploitation (Krishnamurthi, 1969).

The presence of hydroxychavicol and chlorogenic acid in betel vine is reported to kill cancerous cells without affecting normal cells, unlike the common cancer drugs and other therapeutic means (Amonkar *et al.*, 1989). The possibility of manufacturing of a new blood cancer drug from betel vine cannot then be ruled out. Interestingly, it is also claimed that the inflorescence of betel vine contains carcinogens, whereas the leaves contain anticarcinogenic agents, so seeming to indicate that different parts of the same plant contain both carcinogenic and anticarcinogenic substances (Wu *et al.*, 2004).

Scanty or obstructed urination

Betel leaf juice is reported to possess diuretic properties. Its juice, when mixed with dilute milk and sweetened slightly, helps in easing the passage of urine (Palaniappan *et al.*, 2012).

Weakness of nerves

Betel leaves are beneficial in the treatment of nervous disorders. The juice of a few betel leaves, with a teaspoon of honey, serves as a good tonic. A teaspoon of this mixture can be taken twice a day (Palaniappan *et al.*, 2012).

Headaches

The betel leaf has analgesic and cooling properties. It can be applied to relieve intense headaches (Palaniappan *et al.*, 2012).

Respiratory disorders

Betel leaves are useful in pulmonary afflictions suffered in childhood and old age. The leaves, soaked in mustard oil and warmed, may be applied to the chest to relieve a cough or difficulty in breathing (Palaniappan *et al.*, 2012).

Constipation

In the case of constipation in children, a suppository (a solid medical preparation in a roughly conical or cylindrical shape, designed to be inserted into the rectum) made of the stalk of betel leaf dipped in castor oil can be introduced in the rectum. This instantly relieves constipation (Palaniappan *et al.*, 2012).

Sore throats

Local application of the leaves is effective in treating sore throats. The crushed fruit or berry should be mixed with honey and taken to relieve an irritating cough (Palaniappan *et al.*, 2012).

Wounds

Betel leaves can be used to heal wounds. The juice of a few leaves should be extracted and applied on to the wound. A betel leaf should then be wrapped over it and bandaged. The wound will heal with just a single application, within 2 days (Palaniappan *et al.*, 2012).

Boils

The herb is also an effective remedy for boils. A leaf is gently warmed until it is softened, and then it is coated with a layer of castor oil. The oiled leaf is spread over the inflamed part and is replaced every few hours. After a few applications, the boil will rupture, draining out all of the purulent matter. The application can be made at night and removed in the morning (Palaniappan *et al.*, 2012).

Problems with breast milk secretion

The application of leaves smeared with oil is said to promote the secretion of milk when applied on to the breasts of feeding mothers during lactation (Palaniappan *et al.*, 2012).

Antimicrobial activity

The leaf has significant antimicrobial activity against a broad spectrum of microorganisms (Jesonbabu *et al.*, 2012). It has antimicrobial activity against *Streptococcus pyogenes*, *Staphylococcus aureus*, *Proteus vulgaris*, *Escherichia coli*, *Pseudomonas aeruginosa*, etc. The leaf extract also shows bactericidal activity against urinary tract pathogenic bacteria such as *Enterocococcus faecalis*, *Citrobacter koseri*, *C. fruendi*, *Klebsiella pneumoniae*, etc. (Chakraborty and Shah, 2011; Agarwal and Singh, 2012). The bioactive molecule thought to be responsible for antibacterial activity is a sterol, which has been obtained in large quantities in betel leaf extracts (Chakraborty and Shah, 2011).

Gastroprotective activity

A hot water betel leaf extract can protect against indomethacin-induced gastric ulceration owing to its antioxidant and mucin-protecting properties (Pradhan *et al.*, 2013). The allylpyrocatacol that is present in the betel leaf has been shown to have powerful antioxidant potential. Treatment with allylpyrocatacol significantly accelerates the ulcer-healing process by enhancing mucus production rather than decreasing acid production. This assists in the healing process of ulcers by protecting the ulcer crater against irritant stomach secretions (HCl and pepsin), which enhances the rate of the local healing process (Bhattacharya *et al.*, 2007).

Antioxidant activity

Oxidative damage is an important effect of ionizing radiation on biological membranes. The presence of phenolic compounds such as catechol and allylpyrocatechol in betel leaf extract effectively inhibits the radiation-induced lipid peroxidation process in biological membranes. This can be attributed to the ability of phenolics to scavenge the free radicals involved in the initiation and propagation steps of peroxidation. Betel leaf extracts had a strong reductive ability and reduced most of the Fe^{3+} ions (Manigauha *et al.*, 2009). The extract also shows strong hydroxyl radical and superoxide anion radical scavenging properties in comparison with standards such as ascorbic acid and BHT (butylated hydroxytoluene) (Arambewela *et al.*, 2006; Pin *et al.*, 2010).

Antidiabetic activity

In glucose tolerance tests in streptozocin (STZ)-induced diabetic rats, hot and cold aqueous extracts of betel leaves markedly reduced the external glucose load. In tests with a leaf suspension, blood glucose and glycosylated haemoglobin were significantly reduced, and the activities of liver glucose-6-phosphatase and fructose-1,6-bisphosphatase decreased, whereas liver hexokinase increased in a streptozocin (STZ) rat diabetic model compared with untreated diabetic models (Arambewela *et al.*, 2005; Santhakumari *et al.*, 2006).

Anti-fertility activity

Any small changes in oestrogen level may lead to altered structural and functional activity of the reproductive organs. Earlier studies with betel leaf extract suggest that it has reversible antifertility and anti-oestrogenic effects in female rats. The action of the betel leaf extract may be via the pituitary-gonadal axis, resulting in diminished release of gonadotrophin, which in turn reduces reproductive organ weights and oestrogen levels, thus affecting ovarian cysts. Serum biochemical analyses show that treatment lowers glucose but raises cholesterol and vitamin C concentrations, indicating the non-utilization of cholesterol by the system, and hence a decrease in oestrogen (Pradhan *et al.*, 2013).

Oral care agent

Plaque and dietary carbohydrates play roles in the initiation of dental caries. Certain cariogenic and highly acidogenic strains of streptococci, especially *S. mutans*, have the ability to metabolize dietary sucrose and synthesize glucan by cell-surface and extracellular glucosyl transferase, an enzyme that is considered to be of special importance in the establishment of *S. mutans* in dental plaque. Aqueous extracts of betel leaves have been reported to show anticaries properties (Zain, 2011). Razak and Rahim (2003) described the anti-adherence effects of aqueous betel extracts on the adhesion of early plaque settlers such as *S. mitis*, *S. sanguinis* and *Actinomyces* sp. Aqueous extracts of betel vine have also been shown to inhibit various acid-producing oral pathogens that change the ultrastructure of the enamel and its properties, including streptococci, lactobacilli, staphylococci, corynebacteria, *Porphyromonas gingivalis* and *Treponema denticola*. Bissa *et al.* (2007) demonstrated the advantages of chewing betel leaves for oral hygiene, and in particular the effectiveness of the mixture used in betel quids against the bacterial population of mouth cavity. Betel is viewed as the best natural substance and rated the as second most popular daily consumed item in Asia that contributes to oral hygiene.

Drawbacks

The harmful effects of pan as described in Ayurveda are that it weakens teeth, impairs health and deadens the taste buds of the tongue. In the Indian subcontinent, where chewing tobacco with pan is a common habit, cancer of the mouth is very common. However, educated Indians are of the opinion that moderate use of betel leaf is not only innocuous but may even be conducive to good health. As classically attributed to Paracelsus, the saying is that: 'All things are poison and nothing is without poison; only the dose makes a thing not a poison'.

Neuropharmacological profile

A hydroalcoholic extract of betel leaves showed improvement in the discrimination index, potentiated haloperidol-induced catalepsy, reduced basal as well as amphetamine-induced increased locomotor activity and delayed sodium nitrite-induced respiratory arrest. Vyawahare and Bodhankar (2007) suggest a possible facilitation of cholinergic transmission and the inhibition of dopaminergic as well as noradrenergic transmission by the extract.

5.5 Summary

Betel leaf contains various biologically active compounds that are responsible for its antioxidant and chemopreventive activities, etc. The leaf has a great potency in acting as a natural antioxidant. Further research needs to be performed to study the detailed mechanism of action of betel leaves in various metabolic activities in humans, which may be beneficial to mankind. No systematic efforts have been made so far to improve the processing, packaging and marketing of this potentially valuable cash crop, although a large number of villagers/stakeholders exclusively engage themselves traditionally in betel vine cultivation. Such businesses mostly operate based on personal experience. Scientific studies on the optimization of all of the above-mentioned parameters will no doubt lead to the minimization of losses. The medicinal importance of the herb that has been discussed suggests that betel leaf is one of the most promising medicinal herbs and that it has many therapeutic values.

References

Agarwal, T. and Singh, R. (2012) Evaluation of antimicrobial activity of *Piper betel* cultivars. *Novus International Journal of Pharmaceutical Technology* 1(1), 50–58.

Ahuja, S.C. and Uma, A. (2011) Betel leaf and betel nut in India: history and uses. *Asian Agri-History* 15, 13–35.

Amonkar, A.J., Padma, P.R. and Bhide, S.V. (1989) Protective effect of hydroxychavicol, a phenolic component of betel leaf, against the tobacco-specific carcinogens. *Mutation Research* 210, 249–253.

Arambewela, L.S., Arawwawala, L.D. and Ratnasooriya, W.D. (2005) Antidiabetic activity of aqueous and ethanolic extract of *Piper betle*. *Journal of Ethnopharmacology* 102, 239–245.

Arambewela, L., Arawwawala, M. and Rajapaksa, D. (2006) *Piper betle*: a potential natural antioxidant. *International Journal of Food Science and Technology* 41(Supplement s1), 10–14.

Bhattacharya, S., Banerjee, D., Bauri, A.K., Chattopadhyay, S. and Bandyopadhyay, S.K. (2007) Healing property of the *Piper betel* phenol, allylpyrocatechol against indomethacin induced stomach ulceration and mechanism of action. *World Journal of Gastroenterology* 13, 3705–3713.

Bissa, S., Songara, D. and Bohra, A. (2007) Traditions in oral hygiene: chewing of betel (*Piper betle* L.) leaves. *Current Science* 92, 26–28.

Chakraborty, D. and Shah, B. (2011) Antimicrobial, antioxidative and antihemolytic activity of *Piper betel* leaf extracts. *International Journal of Pharmaceutical Sciences* 3(Supplement 3), 192–199.

Chopra, R.N., Nayar, S.L. and Chopra, I.C. (1956) *Glossary of Indian Medicinal Plants*. Council of Scientific and Industrial Research, New Delhi.

Chopra, R.N., Chopra, I.C., Handa, K.L. and Kapur, L.D. (1958) *Chopra's Indigenous Drugs of India*, 2nd edn. Academic Publishers, Kolkata, India.

Guha, P. (1997) *Exploring Betel Leaves for Cottage Industry*. Agricultural and Food Engineering Department, Indian Institute of Technology, Kharagpur, India, pp. 15–19.

Guha, P. (2006) Betel leaf: the neglected green gold of India. *Journal of Human Ecology* 19, 87–93.

Guha, P. (2007) Extraction of essential oil: an appropriate rural technology for minimizing wastage of surplus betel leaves. *Agricultural Mechanization in Asia, Africa, and Latin America* 38(4), 47–50.

Jesonbabu, J., Spandana, N. and Aruna Lakshmi, K. (2012) *In vitro* antimicrobial potentialities of chloroform extracts of ethnomedicinal plant against clinically isolated human pathogens. *International Journal of Pharmacy and Pharmaceutical Science* 4(Supplement 3), 624–626.

Kaleeswari, V. and Sridhar, T. (2013) A study on betel vine cultivation and market crisis in Karur District. *Indian Journal of Applied Research* 3(10), 1–3.

Khanra, S. (1997) Betel leaf based industry. *Nabanna Bharati* 30(2), 169.

Krishnamurthi, A. (ed.) (1969) *The Wealth of India – Raw Material Series, Vol. 8, Ph–Re*. Council of Scientific and Industrial Research, New Delhi, pp. 84–94.

Maity, P. (1989) *The Betel Vine*. Extension Bulletin, All India Coordinated Research Project of Betel Vine, Indian Institute for Horticultural Research, Hessarghatta, Bangalore, India.

Manigauha, A., Ali, H. and Maheshwari, M.U. (2009) Antioxidant activity of ethanolic extract of *Piper betel* leaves. *Journal of Pharmacy Research* 2, 194–195.

Mehrotra, R.S. (1981) Fungal diseases of betelvine and their control. In: Khanduja, S.D. and Balasubrahmanyam, V.R. (eds) *Proceedings of Group Discussion on Improvement of Betelvine Cultivation*. National Botanical Research Institute, Lucknow, India, pp. 3–12.

Palaniappan, G., Sengottiyan, A. and Saravanan, T. (2012) Betel leaf: the green gold of India. *Facts for You, April* 2012, pp. 21–24.

Pin, K.Y., Chuah, A.L., Rashih, A.A., Mazura, M.P., Fadzureena, J., Vimala, S. and Rasadah, M.A. (2010) Antioxidant and anti-inflammatory activities of extracts of betel leaves (*Piper betle*) from solvents with different polarities. *Journal of Tropical Forest Science* 22, 448–455.

Pradhan, D., Suri, K.A., Pradhan, D.K. and Biswasroy, P. (2013) Golden heart of the nature: *Piper betle* L. *Journal of Pharmacognosy and Phytochemistry* 1(6), 147–167.

Ramamurthi, K. and Usha Rani, O. (2012) Betel leaf: nature's green medicine. *Facts for You*, September 2012, 8–10.

Rathnasoma, H.A. and Senevirathna, J.M. (2002) *Technical Bulletin on Betel Cultivation*. Department of Export Agriculture, Peradeniya, Sri Lanka.

Rayaguru, K., Pal, U.S., Khan, M.K., Sahoo, G.R., Panda, M.K. and Sahoo, N.R. (1999) *Post Harvest Profile of Betel Leaves*. Technical Bulletin OUAT/CAET/PHT5/99/2, Indian Council of Agricultural Research, New Delhi.

Rayaguru, K., Khan, K., Sahoo, G. and Pal, U.S. (2007) Post-harvest practices of betel leaves in Orissa, India. *Agricultural Mechanization in Asia, Africa, and Latin America* 38(3), 33–37.

Razak, F.A. and Rahim, Z.H.A. (2003) The anti-adherence effect of *Piper betle* and *Psidium guajava* extracts on the adhesion of early settlers in dental plaque to saliva-coated glass surfaces. *Journal of Oral Science* 45, 201–206.

Rekha, V.P.B., Kollipara, M., Gupta, B.S., Bharath, Y. and Pulicherla, K.K. (2014) A review on *Piper betle* L.: nature's promising medicinal reservoir. *American Journal of Ethnomedicine* 1, 276–289.

Santhakumari, P., Prakasam, A. and Pugalendi, K.V. (2006) Antihyperglycemic activity of *Piper betle* leaf on streptozotocin-induced diabetic rats. *Journal of Medicinal Food* 9, 108–112.

Vyawahare, N.S. and Bodhankar, S.L. (2007) Neuropharmacological profile of *Piper betel* leaves extract in mice. *Pharmacology Online* 2, 146–162.

Wu, M.T., Chen, M.C. and Wu, D.C. (2004) Influences of life-style habits and p53 codon 72 and p21 codon 31 poly-morphisms on gastric cancer risk in Taiwan. *Cancer Letters* 205, 61–68.

Zain, N.B.M. (2011) Differential expression of gene of *Streptococcus mutans* in response to treatment with *Piper betle* aqueous extract – A research framework. In: *2011 International Conference on Bioscience, Biochemistry and Bioinformatics (IPCBEE), Vol. 5*. IACSIT Press, Singapore, pp. 467–469.

6 Celery

Svein Øivind Solberg*
Nordic Genetic Resource Center, Alnarp, Sweden

6.1 Botany

6.1.1 Introduction

Celery (*Apium graveolens* L.) is an important vegetable but also a spice and medicinal plant (Fig. 6.1). All parts of the plant are used. The crop is grown in all continents, with the largest production in the USA, Europe, China and India. The common name in English is celery; in French it is *céleri*, in Italian *seleri*, in Hindi *ajavaina*, in Urdu *kharasanior ajwain* and in Chinese *qíncài*.

Taxonomy

Celery belongs to the Apiaceae family and to the genus *Apium*, which contains around 30 species. The wild form of celery grows on the coastlines of Europe, West Asia and North Africa. The global database GBIF (2014) reports more than 7000 georeferenced records, most of them from Europe (from Sweden in the north to Spain, Italy and Greece in the south). Records from Southern Africa, from North America and from Asia and Australia can also be found, most likely owing to the occurrence of naturalized plants from earlier cultivation. According to Vavilov (1926), the centre of origin is in the Mediterranean.

Cultivated celery can be divided into three subtypes:

- Leaf celery (*A. g.* var. *secalinum* (Alef.) Mansf.), also termed smallage, Indian celery or Chinese celery, cultivated for its aromatic leaves and seeds. The slender petioles are not as thick as in stalk celery. The leaves are dark green and smaller than in stalk celery. Leaf celery is closest to the wild type and is cultivated in Asia.
- Stalk celery (*A. g.* var. *dulce* (Mill.) Pers.), which is cultivated for its fleshy aromatic petioles. The leaves are more yellowish green than in leaf and root celery.
- Root celery (*A. g.* var. *rapaceum* (Mill.) Gaud.), which is cultivated for its aromatic tubers (celeriac) derived from the hypocotyl and upper part of the taproot. The tubers have a creamy white and firm flesh.

A recent study on Indian celery, based on DNA and phytochemical studies of seeds sampled from markets in Pakistan, reported the species to be *Seseli diffusum*, even though workers said it was *A. graveolens* (Maruyama *et al.*, 2009).

*Corresponding author, e-mail: svein.solberg63@gmail.com

© CAB International 2016. *Leafy Medicinal Herbs: Botany, Chemistry, Postharvest Technology and Uses* (eds D.C.P. Ambrose *et al.*)

Fig. 6.1. Celery stalks. Photo courtesy of A.B. Tskhovrebova and S.Ø. Solberg.

6.1.2 History/origin

The first historical record of celery is from Egypt (1200–600 BC), where the plant was used as a medicine. In ancient Greece, due to their strong odour, the leaves were used in ceremonies. The domestication of celery is believed to have taken place in Roman times, most likely during the first centuries AD. Since the 17th century, celery has been extensively cultivated in Europe. It also has a long medicinal tradition in Ayurvedic medicine in India, in Persian medicine and in Chinese medicine (Mathias, 1994).

6.1.3 Location

Celery benefits from humid and mild growth conditions, and has an optimal temperature range of 15–22°C (59–72°F). At lower temperatures, growth is slow, and bolting can destroy the crop, producing flower stalks instead of the marketable produce. At 6–10°C (43–50°F), bolting is induced in a couple of weeks. Celery is sensitive to both frost and very high temperatures; frost will damage the plant and very high temperatures will increase stress reactions, with a risk of physiological calcium deficit. The soils used for celery production are deep and rich, with good drainage and irrigation. A crop rotation system of a minimum of 4–5 years is used in order to keep soil-borne fungal diseases (such as *Septoria apiicola* and *Phoma apiicola*) and nematodes (such as *Ditylenchus dipsaci*) under control (Krug, 1991).

In the USA, celery is produced on 12,000 ha annually and California has about 75% of the production, followed by Michigan, Florida and Texas (USDA, 2014). In Europe, the most extensive production areas can be found in Italy (approx. 5000 ha), but Spain and France also have extensive production. In Central and Eastern Europe, root celery is the most common. In parts of Europe, such as France, a stalk celery type with greenish white tender petioles is also grown (Krug, 1991). The petioles acquire a light colour when straw or mulch are used to protect them from direct sunlight.

In India, celery is grown during the winter season and used as a vegetable, but the seeds are used as well. The main production is in the north, in Punjab and Uttar Pradesh (Sowbhagya, 2014), over a total area of approx. 5000 ha (Fazal and Singla, 2012). The seeds are used directly as a spice or processed into celery powder, essential oils or other products. India exports significant amounts of celery seeds for processing.

Celery is very popular in China (Yuman *et al.*, 2004). The leaf type is most commonly used and the term 'Chinese celery' is a synonym for this type, which has small and aromatic leaves. In Africa, celery is cultivated in highland regions and for market gardening (Schippers, 2004), but more recently also for processing and the food industry.

6.1.4 Morphology

Celery is a biannual herb plant. In the first year, a tight rosette of leaves develops on a compressed stem. The leaves alternate in direction, and are glossy with long petioles and distinct toothed sheets. The leaflets are triangular to rhombic, 2–5 cm × 1.5–3 cm, and often deeply lobed. The leaf type of celery has a higher number of leaves that are darker green in colour and with thinner petioles than other cultivated types (Rożek, 2007). The leaf petioles are crescentic in cross section, with a ribbed surface made of separate collenchyma bundles (Rubatzky and Yamaguchi, 1997). The roots are deep and strongly branched.

The root type of celery forms 10–20 cm wide tubers from the hypocotyl part of the stem and the upper part of the taproot. Flowers and seeds develop in the second year. The stems are strongly grooved and 70–100 cm tall. The flowers are greenish white and bisexual. The male anthers shed pollen before the female stigma is ready, thus preventing self-pollination. The assistance of flies, honeybees and other hairy insects is needed for pollination and the production of seeds. The fruits (seeds) are broadly ovoid, distinctly ribbed and small (up to 1.5 mm long).

6.2 Chemistry

6.2.1 Chemical composition

Celery contains vitamins, minerals and aromatic compounds, and is regarded as a healthy, low-energy vegetable. Every edible 100 g gives only approx. 20 kcal, but 32 mg vitamin C and 0.2 mg vitamin A, and is rich in potassium (280 mg) (Rizzo and Muratore, 2009). Variation in nutritional composition is due to the soil and fertilizers used, genotype, harvesting and storage conditions. After harvest, the contents of sugars, minerals and other compounds in leaves are reduced owing to translocation to the storage organs (stalks or tubers). Ninfali and Bacchiocca (2003) reported that plant genotype is an important source of variation in the antioxidant capacity of the vegetable. The leaves are normally higher in vitamins and minerals than the stalks and tubers. Celery contains many aromatic compounds, and certain flavone glycosides and volatile oils provide its typical aroma.

Antioxidant activity can be expressed in µmol vitamin C equiv./g. Compared with some other vegetables, the total antioxidant activity of celery is, however, relatively low: 5 µmol vitamin C equiv./g compared with values of over 40 for pepper, broccoli, carrot and spinach (Chu *et al.* 2002). The content of free phenolic compounds is also low compared with that of other vegetables. However, celery is still regarded as a valuable and healthy plant, mainly because of its content of flavonoids and other secondary metabolites.

Flavonoids

Flavonoids are secondary plant metabolites comprising several thousand compounds, including anthocyanins, flavonols, flavones, catechins and flavonones (Harborne, 1994). Studies have shown that celery has a high content of flavone glycosides, particularly apigenin-7-*O*-apiosylglucoside (apiin) (Lin *et al.*, 2007) and malonylapiin. The latter can be transformed to apiin (Hostetler *et al.*, 2012). A third flavonoid, luteolin, is also found in celery. Crozier *et al.* (1997) reported that flavonoid content could vary from very low levels up to 40 µg luteolin and 191 µg apigenin/g fresh weight.

Apiin is regarded as one of the main bioactive components in celery (Rithidech *et al.*, 2005). This has been further confirmed in recent studies by Li *et al.* (2014), which showed that the apiin could have free radical-scavenging activities as well as antioxidant activity in mice, thus protecting important organs from oxidative stress. The results highlight the potential of using celery to produce medicines.

Volatile and fatty oils

The leaves, roots and seeds of celery contain oils, but the seeds have the largest concentration. The seeds contain 2–3% essential, volatile oil and up to 15% fatty oil (Sowbhagya, 2014). The volatile oils include D-limonene (60%), β-selinene (10–12%), sedanoic acid anhydride (0.5%) and sedanolide (2.5–3.0%). The latter two components especially contribute to the characteristic odour of the seeds (Tang *et al.*, 1990). Also present are chlorogenic acid, caffeic acid, bergapten, niacin and inositol.

The essential oil content of celery leaves is 0.4–0.7% (Rożek *et al.*, 2012). 3-Butylphthalide is one of the compounds present in the essential oil that is one of the well-known flavour compounds of celery (Gold and Wilson, 1963; Bjeldanes and Kim, 1977). Bartschat *et al.* (1997) reported the analysis of such compounds for testing the quality of natural products.

6.2.2 Phytochemistry

Celery represents an important plant allergen source, causing reactions from mild oral allergy and skin symptoms to more severe respiratory symptoms and life-threatening anaphylactic reactions (Wuthrich and Dietschi, 1985; Ballmer-Weber *et al.*, 2000). Due to phototoxic reactions, skin reactions can occur when handling the leaves in light. Furocoumarins are present in the leaves at 2.1–4.5 mg/kg fresh weight (Järvenpää *et al.*, 1997). Reactive molecules have been found in tubers as well (Andre *et al.*, 1994; Wuthrich, 2005), and two of these are isoforms of the major allergen Api g 1 (Bauermeister *et al.*, 2009). Recently, the function of another allergen, Api g 6, has been described (Vejvar *et al.*, 2013).

The labelling of allergenic foods is essential for consumers for the purpose of avoidance diets (Sheth *et al.*, 2010). For celery, reference doses have so far not been established owing to poor model fits with existing data (Taylor *et al.*, 2014). This should be solved within the near future.

6.3 Postharvest Technology

6.3.1 Processing

Shelf life and storage

Fresh celery, and especially leaves and stalks (petioles), have a short shelf life, even under chilled conditions. Fresh celery should be taken from the field as soon as possible after harvesting, washed, sorted, packed and taken to the market. Stalk celery can keep its quality for about a month at low temperatures (2–4°C or 24–36°F) and high air humidity (up to 95% RH). In stalk and leaf celery, chilling injuries can be found at temperatures slightly above 0°C (32°F), but tubers can be stored at temperatures down to 0°C, and for a longer time than stalks and leaves. Controlled atmosphere storage of celery has not been successful and it is very sensitive to high concentrations of CO_2 if stored under airtight conditions or in non-perforated plastic bags (Ryall and Lipton, 1978). Ethylene has a negative impact on shelf life, as it increases respiration and ageing. Celery can also absorb flavour from other products kept in the same storage area.

Minimally processed fresh celery

Celery is included in several ready-to-eat dishes such as ready-packed chilled salads. This type of processed food is becoming increasingly popular and fits into the healthy and busy lifestyle of many people. Despite minimal mechanical treatment, the plant products have undergone some kind of peeling, cutting and washing. The main challenge is related to the shelf life of processed fresh products, as fresh celery in particular has a short shelf life. Even under chilled conditions and with proper packing, a week is a long time. To prolong shelf life, the application of citric acid (or other compounds) in preprocessing is used in combination with modified atmosphere packing. Tamer *et al.* (2012) showed positive effects on shelf life of minimally processed celery leaf stalks both with the use of citric acid (1.5%) and L-cysteine (0.5%) in combination with modified atmosphere packing. After 20 days at 4°C (36°F) antioxidant activity decreased by 26% using this method. Generally, citric acid was more effective than L-cysteine for the preservation of phenolic compounds.

One of the main problems in the processing of root celery is browning of the flesh. The mechanism is connected with the activity of enzymes catalysing phenolic oxidation. A proper selection of varieties and of raw material is important here (Radziejewska-Kubzdela and Czapski, 2004). Vina and Chaves (2006) studied the influence of storage temperature (0, 4 and 10°C) and storage time on the antioxidant capacity of fresh-cut celery roots disinfected by chlorinated water (100 ppm for 3 min) before packing. The roots retained their initial antioxidant capacity for 3 weeks at 0°C, and browning was lowest at this temperature.

Peeling will influence the content of vitamins and antioxidants, as such compounds are more concentrated in the outer surface than in the central parts. However, consumers prefer using peeled or half-prepared roots for cooking. For leaf and stalk types of celery this consideration is not relevant.

Thermal processing

Blanching, steam blanching, boiling, roasting or other temperature treatments are applied to celery used for ready-made dishes, pickles or other canned or vacuum-packed storage products (Salunkhe and Kadam, 1998).

The aromatic compounds of celery vary in their thermal stability. Hostetler *et al.* (2012) examined the effects of thermal processing of celery leaves (5 h at 100°C). In general, the flavones were resistant to degradation, and both heat and low pH were necessary for the degradation of apiin. Under such conditions, apiin is hydrolysed to apigenin-7-*O*-glucoside. Furthermore, apigenin-7-*O*-glucoside can be converted to apigenin, a compound that is of interest to the functional food industry. The aim is to use celery to make a product with high flavone content and a good bioavailability.

Husband *et al.* (2011) examined the impact of thermal and high-pressure processing of celeriac on the immunoreactivity of Api g 1 and other allergens. They concluded that the use of high pressure and high temperature in combination was the most effective method for reducing allergenicity.

Freezing

Freezing is a common conservation method and diced and frozen celery is a trade commodity used in the food industry. Raw material is blanched before freezing to reduce enzymatic activity. After blanching, the material should be immediately cooled in cold water and left to drip on sieves before freezing. Experiments have shown that freezing and drying caused decreasing content of both assimilation pigments and vitamin C in investigated raw material. Freezing also decreased the content of essential oil and flavonoids while drying increased the content of these compounds in comparison with fresh raw material (Roslon *et al.*, 2010).

When comparing different freezing methods, Lisiewska *et al.* (2006) showed that carrot, root celery, and parsnip previously cooked (blanched) and then frozen had a significant higher content of sodium and manganese than vegetables that had not been blanched, while the chromium content did not change (Fig. 6.2).

Dehydration

Drying is probably the oldest method of food preservation and conventional air drying is still the most frequently used method for the dehydration of vegetables in the food industry. The basic objective is to remove water to a level at which microbial breakdown and chemical reactions are minimized. Dehydration operations are commonly used for celery leaves and seeds. Tubers can also be cut, dried and used in processed foods such as instant soups. The different dehydration methods affect the quality. However, there have been only a few studies carried out on celery and other Apiaceae, and these have mainly been done on carrot (Rawson *et al.*, 2013). Nevertheless, the results that have been obtained from other herbs and medicinal plants are relevant to increasing the understanding of how to handle celery in an optimal way (see other chapters). Blanching before dehydration keeps the green leaf colour better than no

Fig. 6.2. Diced celery before freezing. Photo courtesy of A.B. Tskhovrebova and S.Ø. Solberg.

blanching. This will reduce the enzymatic activity of the leaves and is carried out by brief boiling or steaming. Dehydrated celery leaves are used as a spice in households and within the food industry in sauces, in tomato juice and to flavour various dishes. The seeds are also processed by grinding and mixing with salt to give celery salt.

Canning

In general, canned vegetables show a more pronounced loss of antioxidant activity than frozen vegetables. Murcia *et al.* (2009) showed that canned celery could totally lose its antioxidant activity, probably because of the filling medium used. They detected a significant loss of OH radical scavenging activity in canned compared with raw celery.

Extraction of oils

Both volatile oils and fatty oils are interesting to industry. Celery oil is pale yellow in colour, with an odour. The essential oils are isolated by steam distillation. The seeds are ground and immediately steam distilled to avoid any oil loss by evaporation. The ground seeds are spread out evenly upon perforated grids for the steam distillation process. The oil that is obtained is stored in airtight containers and kept cool, dry and dark to maintain quality.

Extraction of flavonoids

The extraction of flavonoids is used as a method to concentrate compounds with health benefits, which must then be isolated and identified. Several methods are employed to extract active substances from plants, most utilizing alcohol or other organic solvents. New methods, such as supercritical water extraction using high temperatures (110–200°C) for a short period (5–15 min) and under high pressure (about 10 MPa) have shown better results than traditional methods using alcohol (Cheigh *et al.* 2015). Another new method is to use enzymatic hydrolysis. Wu *et al.* (2012) found that 60 IU/ml cellulase/hemicellulase and 4 h hydrolysis time at pH 3.0 and a temperature 45°C were the optimum conditions for the extraction of flavonoids from celery stalks. Bartusch *et al.* (2013) examined the effect of ethanol concentration, temperature and pH on the extraction of phenolic compounds from a range of vegetable and fruit plants. For most plants, including celery, 50% ethanol was found to be the best solvent. Overall, 60°C was the optimal temperature, but for celery leaves, the total extractable phenolic content was highest at 40°C. The pH did not have any significant effect.

6.3.2 Value addition

In general, processing adds value to produce. For farmers, the margins for delivery to the industry are low compared with production for local trade, but deliveries to the industry give opportunities to expand in scale and produce a higher volume. A limiting factor is the presence of a food industry. Clusters of production are built around the industry, but without an industry, farmers must rely on the local market or retailers, or produce for small-scale processing.

6.4 Uses

6.4.1 General uses

Celery is used as a fresh vegetable, as a processed food or spice, or in industry as medicine or as perfume. All parts of the plant can be used. The leaves and stalks are used fresh, as in salads. A classic salad of celery is combined with apples, nuts and cheese. The tubers (celeriac) are usually used cooked, but are also be eaten raw in salads. As described in Section 6.3.1, celery is frequently processed and used raw, as in minimally processed salad mixtures. Dehydrated celery would not be the best option for salads, but is good for any cooked dish that needs celery flavour. Furthermore, celery is a basic ingredient in vegetable stock and it is used as a spice in both meat and vegetable dishes. Celery has been examined as an alternative to nitrite as an additive to meat products (Mandrean and Tita, 2011). The seeds are used to flavour food and liquids, but also for oil extraction.

6.4.2 Traditional and pharmacological uses

Celery was, and still is, used against rheumatism and colds, but also to improve digestion and to reduce blood pressure. Attempts have been made to provide overviews of the use of plants in traditional medicine. One such recent overview (in which celery is included) was published by Al-Asmari *et al.* (2014), with the emphasis on Saudi Arabian traditional medicine. Another examined the hypoglycaemic effect of fumitory leaf, celery leaf and lemon, which are recommended in Persian folklore medicine as beneficial to diabetes and for their effect(s) on pancreatic tissue (Jelodar *et al.*, 2007). This particular study was done on rats and it was concluded that fumitory leaf and celery had a potential hypoglycaemic effect. As such, the results support the traditional claims on the use of these plants in diabetes. In traditional Chinese medicine, it is estimated that more than 200 plant species exhibit hypoglycaemic properties, including many that are common – pumpkin, celery, lotus root and bitter melon (Jia *et al.*, 2003). Evaluations reported in this review indicated that the use of these plants in combination with conventional medical treatments permitted the use of lower drug doses. A survey among Latinos in the USA showed that approx. 70% of them used herbal remedies together with prescribed medications for diabetes, and that celery was frequently used (Amirehsani and Wallace, 2013).

In a recent review of claims reported on the Internet of the beneficial effects of herbal supplements, celery was among the ten most common plant ingredients mentioned (Vamenta-Morris *et al.*, 2014). However, in contrast to the many beneficial claims, the authors concluded that the substances were not adequately studied in humans and that one needs to be cognizant of the lack of substantiated proven benefits of these substances. Nevertheless, a few reports on celery do support positive health effects, but so far proven only in animals. Celery extract decreased the blood pressure and increased the heart rate in hypertensive rats (Moghadam *et al.*, 2013), an action that the authors related to the *n*-butylphthalide present in the extract. In general, flavonoids have antioxidant and free radical scavenging activities in foods (Shahidi and Wanasundara, 1992). Epidemiological studies have indicated that their consumption is associated with a reduced cancer risk (Verma *et al.*, 1988; Wattenburg, 1990; Wei *et al.*, 1990) and a reduced cardiovascular disease risk (Gregory *et al.*, 1990; Hertog, 1994).

The potential use of flavonoids as a protector for radiation-induced chromosomal damage is of increasing interest due to their high antioxidant activity (Rithidech *et al.*, 2005). This has been shown in experiments where blood samples have been taken from healthy volunteers for lymphocyte cultures. No effect was found with 2.5, 5.0 or 10 μg/ml apigenin, but 25 μg/ml apigenin gave a positive effect and indicated the potential use of celery as a radioprotector. Hostetler *et al.* (2012) have worked on functional foods from celery, emphasizing improvements in the bioavailability of important compounds. They say that to fully understand and utilize the potential benefits of flavone-rich food, research should place an emphasis on the absorption of flavones in humans and the stability of the active substances during processing. As already mentioned, they also note that apigenin-7-*O*-glucoside can be converted to apigenin, which potentially has a very high bioavailability. Another interesting compound is sedanenaloide (3-*n*-butyl-4-5-dihydrophthalide), which is one of the constituents responsible for the typical aroma of celery seed oil. This is also claimed to have positive health benefits (Rawson *et al.*, 2013).

For pharmacological use, the seeds have so far been the most important raw material from celery, although other parts of the plant can be used. Fatty oil extracted from the seeds is used as an antispasmodic and nerve stimulant. The seeds are mainly processed for essential oils and used as perfume or as medicine. Sowbhagya and Srinivas (2013) examined ways to process celery seed oil into a phthalide-enriched product with high nutraceutical potential. Fractional distillation afforded a limonene-rich fraction (7.6 g, 97% purity), a fraction containing β-selinene (2.8 g, 90% purity) and a fraction containing phthalides (2.9 g, 90% purity). Solvent–solvent partition of celery seed oil gave limonene (87%) and a fraction containing phthalides (49%), which on further fractionation afforded a phthalide-enriched fraction (90%). By conventional silica gel column chromatography, a product rich in phthalides (53–74%) could be obtained. Huang *et al.* (2013) examined celery in combination with pine needles and as a raw material in processing a health beverage rich in flavonoids. The results showed that the best formula was 20% pine needles extract, 20% celery extract, 2% honey and 0.1% citric acid. The beverage was a light yellow clear liquid that according to the authors tasted good.

6.5 Summary

Celery is an important crop and all parts of the plant can be used. The leaves and stalks are commonly used as fresh herbs. The tubers (celeriac) are used cooked, but also fresh. For the food industry, celery is an important crop. All parts of the plant can be processed and celery is a major ingredient in vegetable stock and other spices used for flavouring food. The seeds are processed for essential oils that are used in perfume or medicine. The crop contains important minerals and vitamins, but also bioactive flavonoids useful for human health. Celery might cause allergic reactions and there is an increasing awareness of how to label and set reference doses.

References

Al-Asmari, A.K., Al-Elaiwi, A., Athar, M.T., Tariq, M., Al Eid, A. and Al-Asmary, S.M. (2014) A review of hepatoprotective plants used in Saudi traditional medicine. *Evidence-Based Complementary and Alternative Medicine* 2014: Article ID 890842.

Amirehsani, K.A. and Wallace, D.C. (2013) Tés, Licuados, and Cápsulas. Herbal self-care remedies of Latino/Hispanic immigrants for type 2 diabetes. *The Diabetes Educator* 39, 828–840.

Andre, F., Andre, C., Colin, L. and Cacaraci, F. (1994) Role of new allergens and of allergens consumption in the increased incidence of food sensitizations in France. *Toxicology* 93, 77–83.

Ballmer-Weber, B.K., Vieths, S., Luttkopf, D. and Heuschmann, P. (2000) Celery allergy confirmed by double-blind, placebo-controlled food challenge: a clinical study in 32 subjects with a history of adverse reactions to celery root. *Journal of Allergy and Clinical Immunology* 106, 373–378.

Bartschat, D., Beck, T. and Mosandl, A. (1997) Stereoisomeric flavor compounds. 79. Simultaneous enantioselective analysis of 3-butylphthalide and 3-butylhexahydrophthalide stereoisomers in celery, celeriac, and fennel. *Journal of Agricultural and Food Chemistry* 45, 4554–4557.

Bartusch, J., Zhou, J. and Saleh, Z. (2013) Recovery of functional horticultural ingredients using cost effective and commercially viable methods. *International Proceedings of Chemical, Biological and Environmental Engineering* 50, 103–107.

Bauermeister, K., Ballmer-Weber, B.K., Bublin, M. and Fritsche, P. (2009) Assessment of component-resolved *in vitro* diagnosis of celeriac allergy. *Journal of Allergy and Clinical Immunology* 124, 1273–1281.

Bjeldanes, L.F. and Kim, I.S. (1977) Phthalide components of celery essential oil. *The Journal of Organic Chemistry* 42, 2333–2335.

Cheigh, C.I., Yoo, S.Y., Ko, M.J., Chang, P.S. and Chung, M.S. (2015) Extraction characteristics of subcritical water depending on the number of hydroxyl group in flavonols. *Food Chemistry* 168, 21–26.

Chu, Y.F., Sun, J., Wu, X. and Liu, R.H. (2002) Antioxidant and antiproliferative activities of common vegetables. *Journal of Agricultural and Food Chemistry* 50, 6910–6916.

Crozier, A., Lean, M.E.J., McDonald, M.S. and Black, C. (1997) Quantitative analysis of the flavonoid content of commercial tomatoes, onions, lettuce, and celery. *Journal of Agricultural and Food Chemistry* 45, 590–595.

Fazal, S.S. and Singla, R.K. (2012) Review on the pharmacognostical and pharmacological characterization of *Apium graveolens*. *Indo Global Journal of Pharmaceutical Sciences* 2, 36–42.
GBIF (2014) *Apium graveolens* L. Global Biodiversity Information Facility, Copenhagen. Available at: http://www.gbif.org/species/5371879 (accessed 26 June 2014).
Gold, H.J. and Wilson, C.W. (1963) Alkylidenephthalides and dihydrophthalides from celery. *The Journal of Organic Chemistry* 28, 985–987.
Gregory, J., Foster, K., Tyler, H. and Wiseman, N. (1990) *The Dietary and Nutritional Survey of British Adults*. Her Majesty's Stationery Office, London.
Harborne, J.B. (1994) *The Flavonoids: Advances in Research Since 1986*. Chapman and Hall, London.
Hertog, M.G.L. (1994) Flavonols and flavones in foods and their relation with cancer and coronary heart disease risk. PhD thesis, Agricultural University Wageningen, The Netherlands.
Hostetler, G., Riedl, K. and Schwartz, S. (2012) Endogenous enzymes, heat, and pH affect flavone profiles in parsley (*Petroselinum crispum* var. *neapolitanum*) and celery (*Apium graveolens*) during juice processing. *Journal of Agricultural and Food Chemistry* 60, 202–208.
Huang, J.-L., Deng, L.-Y., Lu, M.-Y. and Chen, X.-Y. (2013) Study on processing technology of compound beverage of pine needle and celery. *Food Research and Development* 34(5), 47–50. [In Chinese.]
Husband, F.A., Aldick, T., van der Plancken, I., Grauwet, T., Hendrickx, M., Skypala, I. and Mackie, A.R. (2011) High-pressure treatment reduces the immunoreactivity of the major allergens in apple and celeriac. *Molecular Nutrition and Food Research* 55, 1087–1095.
Järvenpää, E.P., Jestoi, M.N. and Huopalahti, R. (1997) Quantitative determination of phototoxic furocoumarins in celeriac (*Apium graveolens* L. var. *rapaceum*) using supercritical fluid extraction and high performance liquid chromatography. *Phytochemical Analysis* 8, 250–256.
Jelodar, G., Maleki, M. and Sirus, S. (2007) Effect of fumitory, celery and lemon on blood glucose and histopathology of pancreas of alloxan diabetic rats. *Journal of Applied Animal Research* 31, 101–104.
Jia, W., Gao, W.Y. and Tang, L. (2003) Antidiabetic herbal drugs officially approved in China. *Phytotherapy Research* 17, 1127–1134.
Krug, H. (1991) *Gemüseproduktion*. Verlag Paul Parey, Berlin/Hamburg, Germany, pp. 335–341.
Li, P., Jia, J., Zhang, D., Xie, J., Xu, X. and Wei, D. (2014) *In vitro* and *in vivo* antioxidant activities of a flavonoid isolated from celery (*Apium graveolens* L. var. *dulce*). *Food and Function* 5, 50–56.
Lin, L.Z., Lu, S. and Harnly, J.M. (2007) Detection and quantification of glycosylated flavonoid malonates in celery, Chinese celery, and celery seed by LC-DAD-ESI/MS. *Journal of Agricultural and Food Chemistry* 55, 1321–1326.
Lisiewska, Z., Kmiecik, W. and Gebczynski, P. (2006) Effects on mineral content of different methods of preparing frozen root vegetables. *Food Science and Technology International* 12, 497–503.
Mandrean, N.L. and Tita, O. (2011) Celery, a natural alternative to chemical nitrite added to meat products. *Bulletin of University of Agricultural Sciences and Veterinary Medicine (Cluj-Napoca)* 68, 317–320.
Maruyama, T., Abbaskhan, A., Choudhary, M.I., Tsuda, Y., Goda, Y., Farille, M. and Reduron, J.-P. (2009) Botanical origin of Indian celery seed (fruit). *Journal of Natural Medicines* 63, 248–253.
Mathias, M.E. (1994) Magic, myth and medicine. *Economic Botany* 48, 3–7.
Moghadam, M.H., Imenshahidi, M. and Mohajeri, S.A. (2013) Antihypertensive effect of celery seed on rat blood pressure in chronic administration. *Journal of Medicinal Food* 16, 558–563.
Murcia, M.A., Jiménez, A.M. and Martínez-Tomé, M. (2009) Vegetables antioxidant losses during industrial processing and refrigerated storage. *Food Research International* 42, 1046–1052.
Ninfali, P. and Bacchiocca, M. (2003) Polyphenols and antioxidant capacity of vegetables under fresh and frozen conditions. *Journal of Agricultural and Food Chemistry* 51, 2222–2226.
Radziejewska-Kubzdela, E. and Czapski, J. (2004) A comparison of processability of selected varieties of celeriac for the production of minimally processed shredded celeriac. *Electronic Journal of Polish Agricultural Universities* 7(2), 15. Available at: http://www.ejpau.media.pl/volume7/issue2/food/abs-15.html (accessed 17 July 2015).
Rawson, A., Brunton, N.P., Rai, D.K., McLoughlin, P., Tiwari, B.K. and Tuohy, M.G. (2013) Stability of falcarinol type polyacetylenes during processing of Apiaceae vegetables. *Trends in Food Science and Technology* 30, 133–141.
Rithidech, K.N., Tungjai, M. and Whorton, E.B. (2005) Protective effect of apigenin on radiation-induced chromosomal damage in human lymphocytes. *Mutation Research* 585, 96–104.

Rizzo, V. and Muratore, G. (2009) Effects of packaging on shelf life of fresh celery. *Journal of Food Engineering* 90, 124–128.

Roslon, W., Osinska, E. and Gajc-Wolska, J. (2010) The influence of raw material stabilization on the quality of celery (*Apium graveolens* L.) leaves. *Acta Horticulturae* 877, 201–208.

Rożek, E. (2007) Content of some chemicals and essential oil in leaves of leaf celery *Apium graveolens* L. var. *secalinum* Alef. *Herba Polonica* 53, 213–217.

Rożek, E., Nurzyńska-Wierdak, R. and Dzida, K. (2012) Factors modifying yield quantity and quality, as well as the chemical composition of the leaves of leaf celery *Apium graveolens* L. var. *secalinum* Alef. grown from seedlings. *Acta Scientiarum Polonorum, Hortorum Cultus* 11, 201–210.

Rubatzky, V.E. and Yamaguchi, M. (1997) *World Vegetables: Principles, Production and Nutritive Value*, 2nd edn. Chapman and Hall, New York.

Ryall, A.L. and Lipton, W.J. (1978) *Handling, Transportation and Storage of Fruits and Vegetables, Vol. 1.* Avi Publishing, Westport, Connecticut.

Salunkhe, D.K. and Kadam, S.S. (eds) (1998) *Handbook of Vegetable Science and Technology: Production, Composition, Storage, and Processing.* Marcel Dekker, New York.

Schippers, R.R. (2004) *Apium graveolens* L. [Internet] Record from PROTA4U. In: Grubben, G.J.H. and Denton, O.A. (eds) *PROTA (Plant Resources of Tropical Africa)*, Wageningen, The Netherlands. Available at: http://www.prota4u.org/protav8.asp?h=M4&t=Apium,graveolens&p=Apium+graveolens#Synonyms (accessed 17 July 2015).

Shahidi, F. and Wanasundara, P.K. (1992) Phenolic antioxidants. *Critical Reviews in Food Science and Nutrition* 32, 67–103.

Sheth, S.S., Waserman, S., Kagan, R., Alizadehfar, R., Primeau, M.-N., Elliot, S., St. Pierre, Y., Wickett, R., Joseph, L., Harada, L. *et al.* (2010) Role of food labels in accidental exposures in food-allergic individuals in Canada. *Annals of Allergy, Asthma and Immunology* 104, 60–65.

Sowbhagya, H.B. (2014) Chemistry, technology, and nutraceutical functions of celery (*Apium graveolens* L.): an overview. *Critical Reviews in Food Science and Nutrition* 54, 389–398.

Sowbhagya, H.B. and Srinivas, P. (2013) Enrichment of bio-active phthalides in celery seed oil. *Journal of Pharmacy and Nutrition Sciences* 3, 250–257.

Tamer, C.E., Copur, Ö.U., Incedayi, B. and Vural, H. (2012) Evaluation of some quality parameters of minimally processed celery by quantitative analysis. *Journal of Food Processing and Preservation* 37, 717–726.

Tang, J., Zhang, Y., Hartman, T.G., Rosen, R.T. and Ho, C.T. (1990) Free and glycosidically bound volatile compounds in celery (*Apium graveolens* L.). *Journal of Agricultural and Food Chemistry* 38, 1937–1940.

Taylor, S.L., Baumert, J.L., Kruizinga, A.G., Remington, B.C., Grevel, R. and Brooke-Taylor, S. (2014) Establishment of Reference Doses for residues of allergenic foods: report of the VITAL Expert Panel. *Food and Chemical Toxicology* 63, 9–17.

USDA (2014) *Vegetables 2013 Summary.* National Agricultural Statistics Service, US Department of Agriculture, Washington, DC.

Vamenta-Morris, H., Dreisbach, A., Shoemaker-Moyle, M., Emaad, M. and Abdel-Rahman, E.M. (2014) Internet claims on dietary and herbal supplements in advanced nephropathy: truth or myth. *Journal of Nephrology* 40, 393–398.

Vavilov, N.I. (1926) Studies on the origin of cultivated plants. *Bulletin of Applied Botany and Plant Breeding (Russian)* 14, 1–245.

Vejvar, E., Himly, M., Briza, P., Eichhorn, S., Ebner, C., Hemmer, W., Ferreira, F. and Gadermaier, G. (2013) Allergenic relevance of nonspecific lipid transfer proteins 2: identification and characterization of Api g 6 from celery tuber as representative of a novel IgE-binding protein family. *Molecular Nutrition and Food Research* 57, 2061–2070.

Verma, A.K., Johnson, J.A., Gould, M.N. and Tanner, M.A. (1988) Inhibition of 7,12-dimethylbenz(a) anthracene and *N*-nitromethylurea induced rat mammary cancer by dietary flavonol quercetin. *Cancer Research* 48, 5754–5758.

Vina, S.Z. and Chaves, A.R. (2006) Antioxidant responses in minimally processed celery during refrigerated storage. *Food Chemistry* 94, 68–74.

Wattenburg, L. (1990) Inhibition of carcinogenesis by minor nutrient constituents of the diet. *Proceedings of the Nutrition Society* 49, 173–183.

Wei, H., Tye, L., Bresnick, E. and Birt, D.F. (1990) Inhibitory effect of apigenin, a plant flavonoid, on epidermal ornithine decarboxylase and skin tumour promotion in mice. *Cancer Research* 50, 499–502.
Wu, J.C., Huang, G.R. and Cheng, J. (2012) Optimization of enzymatic hydrolysis for extraction of flavonoids from *Apium graveolens* L. stalks by entropy weight method. *Journal of Food, Agriculture and Environment* 10, 182–185.
Wuthrich, B. (2005) Frequency of food allergies over time – longitudinal statistics from 1978–1988. *Allergologie* 28, 355–358.
Wuthrich, B. and Dietschi, R. (1985) The celery–carrot–mugwort condiment syndrome: skin test and RAST results. *Schweizerischemedizinische Wochenschrift* 115, 258–264.
Yuman, L., Jinsong, C., Zhang, X. and Kamphuis, B. (2004) *The Vegetable Industry in China. Developments in Policies, Production, Marketing and International Trade.* Report 6.04.14, Agricultural Economics Research Institute, The Hague, The Netherlands.

7 *Centella*

Terrence Madhujith[1*] and Subajiny Sivakanthan[2]
[1]*University of Peradeniya, Sri Lanka;* [2]*University of Jaffna, Sri Lanka*

7.1 Botany

7.1.1 Introduction

Taxonomy

Centella asiatica (L.) Urban (or *Centella asiatica* (L.) Urb.), a tropical plant belonging to the family Apiaceae (Umbelliferae), is commonly known as 'gotu kola', Asiatic pennywort, Indian pennywort, Indian water navelwort, wild violet, tiger herb and spadeleaf, and also just as *Centella* or 'centella' (Yu *et al.*, 2006; James and Dubery, 2009; Orhan, 2012). Its taxonomic hierarchy is as follows:

Kingdom: Plantae
Division: Tracheophyta
Subdivision: Spermatophytina
Class: Magnoliopsida
Order: Apiales
Family: Apiaceae
Genus: *Centella*
Species: *asiatica*

The species is an ancient medicinal herb used in Ayurvedic medicine in India and in herbal medicine in Malaysia and China, and some other Asian countries. Moreover, it has been used in folk and alternative medicines for treating numerous human ailments, including wounds, psoriasis and leprosy, and improving brain function. In contrast to other medicinal herbs, *C. asiatica* has been investigated experimentally and clinically rather extensively (Brinkhaus *et al.*, 2000).

The use of *C. asiatica* in food and beverages has become popular owing to its health benefits, including its antihypertensive (Intharachatorn and Srisawat, 2013), antioxidant (Pittella *et al.*, 2009), immunomodulatory (Wang *et al.*, 2003), antimicrobial, anticancer (Roy *et al.*, 2013), anti-inflammatory (Somchit *et al.*, 2004), wound healing (Suguna *et al.*, 1996), memory-enhancing (Meena *et al.*, 2012) and anti-ulcer (Abdulla *et al.*, 2010) properties. Currently, the resources of *C. asiatica* are gradually being depleted and thus the species is included in the IUCN (International Union for Conservation of Nature) list of threatened species (Singh *et al.*, 2010; IUCN, 2013).

C. asiatica has various synonyms (for example, *C. coriacea*, *C. asiatica* var. *floridana*, *Hydrocotyle asiatica*, *H. asiatica* var. *floridana*, *H. lunata* and *Trisanthus cochinchinensis*), a detailed list of which is given by the Institute for Systematic Botany of the University of South Florida (Wunderlin and

*Corresponding author, e-mail: madujith@yahoo.com

© CAB International 2016. *Leafy Medicinal Herbs: Botany, Chemistry, Postharvest Technology and Uses* (eds D.C.P. Ambrose *et al.*)

Hansen, 2008). The herb is known as 'Brahmi' in Unani medicine and 'Mandookaparni' in Ayurvedic medicine (Jamil *et al.*, 2007). There is an array of vernacular names used for *C. asiatica* worldwide, many of which are listed in Table 7.1.

7.1.2 History/origin

Centella is native to South-east Asian countries such as China, India, Indonesia, Japan, Malaysia and Sri Lanka, as well as Madagascar and South Africa (Jamil *et al.*, 2007). It is an ethnomedical plant used by diverse ethnic groups in various parts of the world. It was referred to in the ancient Chinese Shennong Herbal and in Indian Ayurvedic medicine about 2000 and 3000 years ago, respectively (European Medicines Agency, 2010). It has been listed in the *Susruta Samhita*, an ancient and historic Indian medical text. In Ayurvedic medicine, *Centella* is reputed as a mental rejuvenator and as a local stimulant of skin, and thus is used to treat skin diseases such as herpes, eczema, psoriasis and wounds (Premila, 2006).

In India and Indonesia, *Centella* was used to treat wounds and leprosy, enhance memory and prolong lifespan. In China, it is used for boils, contusions, fractures, snakebites, strains and turbid leucorrhoea. The use of *Centella* as a drug was first accepted in France in the 1980s. In India, *Centella* and its extracts were incorporated into the

Table 7.1. Vernacular names of *Centella asiatica*. From WHO, 1999; EMEA, 2010; Singh *et al.*, 2010.

Region/language	Vernacular name
South Asia	
Assam	Manimuni
Bengal	Thankuni, tholkuri
Bihar	Chokiora
Deccan	Vallarai
Gujarati	Barmi, moti brahmi
Hindi	Bemgsag, brahma-manduki, gotukola (or gotu kola), khulakhudi, mandookaparni
Kanarese	Brahmisoppu, urage, vandelagaillikiwigidda, vondelaga
Malayalam	Kodagam, kodangal, kutakm, kutannal, muthal, muttil, muyalchevi
Marathi	Karinga, karivana
Meghalaya	Bat-maina
Nepal	Ghod tapre
Oriya	Thalkudi
Sanskrit	Bhekaparni, bheki, brahmamanduki, darduchhada, divya, mahaushadhi, mandukaprnika, manduki, mutthil, supriya, tvasthi
Sinhalese	Hingotukola
Tamil	Babassa, vallarai
Telugu	Bekaparnamu, bokkudu, saraswataku, mandukbrahmmi, saraswati (plant)
Tripura	Thankuni, thunimankuni
Urdu	Brahmi
Other regions	
China	Fo-ti-tieng, chi-hsueuh-ts'ao, tungchian, luei gong gen
Cook Islands	Kapukapu
Fiji	Totodro
Hawaii	Pohe kula
Japan	Tsubo-kusa
Italian	Idrocotile
Samoa, Tonga	Tono
Spanish	Blasteostimulina
Tahiti	Tohetupou
USA	Indian pennywort, marsh pennywort

Indian Pharmacopoeia in the 19th century (Murray, 2012).

In 1884, *Centella* was referred to in the French Pharmacopoeia, but regardless of its long history of conventional use, it only appeared in the Codex in 1884. The first extract of *Centella* in dried form was created after 1941, and 3 years later, the French scientist, Boiteau isolated its triterpenoid molecules. The use of the aerial part of *Centella* in rheumatism and its topical application for leprous ulcers, indolent wounds, and post-surgical wound healing (cicatrization) are reported in the British Herbal Pharmacopoeia. In Europe, therapeutic products containing the triterpenic fractions extracted from *Centella* have been legitimized for cutaneous use (as a powder, cream and ointment) since 1968. Oral tablets containing Madecassol® have been permitted in Europe since 1969, but are only allowed in Belgium, France, Greece, Italy and Portugal (EMEA, 2010).

7.1.3 Location

Centella is native to South-east Asian countries, South Africa and Madagascar. However, it is now widely grown in the West as well (Roy *et al.*, 2013). The species is widespread throughout tropical and subtropical countries worldwide, including: Angola, Australia, Bhutan, Botswana, Cambodia, Cameroon, Central America, China, Congo, Côte d'Ivoire, Ethiopia, Gabon, Gambia, Ghana, Guinea, India, Indonesia, Iran, Kenya, Korea, Laos, Liberia, Madagascar, Malawi, Malaysia, Mali, Mauritius, Mozambique, Myanmar, Nepal, Nigeria, the Pacific Islands, Pakistan, Sao Tomé and Principe, Saudi Arabia, Senegal, Somalia, South Africa, South America, Sri Lanka, Sudan, Taiwan, Tanzania, Thailand, Vietnam, Yemen, Zambia and Zimbabwe (Alternative Medicine Review, 2007; Satake *et al.*, 2007; Zheng and Qin, 2007; IUCN, 2013).

7.1.4 Morphology

Centella is a perennial, slender, prostrate, faintly aromatic, stoloniferous, creeper herb. It attains a height of up to 15 cm (see Fig. 7.1). The stem is long, prostrate and emerges from the leaf axils of a vertical rootstock; it is glabrous, striated, filiform, often reddish, with long internodes and rooting at the nodes. Leaves emerge alternately in clusters at stem nodes; they are 1.3–6.3 cm across, thin, long petioled, 2–6 cm long and 1.5 to 5 cm wide; there are several from each rootstock, one to three from each node of the stems, and they are orbicular–reniform, more or less cupped, entire, crenate or lobulate, and glabrous. The petioles are variable in length, 7.5 to 15 cm long or more and channelled. The stipules are short and adnate to the petioles forming a sheathing base. The peduncle is about 6 mm long with no pedicels; small bracts embrace the flowers. The inflorescence is in a single umbel, bearing one to five flowers, sessile, white or reddish. The fruits are small, compressed, 8 mm long, the mericarps are longer than broad, curved, rounded at the top, seven to nine-ridged, with secondary ridges as prominent as the primary, and reticulate between them. The pericarp is much thickened. The seeds have a pendulous embryo that is laterally

Fig. 7.1. A plant of *Centella asiatica*.

compressed. The plant develops quickly and extensively in shady, marshy, damp and wet places. Sandy loam soil is the most fertile soil for regeneration (WHO, 1999; Singh *et al.*, 2010; Tiwari *et al.*, 2011). The plant grows along stone walls or other rocky areas up to an elevation of 2000 ft in India and Sri Lanka (Murray, 2012). Propagation is by seeds and stolons (Jayasinha *et al.*, 1999).

7.2 Chemistry

7.2.1 Chemical and nutritional composition

Centella has been reputed to contain a huge number of constituents belonging to various chemical classes. The types of phytochemicals present depend on the location and environmental conditions (James and Dubery, 2009). Active constituents of *Centella* differed significantly between samples collected from India, Nepal and Madagascar (Randriamampionona *et al.*, 2007; Devkota *et al.*, 2010), so that geographical and environmental factors can influence the type and amount of constituents present. However, the method of extraction could also have an influence on the diversity of compounds identified in *Centella* from various locations (Booncong, 1989). The chemical constituents identified include terpenes, phenolics, alkaloids, polyacetylenes, carbohydrates, amino acids, vitamins and minerals. Among these, the main active components are the triterpenoid glycosides asiaticoside (40%) and madecassoside (1%), and their respective aglycones (asiatic acid, 29–30%, and madecassic acid, 29–30%) (EMEA, 2010), which are used in the standardization of *Centella* as explained in the *European Pharmacopoeia* (Bosse *et al.*, 1979; Randriamampionona *et al.*, 2007). The leaves contain higher concentrations of these phytochemicals relative to the petioles and roots (Zainol *et al.*, 2008). The various classes of chemical constituents found in centella are listed in Table 7.2.

The nutritional value of *Centella* is promising as the herb serves as a good source of micronutrients and macronutrients. The total energy present in *Centella* is 37 kcal/100 g. The herb contains moisture (87.7%), carbohydrate (6.7%), protein (2%), crude fibre (1.8%), fat (0.2%) and ash (1.6%). *Centella* is rich in minerals such as potassium and calcium (391 and 171 mg/100g, respectively). It is also rich in vitamin C (48.5 mg/100g) (Tee *et al.*, 1997). The nutrient composition of *Centella* is influenced by its stage of maturity (Rosalizan *et al.*, 2008), morphotype (Chandrika *et al.*, 2011) and location (Upadhyaya and Saikia, 2012).

7.2.2 Phytochemistry

Terpenes

Centella contains monoterpenes and sesquiterpenes (essential oils), diterpenes, triterpenes, tetraterpenes and glycosides. The major and most important component of the herb is the triterpenes, which are regarded as the marker constituents in terms of quality control. The triterpenes are composed of many compounds, including asiatic acid, madecassic acid, asiaticoside, madecassoside, brahmoside, brahmic acid, brahminoside, brahmoside, thankuniside, isothankunisode, centelloside, madasiatic acid, centic acid and centellic acid (Zheng and Qin, 2007). Among all these constituents, asiaticoside has been subjected to extensive investigations and is reported to be responsible for most of the pharmacological activities of *Centella*. The second most isolated compound is madecassoside, followed by asiatic acid and madecassic acid (Chong and Aziz, 2011). The brahmoside in the plant is known to possess tranquilizing and anabolic activity. Asiaticoside and centelloside are used for the treatment of leprosy. Animal studies have demonstrated that asiaticoside reduces the number of tubercular lesions in the nerve ganglia, lungs, liver and spleen. Oxyasiaticoside (a derivative of asiaticoside) is known to inhibit the growth of the tubercle bacillus (*Mycobacterium tuberculosis*). Asiaticosides also possess hyperglycaemic activity (Khare, 2007). Depending on the raw material, the amounts of triterpenoids can range from 1 to

Table 7.2. Chemical constituents of *Centella asiatica*.

Main class of chemical constituents	Chemical constituents	References
Alkaloid	Hydrocotylin, vallarine	Chong and Aziz, 2011; Roy *et al*., 2013
Amino acids	Alanine and serine (major components), aminobutyrate, aspartate, glutamate, histidine, lysine, threonine, arginine, leucine, isoleucine, valine, methionine, tyrosine, phenylalanine, proline, cysteine, glycine	Chong and Aziz, 2011; Roy *et al*., 2013
Carbohydrates	Monosaccharides: glucose, mesoinositol Oligosaccharides: centellose Polysaccharides: pectin, arabinogalactan	James and Dubery, 2009; Chong and Aziz, 2011; Roy *et al*., 2013
Diterpenes	Neophytadiene	Chong and Aziz, 2011
Essential oil (0.1% of the plant)	Monoterpenes: α-pinene, β-pinene, myrcene, γ-terpinene, α-thjuene, bornyl acetate, linalool Sesquiterpenes: α-copaene , β-elemene, β-caryophyllene, *trans*-β-farnesene, germacrene D, bicycloelemene, α-humulene, (*E*)-β-farnesene, δ-cadinene, *epi*-bicyclosesquiphellandrene, epiglobulol	Chong and Aziz, 2011; Roy *et al*., 2013
Minerals	Calcium, phosphorus, iron, potassium, magnesium, manganese, zinc, sodium, copper	Chong and Aziz, 2011; Roy *et al*., 2013
Phenols	Flavonoids: kaempferol, kaempferol-3-*O*-β-D glucuronide, castilliferol, catechin, epicatechin, quercetin, quercetin-3-*O*-β-D-glucuronide, castillicetin, apigenin, rutin, luteolin, naringin, patuletin, myricetin Tannins: tannin, phlobatannin Phenylpropanoids: rosmarinic acid, 3,5-di-*O*-caffeoylquinic acid, 1,5-di-*O*-caffeoylquinic acid, 3,4-di-*O*-caffeoylquinic acid, 4,5-di-*O*-caffeoylquinic acid, ettacrynic acid, chlorogenic acid, isochlorogenic acid, *p*-hydroxybenzoic acid, vanillic acid, *p*-coumaric acid, *o*-coumaric acid, *trans* cinnamic acid	Suntornsuk and Anurukvorakun, 2005; Yu *et al*., 2006; Subban *et al*., 2008; Hussin *et al*., 2009; Chong and Aziz, 2011; Orhan, 2012; Roy *et al*., 2013
Polyacetylenes	Cadiyenol, centellin, centellicin, asiaticin, 8-acetoxycentellynol, dotriacont-8-en-1-oic acid, 11-oxoheneicosanyl	Chong and Aziz, 2011; Roy *et al*., 2013
Triterpenic acid sugar esters (glycosides or triterpenic saponins or pseudosaponins)	Asiaticoside (major component), asiaticoside A, B, C, D, E and F, sceffoleoside A, madecassoside, centelloside, indocentelloside, brahmoside, brahminoside, thankuniside, isothankuniside	Kuroda *et al*., 2001; Matsuda *et al*., 2001; Schaneberg *et al*., 2003; Zheng and Qin, 2007; James and Dubery, 2009; Chong and Aziz, 2011; Orhan, 2012; Roy *et al*., 2013
Triterpenic acids	Asiatic acid, madecassic acid (6-β-hydroxyasiatic acid), madasiatic acid, thankunic acid, isothankunic acid, centic acid, centellic acid, centoic acid, indocentoic acid, terminolic acid, betulinic acid, brahmic acid, isobrahmic acid, pomolic acid, corosolic acid, ursolic acid	Schaneberg *et al*., 2003; Yoshida *et al*., 2005; Yu *et al*., 2006; Zheng and Qin, 2007; Chong and Aziz, 2011; James and Dubery, 2009; Orhan, 2012; Roy *et al*., 2013
Triterpenic steroids (phytosterols)	Campesterol, stigmasterol, sitosterol	Chong and Aziz, 2011; Roy *et al*., 2013
Vitamins	Ascorbic acid, nicotinic acid, β-carotene	Chong and Aziz, 2011; Roy *et al*., 2013

8% (ESCOP, 2009). The essential oil present in *Centella* is composed of a wide array of monoterpenes and sesquiterpenes (Chong and Aziz, 2011).

In addition to the classification of terpenes found in *Centella* as indicated in Table 7.2, they can be further classified based on their cyclic structure. The monoterpenes reported to be present exist in acyclic, monocyclic and bicyclic structures, while the sesquiterpenes have tricyclic structures as well. The triterpenes in *Centella* are mainly pentacyclic triterpenic acids. Two types of pentacyclic triterpenes are reported: the ursane type and oleanane type. Triterpenes such as asiatic acid, madecassic acid, brahmic acid (also known as 6β-hydroxyasiatic acid), pomolic acid, corosolic acid and ursolic acid are ursane-type pentacyclic triterpenes. The ursane-type pentacyclic triterpene saponins include asiaticoside and asiaticosides A, C, D, E and F, madecassoside, brahminoside, centellasaponin B and C and scheffuroside B. The oleanane-type pentacyclic triterpenes are 3-*epi*-maslinic acid and terminolic acid. The oleanane-type pentacyclic triterpenes saponins are asiaticoside B, centellasaponin D, scheffoleoside A and 23-*O*-acetylasiaticoside B (Chong and Aziz, 2011).

Various acyclic monoterpenes (3-nonen-2-one), monocyclic monoterpenes (linalool, myrcene, γ-terpinene, terpinolene, limonene, terpinen-4-ol, α-terpinene, α-phellandrene, ρ-cymene, pulegone, menthone, methyl thymol and methyl carvacrol) and bicyclic monoterpenes (α-thujene, α-pinene, β-pinene, camphene, chrysanthenyl acetate and bornyl acetate) have also been reported (Chong and Aziz, 2011).

The sesquiterpenes found in *Centella* include acyclic sesquiterpenes (decan-1-ol, trans-β-farnesene), monocyclic sesquiterpenes (germacrene A, B and D, β-elemene, γ-elemene, γ-curcumene, bicycloelemene, bicyclogermacrene, α-humulene, humulene epoxide), bicyclic sesquiterpenes (epibicyclosesquiphellandrene, α-cadinene, δ-cadinene, β-acoradiene, β-caryophyllene, caryophyllene oxide) and tricyclic sesquiterpenes (epi-globulol, *allo*-aromadendrene, spathulenol, mintsulfide, viridiflorol, α-copaene) (Chong and Aziz, 2011).

Phenols

Centella contains high amounts of phenolic compounds having antioxidant properties (Hussin *et al.*, 2009). The phenolic compounds found include flavonoids, phenylpropanoids and tannins (Zainol *et al.*, 2003; Suntornsuk and Anurukvorakun, 2005; Subban *et al.*, 2008; Chong and Aziz, 2011).

Polyacetylenes

Centella is reported to contain polyacetylenes such as cadiyenol, centellin, centellicin, asiaticin, 8-acetoxycentellynol, dotriacont-8-en-1-oic acid and 11-oxoheneicosanyl (Chong and Aziz, 2011).

Other constituents

In addition to the above constituents, *Centella* contains alkaloids, carbohydrates, amino acids, vitamins and minerals as indicated in Table 7.2.

Chong and Aziz (2011) have tabulated the chemical constituents found in different parts of *Centella* – such as the whole plant, leaves, roots, aerial parts, petioles and stolons – using plant samples from different countries (China, India, Italy, Japan, Korea, Madagascar, Malaysia, Pakistan, South Africa, Spain, Sri Lanka and the USA), based on 49 research findings.

7.3 Postharvest Technology

The value addition and processing techniques for commercially important medicinal plants that are described in two sets of guidelines issued for Indian medicinal plants (Raju and Shastry, 2002; Shastry, 2002) are applicable to *Centella*.

7.3.1 Processing

Drying

The whole plant can be sun and shade dried and stored. The storage moisture content should be 10%.

Semi-processing to value-added products

POWDER. Cleaned and dried plant parts are ground to powder and a homogenous powder of the desired particle size obtained by sieving.

TABLETS/CAPSULES. Tablets are produced by mixing the homogenous powder with a suitable binding agent and compressing into a tablet; capsules are produced by filling a capsule with the homogenous powder at the desired dosage.

EXTRACTS. Dried and clean plant parts are powdered and extracted with a suitable solvent, such as ethyl alcohol or methyl alcohol, in a percolator (cold extraction) or in a Soxhlet extractor (hot extraction). The extract is then concentrated by removing the solvent under reduced pressure at low temperature. The concentrated extract is spray dried. The extract can be standardized to the required strength of active constituent (Khare, 2007).

CRUDE DRUGS. Cleaned and dried (under sun or shade) plant material is packed as the crude drug. The season, time of day and the maturity stage of the plant should be appropriate for the collection of the plant material. For example, leaves should be collected during the flowering season when the plant is highly active.

7.3.2 Value addition

Traditional food preparations

In Thailand, blended leaves of *Centella* are used to make cordial, tea and juice. In Sri Lanka, the juice extracted from *Centella* is used to make porridge known as 'Kolakenda' in Sinhala. Extract of *Centella* is also used in the production of food products such as herbal noodles. In China, *Centella* is used as a cooling drink. Even though *Centella* is widely used in the form of tea, soft drinks and syrup, some studies have been carried out on fermentation of the plant material to explore the feasibility of producing fermented herbal tea (Heong *et al.*, 2011).

The leaves of *Centella* are used to make various types of food in various countries, for example 'Sambai oipeugaga', and salad of the Aceh type in Indonesia, drinks and cold rolls in Thailand and Vietnam, the Malay salad called 'Ulam' and 'Malluma', a traditional complement to rice and curry in Malaysia, and chutney in Brazil (Das, 2011). The leaves can be cooked as a part of a soup or as a main vegetable. Due to its mild bitterness, it is cooked and coconut milk and/or shredded coconut and sometimes sweet potatoes or potatoes are added. In Thailand, it is used as a vegetable, tea and juice. *Centella* is also commonly used for making herbal tea prepared either using a mixture of many different herbs or a single herb (Hashim, 2011).

Commercial products

Pharmaceutical products containing refined extracts of *Centella* have been authorized and marketed in Europe in several Member States (Belgium, France, Greece, Italy, Portugal and Spain) for both external and internal medical uses since the late 1960s (EMEA, 2010). Products containing *Centella* are also available in India, Korea and China (Singh *et al.*, 2010).

Extracts of *Centella* are used in the manufacture of various pharmaceutical products. Those reported in the literature include: TECA (titrated extract *C. asiatica*), TTFCA (total triterpenoid fraction of *C. asiatica*), TTF (total triterpenic fraction), ETCA (estratto titolato di *C. asiatica*) and CATTF (*C. asiatica* total triterpenic fraction). According to the information from the literature and on licensed medicinal products, it is obvious that all the above-mentioned acronyms are different names to designate the same extract, which is commercially known as Madecassol®, Centellase® or Blastoestimulina®, and contains 40% asiaticoside and 60% asiatic acid and madecassic acid (EMEA, 2009).

The products are available in various forms, such as oral tablets, cream and ointment and cutaneous powder. Cream 1%, cutaneous powder 2%, sterilized impregnated dressing and ointment 1% are used for healing wounds and as a treatment for ulcerations. Tablets containing TECA extract, tablets containing triterpenic fraction extract and hard capsules containing asiatic acid are some examples of products for internal medical uses, such as the treatment of pre-varicose syndromes, oedema, itching and leg ulcers and pain (EMEA, 2010). Some examples for commercial products containing *Centella* include Mentat, Gertiforte (stress care), Abana (heart care), Menosan, Diamond shiny pearl BB and Blastoestimulina (EMEA, 2010; Singh *et al.*, 2010).

7.4 Uses

The use of *Centella* in food and beverages, and in medical and cosmetic applications has been increasing worldwide over the years.

7.4.1 General uses

Centella has been reputed to have a wide range of pharmacological uses since prehistoric times. The herb is also used as a traditional vegetable in China, India, Indonesia and Sri Lanka, and is being introduced as a cultivated vegetable into developed countries (Zheng and Qin, 2007). The herb is eaten fresh as salad and juice, as a cooked vegetable and is used as a drink – the dried leaves being used to make tea. The whole plant, including the leaves, stem and root, is consumed (Brinkhaus *et al.*, 2000). *Centella* is also used in nutraceutical and cosmetic preparations, and is becoming an important commercial plant worldwide (James and Dubery, 2009; Hashim *et al.*, 2011).

7.4.2 Pharmacological uses

Centella is a popular herb owing to its efficacy and versatility, especially as a wound healing agent and brain stimulant (in promoting brain growth and improving learning and memory) (Zheng and Qin, 2007). In India, it has a reputation as an ethnomedicine. In Ayurvedic and Unani medicine, it is used to treat various diseases. These include respiratory diseases (bronchitis and asthma), skin ailments (wound healing, psoriasis and eczema), the revitalization of connective tissues, burn and scar treatment, and the treatment of leprosy, elephantiasis, leucorrhoea, body aches, gastric catarrh, dropsy, stomach disorders and urethritis. Besides these uses, it is being used in maternal healthcare and to improve memory and as a nervine (nerve) tonic (Singh *et al.*, 2010; Murray, 2012). Further, it acts as a sedative, anti-stress and anti-anxiety agent, and is used for the treatment of depression. Its antibacterial properties are being used to treat periodontal disease, syphilis and hepatitis. *Centella* stimulates lipolysis and blood circulation and is, thus, used in the management of local adiposity or 'cellulite'. It assists in destroying toxin accumulated in the brain as well as in the nerves, while it helps to clear the body from heavy metals and drugs, and is therefore useful as a general health tonic, an aphrodisiac and immune booster (James and Dubery, 2009).

Fresh extracts of *Centella* are used by the people of Java and the Malay Peninsula as topical and internal agents in wound healing (Kartnig, 1986). In China, *Centella* is used for leucorrhoea, dysentery, nosebleeds, jaundice and scabies, leprosy, fractures, tonsillitis, measles, tuberculosis and urinary difficulties (Singh *et al.*, 2010; Murray, 2012). In Nepal, it is used to treat indigestion, rheumatism, poor memory and leprosy. In Malaysia, it is used to treat hypertension, diarrhoea, mental fatigue, anxiety, eczema and urinary tract infections. In addition, *Centella* is used to improve memory and as a detoxicant and diuretic (Ahmad and Ismail, 2003; Singh *et al.*, 2010; Murray, 2012). In Bangladesh, the whole plant is used to treat dog bites, asthma, diabetes, itching, leucorrhoea, malaria, tumours and wounds, and also as a carminative (Rahmatullah *et al.*, 2009; Rahman *et al.*, 2012). In Madagascar, the herb is traditionally used to treat leprosy and tuberculosis (Kartnig, 1986). In traditional African medicine, *Centell*a has been used for the treatment of leprosy (James and Dubery, 2009). The oiled leaves are eaten for urinary tract infections and the unfiltered juice is used to treat scrofula and syphilis (Khare, 2007).

Both human and animal studies have proven the pharmacological activities of *Centella* as an antimicrobial, and as an anticancer, wound-healing, neuroprotective, immunomodulatory, anti-inflammatory, hepatoprotective and insecticidal, antioxidant (Roy *et al.*, 2013) and antithrombotic (Satake *et al.*, 2007) agent.

According to WHO (1999), the medicinal uses of *Centella* supported by clinical data include the treatment of wounds, burns and ulcerous skin diseases and the prevention of keloid and hypertrophic scars. Extracts of the plant have been used to treat second- and third-degree burns. Topical application of extracts is used to improve the healing of chronic wounds. Oral administration of extracts is used to treat stress-induced stomach and duodenal ulcers. Uses reported in folk medicine, but not supported by experimental or clinical data, include treatment for therapy for anaemia, asthma, albinism, measles, cholera, bronchitis, cellulite, constipation, epistaxis, dermatitis, diarrhoea, dysentery, dysmenorrhoea, dysuria, dizziness, nervous disorders, epilepsy, haematemesis, hepatitis, hypertension, haemorrhoids, jaundice, leucorrhoea, nephritis, neuralgia, rheumatism, smallpox, syphilis, toothache, urethritis and varices, and use as an antipyretic, analgesic, anti-inflammatory and 'brain tonic' (WHO, 1999).

The chemical constituents of *Centella* are responsible for its medicinal and nutraceutical applications. Its major bioactive ingredients, triterpenoid derivatives, play an important role in its medicinal application, and several of its traditional pharmacological uses have been validated scientifically in terms of these bioactive compounds (Zheng and Qin, 2007). Boiteau *et al.* (1949), in a study of derivatives of *Centella* used against leprosy, first reported the chemical constitution of asiaticoside, which is responsible for most of the pharmacological activities of *Centella*. Animal studies have proven that asiaticoside exhibits protective effects on the central nervous system such as an antidepressant-like effect and an anxiolytic-like effect, and also has wound-healing and anti-ulcer activities (Chong and Aziz, 2011). Table 7.3 gives the structure and summarizes the biological activities of the major constituents of *Centella*.

Wound healing

For universal use in wound healing and various skin disorders, the effectiveness of the extracts and monomers from *Centella* have been subjected to and justified by both experimental and clinical evaluations. Wound healing is a complex physiological process that includes a series of steps. Extracts of *Centella* have been shown to produce different actions on various phases of wound healing (Zheng and Quin, 2007); they accelerate wound healing when administered both topically and orally (systemically). The mechanisms involved in wound healing activity are probably by the enhancement of collagen synthesis, and the stimulation of angiogenesis and antioxidant activity (EMEA, 2010). Research has demonstrated that asiatic acid is the only component responsible for the stimulation of collagen synthesis, while madecassoside is able to increase type III collagen secretion and asiaticoside to induce type I collagen synthesis (Zheng and Quin, 2007). Topical application of *Centella* enhances wound healing, particularly in chronic post-surgical and post-trauma wounds (Khare, 2007).

Based on experimental studies, the triterpenic constituents of *Centella* are believed to have the ability to stimulate collagen synthesis in skin fibroblasts, thus *Centella* has been used in skincare products for restoring skin firmness and elasticity, and improving skin appearance (Hashim *et al.*, 2011). Drugs and cosmetic preparations containing active constituents of *Centella* for skincare are available worldwide (Randriamampionona *et al.*, 2007).

Antioxidative properties

Generation of free radicals or ROS more than the antioxidant capacity of a biological system creates oxidative stress. Excess free radicals and ROS attack biological molecules such as lipids, proteins and nucleic acids, causing tissue or cellular injury (Halliwell and Gutteridge, 1999; Stadtman, 2004).

Table 7.3. Structure and biological activities of major bioactive compounds isolated from *C. asiatica*.

Name of compound	Structure	Biological activity	Reference
Apigenin	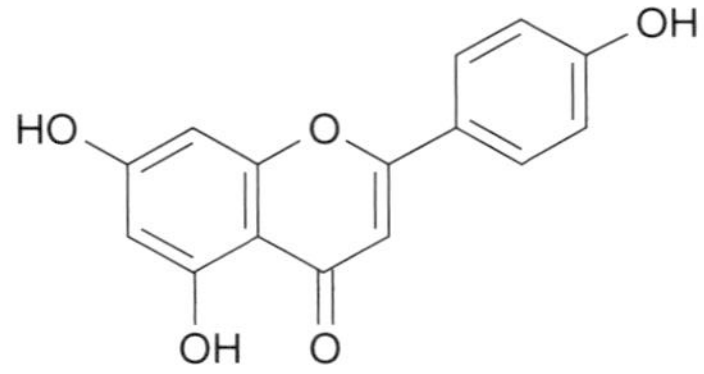	Antibacterial, anti-ulcerative, diuretic, aldose reductase inhibitor, antihypertensive, anti-inflammatory, antioxidant	Intharachatorn and Srisawat, 2013; Roy *et al*., 2013
Ascorbic acid	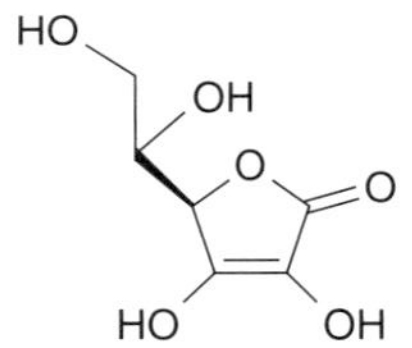	Antioxidant, antibacterial, anti-infective, antidote, antihypercholesterolaemic, inhibits production of carcinogen, induces tissue to produce collagen, hematopoietic activity	Chanwitheesuk *et al*., 2005; Roy *et al*., 2013
Asiatic acid	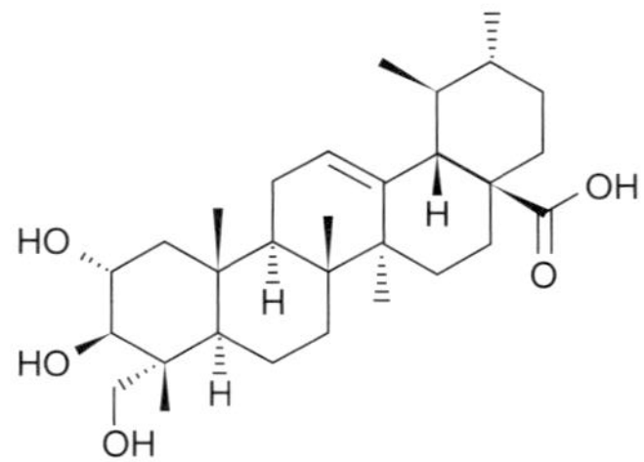	Neuroprotective, wound healing, enhancing learning and memory, anti-inflammatory, anti-apoptotic, anti-hyperglycaemic, neuroprotective, anticancer, antimicrobial	Bunpo *et al*., 2004; Park *et al*., 2005; Khare, 2007; Nasir *et al*., 2011; Xu *et al*., 2012b; Ramachandran and Saravanan, 2013; Roy *et al*., 2013
Asiaticoside	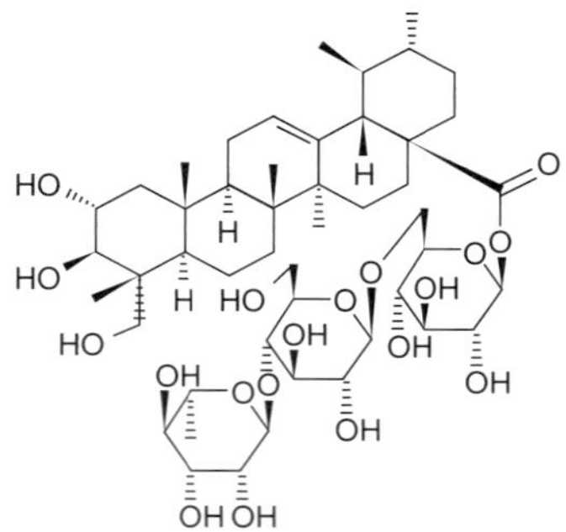	Anti-inflammatory; anticancer, antioxidant, wound healing, antidepressant, neuroprotective, anxiolytic, hepatoprotective, burn cure, antimicrobial, antiulcer, antioxidant, skin protective	Shukla *et al*., 1999; Huang *et al*., 2004; Yoshida *et al*., 2005; Wijeweera *et al*., 2006; Zheng and Qin, 2007; Kimura *et al*., 2008; Zhang *et al*., 2010; Chong and Aziz, 2011; Nhiem *et al*., 2011; Lee *et al*., 2012; Wanasuntronwong *et al*., 2012; Roy *et al*., 2013; Wan *et al*., 2013

Betulic acid (also known as betulinic acid)	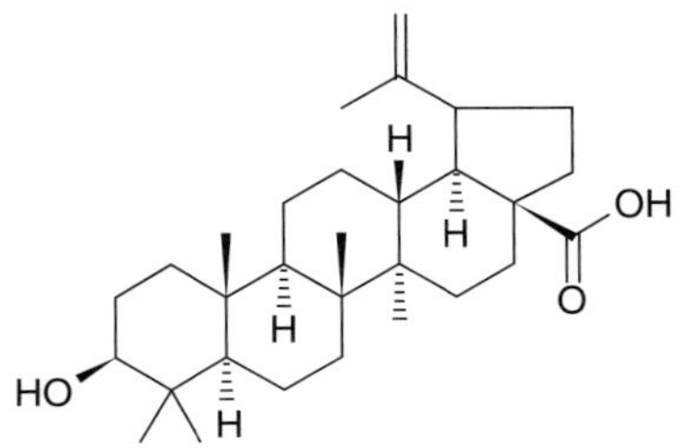	Antineoplastic, cytotoxic, antitubercular, antibacterial	Roy *et al*., 2013
β-Carotene	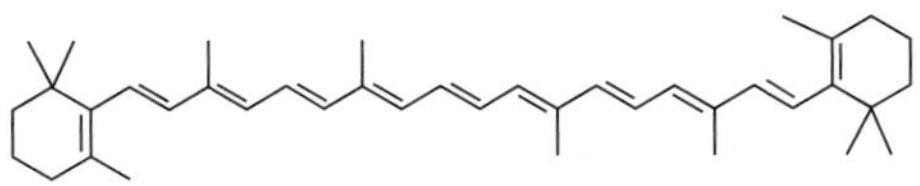	Antioxidant, antitumour	Chanwitheesuk *et al*., 2005; Roy *et al*., 2013
Catechin	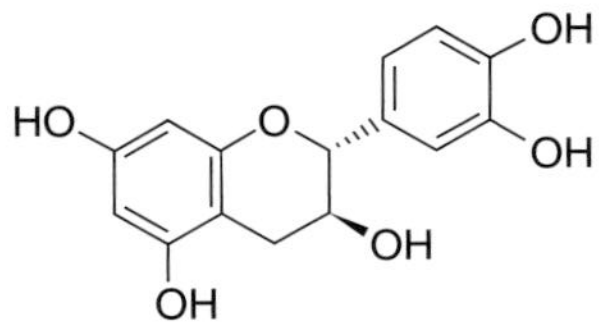	Antihypertensive	Intharachatorn and Srisawat, 2013
Chlorogenic acid	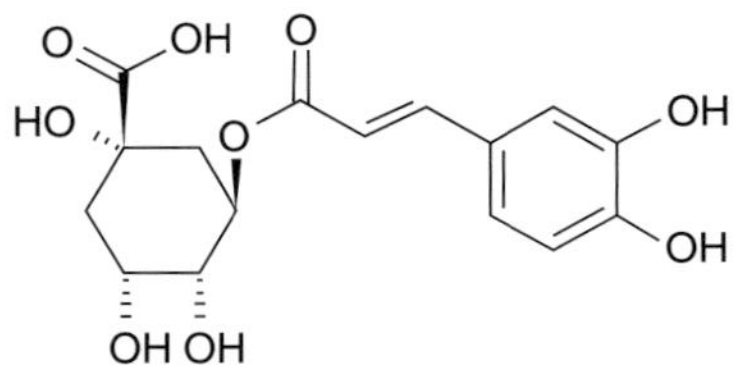	Antioxidant, antineoplastic, cytotoxic, antimutagenic, antiviral, choleretic, haemostatic, leucopoietic, antimalarial	Roy *et al*., 2013
Kaempferol	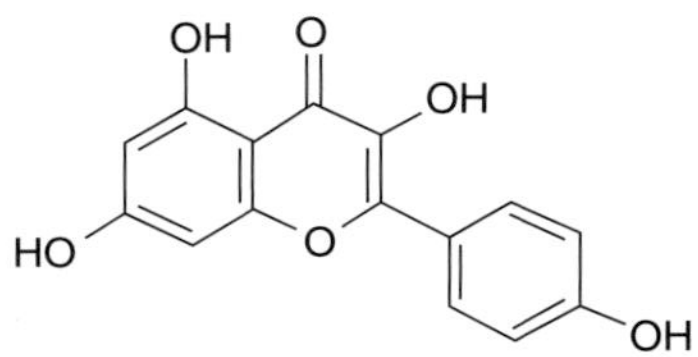	Anti-HIV-1, antibacterial, anti-inflammatory, antioxidant, antihypertensive	Intharachatorn and Srisawat, 2013; Roy *et al*., 2013

Continued

Table 7.3. Continued

Name of compound	Structure	Biological activity	Reference
Madecassic acid	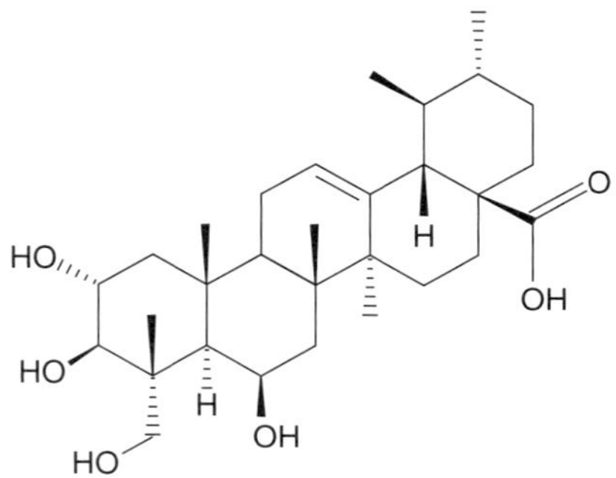	Wound healing	Maquart *et al.*, 1999
Madecassoside	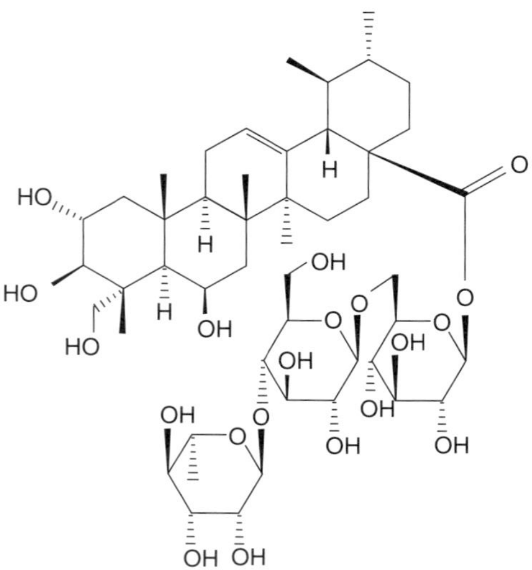	Anti-inflammatory, anti-arthritic, cardioprotective, wound healing	Li *et al.*, 2007; Zheng and Quin, 2007; Bian *et al.*, 2008; Liu *et al.*, 2008a,b; Song *et al.*, 2012; Roy *et al.*, 2013

Naringin	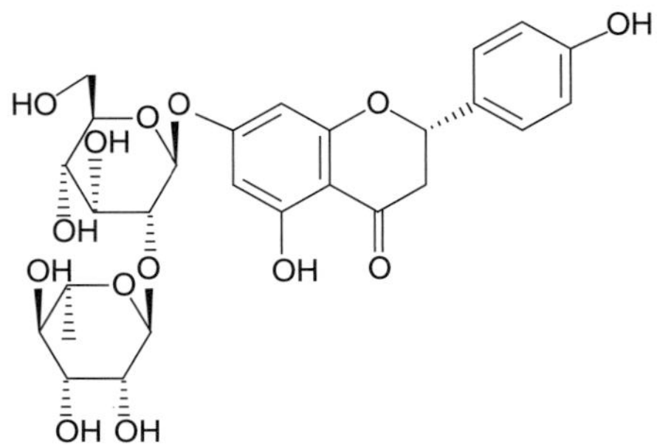	Antibacterial, anti-inflammatory, antiviral	Roy *et al.*, 2013
Quercetin	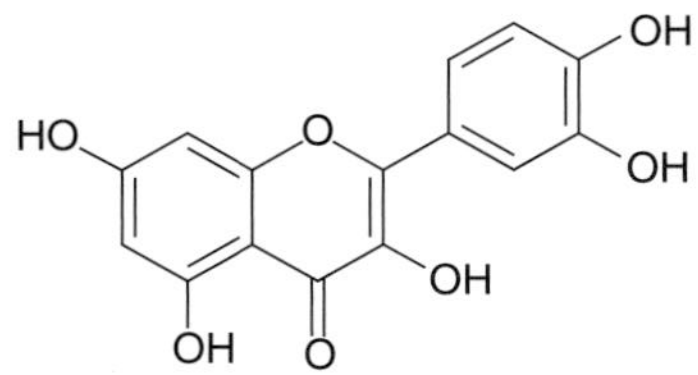	Anti-HIV-1, antiasthmatic, antibacterial, anti-hepatotoxin, antihypertensive, anti-inflammatory, antiviral, antihypercholesterolaemic, platelet aggregation inhibitor, antioxidant	Intharachatorn and Srisawat, 2013; Roy *et al.*, 2013
Rutin	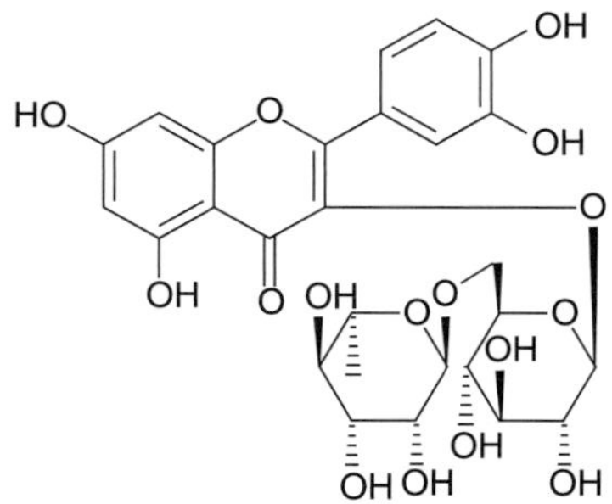	Anti-inflammatory, antiviral, antihypertensive	Intharachatorn and Srisawat, 2013; Roy *et al.*, 2013
Stigmasterol	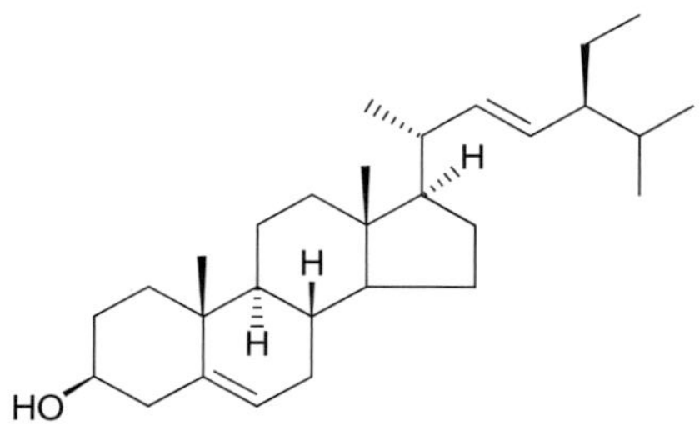	Antihypercholesterolaemic, antimutagenic, cytotoxic inactive, antileishmanial, antimalarial, antitrypanosomal, platelet aggregation inhibitor, antiviral	Roy *et al.*, 2013

Oxidative stress is associated with atherosclerosis, cancer, diabetes, arthritis, reperfusion damage and inflammation (Ames *et al.*, 1993). Antioxidants scavenge free radicals to prevent damage caused by ROS. Many researchers have reported the antioxidative properties of *Centella* (Hamid *et al.*, 2002; Pittella *et al.*, 2009; Rahman *et al.*, 2013). The phenolic compounds with the structural characteristics of free radical scavengers (Bandoniene and Murkovic, 2002), are believed to be the major contributors to these antioxidative properties (EMEA, 2010; Hashim *et al.*, 2011). For instance, the phenolic hydroxyl group in flavonoids has been found to be a strong antioxidant capable of effectively scavenging ROS (Cao *et al.*, 1997). Madecassoside also has antioxidant properties (Chong and Aziz, 2011).

Antihypertensive effects

Intharachatorn and Srisawat (2013) have reported that *Centella* extract supplement had antihypertensive effects in hypertensive rats, and quercetin may be in part responsible for this effect of the extract. The high total phenolic content of *Centella* that is contributed by the flavonoids quercetin, kaempferol, catechin, rutin, apigenin and naringin is said to have a direct effect in lowering blood pressure and *Centella* is often referred to as a rejuvenating pharmaceutical in the Ayurvedic Pharmacopoeia (Intharachatorn and Srisawat, 2013). Raw leaves or a plant decoction are consumed to treat hypertension (Khare, 2007).

Anti-inflammatory activities

Asiatic acid and madecassic acid have anti-inflammatory effects (Roy *et al.*, 2013). An animal study (in rats) has proven that ethanolic extract of *Centella* (100 mg/kg of body weight) has anti-inflammatory activity similar to that of standard ibuprofen (an anti-inflammatory drug) (George *et al.*, 2009). Madecassoside also has anti-inflammatory properties, thus it may improve type II collagen-induced arthritis (Liu *et al.*, 2008a,b).

Anticancer activity

In alternative healthcare, *Centella* has been used in the treatment of tumours and cancerous growths without suppressing the autoimmune system or creating toxic wastes within the body (James and Dubery, 2009). Studies suggested that various extracts of *Centella* induce apoptosis in cancerous cells. Babu *et al.* (1995) examined the antitumour effect of crude and partially purified fractions by both *in vivo* and *in vitro* studies and found that oral administration of both types of extract suppressed the development of tumours and increased the lifespan of tumour-bearing mice. Further, the partially purified fractions were more effective in inhibition of the proliferation of the transformed cell lines than the crude extract and other solvent fractions, and was dose dependent (Babu *et al.*, 1995). A methanolic extract of *Centella* can induce apoptosis in human breast cancerous cells (Babykutty *et al.*, 2009), and aqueous extracts can induce apoptosis in colonic crypts and have a chemopreventive effect on colon cancer in rats (Bunpo *et al.*, 2004). Park *et al.* (2005) have demonstrated that asiatic acid from *Centella* induced apoptosis in human melanoma cells and so may be useful in the prevention of human skin cancer. Further, it has been found that asiaticoside may induce apoptosis and enhance the antitumour activity of vincristine (a chemotherapy drug) in cancer cells, and thus might be useful in cancer chemotherapy (Huang *et al.*, 2004).

Brain-stimulating effect

As described in both the Indian and Chinese medicine systems, *Centella* has several effects on the central nervous system, such as acting as a stimulatory nervine tonic, rejuvenant, sedative and tranquillizer, and having memory-enhancing properties. Leaves of *Centella* are used to improve the mental ability of mentally retarded children and enhance the memory (Jared, 2010). An aqueous extract of *Centella* has a cognitive enhancement effect through an antioxidant mechanism (Zheng and Qin, 2007). Animal studies have proven the brain-stimulating effect of *Centella*. For example, the intragastric administration of *Centella* has memory effects in male Wistar rats (Jared, 2010) and *Centella* extract promoted the brain function of juvenile and young adult mice (Rao *et al.*, 2005).

Anti-ulcer activity

Several scientific findings have suggested the potential use of *Centella* and its bioactive constituents in the treatment of gastric ulcers. Both *Centella* and its bioactive constituents are used in the form of drugs to treat gastric ulcers. Other studies have suggested *Centella* has a significant healing effect against gastric ulcers induced in rats by ethanol, aspirin, cold restraint stress and pyloric ligation (Sairam *et al.*, 2001; Cheng *et al.*, 2004). Another study has shown that the aqueous extract of *Centella* and asiaticoside have anti-inflammatory properties brought about by the inhibition of nitric oxide synthesis and thus facilitate ulcer healing (Guo *et al.*, 2004).

Anti-anxiety activity

Researchers have demonstrated that pure asiaticoside and methanol and ethyl acetate extracts of *Centella* exhibit anxiolytic activity in rats (Wijeweera *et al.*, 2006; Wanasuntronwong *et al.*, 2012). In a study by Jana *et al.* (2010), the treatment of 33 patients with 500 mg/capsule of concentrated lyophilized 70% hydro-ethanolic extract of *Centella*, twice daily after meals for 60 days, demonstrated that *Centella* not only significantly attenuates anxiety-related disorders but also reduces stress and depression.

Hepatoprotective activity

A total glucosides extract of *Centella* showed a protective effect against liver fibrosis effect in rats (Ming *et al.*, 2004). Pingale (2008) reported that an aqueous slurry of *Centella* plant powder protect from carbon tetrachloride-induced liver damage (Pingale, 2008). Another study on the hepatoprotective effect of *Centella* juice revealed that pretreatment with the juice gave hepatoprotection to the same extent as that given by Liv-52 (a hepatoprotective agent made from several herbs) against paracetamol-induced liver damage in broiler chickens (Kumar *et al.*, 2009). Asiaticoside had a hepatoprotective effect against acute liver injury induced by lipopolysaccharide/D-galactosamine induced liver damage in mice (Zhang *et al.*, 2010).

Cardioprotective activity

Animal studies have proven the cardioprotective activity of *Centella*. Administration of *Centella* showed protective effect on myocardial damage (Gnanapragasam *et al.*, 2004; Pragada *et al.*, 2004; Li *et al.*, 2007; Bian *et al.*, 2008).

Neuroprotective effect

Extract of *Centella* has been used in Ayurvedic medicine as a nerve tonic. The asiatic acid, asiaticoside and micronutrients in the *Centella* have neuroprotective effects (Hashim, 2011; Xu *et al.*, 2012a,b). A chloroform and methanolic extract of *Centella* protected monosodium glutamate-induced neurodegeneration in rats, an action that was attributed to its antioxidant and behavioural properties (Ramanathan *et al.*, 2007). An aqueous extract of *Centella* gave protection of the brain against neurodegenerative disorders such as Parkinson's disease (Haleagrahara and Ponnusamy, 2010). *Centella* may be useful for treating stroke, epilepsy, multiple sclerosis and other neuropsychiatric disorders (Barbosa *et al.*, 2008).

Antimicrobial activity

Centella shows antibacterial activity against both Gram-positive and Gram-negative bacteria which is attributed to asiaticoside, which weakens their membranous tissues (Mamtha *et al.*, 2004). Antibacterial activity against *Mycobacterium tuberculosis* and *M. leprae*, Gram-positive cocci, *Pseudomonas pyocyaneus* [current name *P. aeruginosa*], *Trichoderma mentagrophytes* and *Entamoeba histolytica* has been demonstrated. Both the alcoholic and aqueous extract of *Centella* has an antiviral action against type II herpes simplex virus (Khare, 2007; EMEA, 2010). The ethanolic extract of *Centella* at a concentration of 400 mg/ml showed antimicrobial activity against an array of enteric pathogens, and thus could be used as an antidiarrhoeal drug (Mamtha *et al.*, 2004). Arumugam *et al.* (2011) evaluated the *in vitro* antibacterial activity of the methanol, acetone, chloroform and water extracts of leaf and callus of *Centella* against *Bacillus cereus*, *Escherichia coli*, *Staphylococcus aureus* and *P. aeruginosa*.

All of the extracts showed significant antibacterial activity against the test organisms, but the methanol extracts of both leaf and callus gave the maximum inhibitory effect (Arumugam *et al.*, 2011).

Immunomodulatory activity

The triterpenoid saponins of *Centella* have been reported to possess immunomodulatory activity (Plohmann *et al.*, 1997). *Centella* extracts showed an immunomodulatory effect on both non-specific cellular immune responses and humoral immune responses. An aqueous extract of *Centella* showed an immunostimulating activity on the mitogen-stimulated proliferation of human peripheral blood mononuclear cells. In an *in vivo* study in mice, significant response was shown by the aqueous extract of *Centella* to both primary and secondary antibodies against bovine serum albumin (Punturee *et al.*, 2005).

Antithrombotic effect

Centella promotes blood circulation to remove blood stasis. A methanol extract from *Centella* inhibited shear-induced platelet activation and dynamic coagulation in rats. Consequently, *Centella* is a potential medicinal plant for use in the prevention of lifestyle-related diseases such as hypertension, cardiopathy and cerebral apoplexy caused by arteriosclerosis (Satake *et al.*, 2007).

Antidiabetic activity

The antidiabetic activity of *Centella* has been known to the ancient people of Bangladesh for centuries in Ayurvedic medicine (Rahman *et al.*, 2012). The glycosides such as bhramoside and brahminoside found in *Centella* possess sedative and hypoglycaemic effects. In addition, the polyphenolic polymers in *Centella* act as antioxidants, and improve insulin action, thus the plant is useful in treating glucose intolerance and diabetes (Joshi and Chaturvedi, 2013). Animal studies have proven that ethanolic and methanolic extracts of *Centella* lowered blood glucose levels. The mechanism by which the plant extract exerts a hypoglycaemic effect is unclear, but could be attributed to its ability to restore the function of the pancreatic tissues by causing an increase in insulin secretion or a decrease in the intestinal absorption of glucose (Chauhan *et al.*, 2010; Rahman *et al.*, 2012).

Venous insufficiency

The triterpenoid saponins present in *Centella* can strengthen weakened veins by improving the wall alterations that occur in chronic venous hypertension and thereby protect the venous endothelium (Allegra, 1981).

Skin improvement

Centella has been listed as an anti-ageing plant. It can improve the clinical score for skin wrinkles, suppleness, firmness, roughness and hydration, and induces type-I collagen synthesis (Mukherjee *et al.*, 2011). Asiaticoside present in *Centella* can induce type I collagen synthesis in human dermal fibroblast cells (Lee *et al.*, 2006). An alcoholic extract of *Centella* can improve skin conditions such as pruritis (Gohil *et al.*, 2010). A hydroalcoholic extract of *Centella* was used to make herbal creams for cutaneous use together with four other medicinal herbs (*Areca catechu*, *Cinnamon zeylanicum*, *Curcuma caesia* and *Tamarindus indica*), and this extract improved skin hydration, sebum levels and viscoelasticity, as well as reducing melanin content (Saraf *et al.*, 2012).

Other effects

In addition to the above-mentioned effects, other actions of *Centella* include antifertility (Zheng and Qin, 2007), tranquillizing, antipyretic, insecticidal (Jayasinha *et al.*, 1999), radioprotective, antiepileptic, antidepressant (Gohil *et al.*, 2010) and anti-HIV.

Clinical safety

Centella has no known toxicity at the recommended doses. Side effects are rare, administration of high doses may cause skin allergy and burning sensations (with external use), headache, stomach upset, nausea, dizziness and drowsiness. The fresh plant may have a low potential for skin irritation. Some topical preparations may cause contact dermatitis.

The oral consumption of a high dose of *Centella* can cause headaches and transient unconsciousness (Gohil *et al.*, 2010).

Centella is supposed to be abortifacient and to alter the menstrual cycle, so it is advisable to avoid consumption during pregnancy due to its emmenagogue action (Brinker, 2010; EMEA, 2010), and *Centella* should be used during pregnancy under medical advice (ESCOP, 2009). People taking *Centella* for a prolonged period (up to 6 weeks) should take a 2 week gap before continuing to take the herb (Gohil *et al.*, 2010). Even though asiaticoside has been of concern as a possible skin carcinogen in rodents after recurring topical application, further studies are necessary to prove this claim (WHO, 1999).

7.5 Summary

Centella is one of the herbal remedies with multiple therapeutic effects. It has a history of use in Ayurvedic and traditional medicine systems since prehistoric times for treating various human ailments. The biological effects of *Centella* are mainly attributed to its major triterpene derivatives, including asiatic acid, madecassic acid, asiaticoside, madecassoside and brahmic acid. The herb has been subjected to several experimental, phytochemical and clinical investigations, and the studies suggest that it has several pharmacological activities, such as antihypertensive, antioxidant, immunomodulatory, antimicrobial, anticancer, anti-inflammatory, wound healing, memory enhancing, anti-ulcer, antidiabetic and anti-fertility, as well as several other properties. Medicinal, nutraceutical and cosmetic preparations containing extracts of *Centella* are available in the markets of many countries, accompanied by several medicinal claims. However, some of the proposed therapeutic activities of *Centella* and its mechanism of action need to be proved by further clinical studies.

References

Abdulla, M.A., Al-Bayaty, F.H., Younis, L.T. and Abu Hassan, M.I. (2010) Anti-ulcer activity of *Centella asiatica* leaf extract against ethanol-induced gastric mucosal injury in rats. *Journal of Medicinal Plants Research* 4, 1253–1259.

Ahmad, F.B. and Ismail, G.I. (2003) Medicinal plants used by Kadazandusun communities around Crocker range. *ASEAN Review of Biodiversity and Environmental Conservation (ARBEC)* January–March 2003: art1. Available at: http://kdca.org.my/wp-content/files/medicinal_crange.pdf (accessed 20 July 2015).

Allegra, C. (1981) Comparative capillaroscopic study of certain bioflavonoids and total triterpenic fractions of *Centella asiatica* in venous insufficiency. *Clinical Therapy* 99, 507–513.

Alternative Medicine Review (2007) *Centella asiatica. Alternative Medicine Review* 12, 69–72.

Ames, B.N., Shinenaga, M.K. and Hagen, T.M. (1993) Oxidants, antioxidants and the degenerative diseases of aging. *Proceedings of the National Academy of Sciences of the United States of America* 90, 7915–7922.

Arumugam, T., Ayyanar, M., Pillai, Y.J.K. and Sekar, T. (2011) Phytochemical screening and antibacterial activity of leaf and callus extracts of *Centella asiatica. Bangladesh Journal of Pharmacology* 6, 55–60.

Babu, T.D., Kuttan, G. and Padikkala, J. (1995) Cytotoxic and anti-tumour properties of certain taxa of Umbelliferae with special reference to *Centella asiatica* (L.) Urban. *Journal of Ethnopharmacology* 48, 53–57.

Babykutty, S., Padikkala, J., Prasanna Sathiadevan, P., Vijayakurup, V., Abdul Azis, T.K., Srinivas, P. and Gopala, S. (2009) Apoptosis induction of *Centella asiatica* on human breast cancer cells. *African Journal of Traditional, Complementary and Alternative Medicine* 6, 9–16.

Bandoniene, D. and Murkovic, M. (2002) On-line HPLC-DPPH screening method for evaluation of radical scavenging phenols extracted from apples (*Malus domestica* L.) *Food Chemistry* 50, 2482–2487.

Barbosa, N.R., Pittella, F. and Gattaz, W.F. (2008) *Centella asiatica* water extract inhibits iPLA2 and cPLA2 activities in rat cerebellum. *Phytomedicine* 15, 896–900.

Bian, G.X., Li, G.G., Yang, Y., Liu, R.T., Ren, J.P., Wen, L.Q., Guo, S.M. and Lu, Q.J. (2008) Madecassoside reduces ischemia-reperfusion injury on regional ischemia induced heart infarction in rat. *Biological and Pharmaceutical Bulletin* 31, 458–463.

Boiteau, P., Buzas, A., Lederer, E. and Polonsky, J. (1949) Derivatives of *Centella asiatica* used against leprosy: Chemical constitution of asiaticoside. *Nature* 163, 258.

Booncong, P. (1989) A pharmacognostic and taxonomic study of *Centella asiatica* (Apiaceae). PhD thesis, Miami University, Ohio.

Bosse, J.P., Papillon, J., Frenette, G., Dansereau, J., Cadotte, M. and Le Lorier, J. (1979) Clinical study of a new antikeloid agent. *Annals of Plastic Surgery* 3, 13–21.

Brinker, F. (2010) *Herbal Contraindications and Drug Interactions: Plus Herbal Adjuncts with Medicines*, 4th edn. Eclectic Medical Publications, Sandy, Oregon.

Brinkhaus, B., Lindner, M., Schuppan, D. and Hahn, E.G. (2000) Chemical, pharmacological and clinical profile of the East Asian medical plant *Centella asiatica*. *Phytomedicine* 7, 427–428.

Bunpo, P., Kataoka, K., Arimochi, H., Nakayama, H., Kuwahara, T., Bando, Y., Izumi, K., Vinitketkumnuen, U. and Ohnishi, Y. (2004) Inhibitory effects of *Centella asiatica* on azoxymethane-induced aberrant crypt focus formation and carcinogenesis in the intestines of F344 rats. *Food and Chemical Toxicology* 42, 1987–1997.

Cao, G., Sofic, E. and Prior, R.L. (1997) Antioxidant and prooxidant behavior of flavonoids: structure–activity relationships. *Free Radical Biology and Medicine* 22, 749–760.

Chandrika, U.G., Salim, N., Wijepala, G.D.D.J., Perera, K.S.U. and Goonetilleke, A.K.E. (2011) Carotenoid and mineral content of different morphotypes of *Centella asiatica* L. (Gotukola). *International Journal of Food Sciences and Nutrition* 62, 552–557.

Chanwitheesuk, A., Teerawutgulrag, A. and Rakariyatham, N. (2005) Screening of antioxidant activity and antioxidant compounds of some edible plants of Thailand. *Food Chemistry* 92, 491–497.

Chauhan, P.K., Pandey, I.P. and Dhatwalia, V.K. (2010) Evaluation of the anti-diabetic effect of ethanolic and methanolic extracts of *Centella asiatica* leaves extract on alloxan induced diabetic rats. *Advances in Biological Research* 4, 27–30.

Cheng, C.L., Guo, J.S., Luk, J. and Koo, M.W. (2004) The healing effects of *Centella* extract and asiaticoside on acetic acid induced gastric ulcers in rats. *Life Sciences* 74, 2237–2249.

Chong, N.J. and Aziz, Z. (2011) A systematic review on the chemical constituents of *Centella asiatica*. *Research Journal of Pharmaceutical, Biological and Chemical Sciences* 2, 445–459.

Das, A.J. (2011) Review on nutritional, medicinal and pharmacological properties of *Centella asiatica* (Indian pennywort). *Journal of Biologically Active Products from Nature* 1, 216–228.

Devkota, A., Dall'Acqua, S., Jha, P.K. and Innocenti, G. (2010) Variation in the active constituent contents in *Centella asiatica* grown in different habitats in Nepal. *Journal of Plant Science* 7, 43–47.

EMEA (2009) *Evaluation of Medicines for Human Use. Public Statement on* Centella asiatica *(L.) Urban, Herba.* European Medicines Agency, London. Available at: http://www.ema.europa.eu/docs/en_GB/document_library/Public_statement/2009/12/WC500018176.pdf (accessed 20 July 2015).

EMEA (2010) *Assessment Report on* Centella asiatica *(L.) Urban, Herba.* European Medicines Agency, London. Available at: http://www.ema.europa.eu/docs/en_GB/document_library/Herbal_-_HMPC_assessment_report/2012/06/WC500128144.pdf (accessed 20 July 2015).

ESCOP (2009) *Centellae asiaticae* herba. In: ESCOP Monographs, 2nd edn, Supplement 2009. European Scientific Cooperative on Phytotherapy (ESCOP), Exeter, UK, pp. 36–44.

George, M., Joseph, L. and Ramaswamy (2009) Anti-allergic, anti-pruritic, and anti-inflammatory activities of *Centella asiatica* extracts. *African Journal of Traditional, Complementary and Alternative Medicines* 6, 554–559.

Gnanapragasam, A., Ebenezar, K.K., Sathish, V., Govindaraju, P. and Devaki, T. (2004) Protective effect of *Centella asiatica* on antioxidant tissue defense system against adriamycin induced cardiomyopathy in rats. *Life Sciences* 76, 585–597.

Gohil, K.J., Patel, J.A. and Gajjar, A.K. (2010) Pharmacological review on *Centella asiatica*: a potential herbal cure-all. *Indian Journal of Pharmaceutical Sciences* 72, 546–556.

Guo, J.S., Cheng, C.L. and Koo, M.W. (2004) Inhibitory effects of *Centella asiatica* water extract and asiaticoside on inducible nitric oxide synthase during gastric ulcer healing in rats. *Planta Medica* 70, 1150–1154.

Haleagrahara, N. and Ponnusamy, K. (2010) Neuroprotective effect of *Centella asiatica* extract (CAE) on experimentally induced Parkinsonism in aged Sprague–Dawley rats. *Journal of Toxicology Sciences* 35, 41–47.

Halliwell, B. and Gutteridge J.M. (1999) *Free Radicals in Biology and Medicine.* Oxford University Press, Oxford, U.K.

Hamid, A.A., Shah, Z.M., Muse, R. and Mohamed, S. (2002) Characterisation of antioxidative activities of various extracts of *Centella asiatica* (L) Urban. *Food Chemistry* 77, 465–469.

Hashim, P. (2011) Mini review: *Centella asiatica* in food and beverage applications and its potential antioxidant and neuroprotective effect. *International Food Research Journal* 18, 1215–1222.

Hashim, P., Sidek, H., Helan, M.H.M., Sabery, A., Palanisamy, U.D. and Ilham, M. (2011) Triterpene composition and bioactivities of *Centella asiatica. Molecules* 16, 1310–1322.

Heong, C.S., Kaur, B., Huda, N., Karim, A.A. and Fazilah, A. (2011) Effect of fermentation on the composition of *Centella asiatica* teas. *American Journal of Food Technology* 6, 581–593.

Huang, Y.H., Zhang, S.H., Zhen, R.X., Xu, X.D. and Zhen, Y.S. (2004) Asiaticoside inducing apoptosis of tumor cells and enhancing anti-tumor activity of vincristine. *Ai Zheng* 23, 1599–1604.

Hussin, M., Hamid, A.A., Mohamad, S., Saari, N., Bakar, F. and Dek, S.P. (2009) Modulation of lipid metabolism by *Centella asiatica* in oxidative stress rats. *Journal of Food Science* 74, H72–H78.

Intharachatorn, T. and Srisawat, R. (2013) Antihypertensive effects of *Centella asiatica* extract. In: *2013 International Conference on Food and Agricultural Sciences.* IPCBEE 55, IACSIT Press, Singapore.

IUCN (2013) The IUCN Red List of Threatened Species. International Union for Conservation of Nature, UK Office, Cambridge, UK. Available at: http://www.iucnredlist.org/details/168725/0 (accessed 1 May 2014).

James, J.T. and Dubery, I.A. (2009) Pentacyclic triterpenoids from the medicinal herb, *Centella asiatica* (L.) Urban. *Molecules* 14, 3922–3941.

Jamil, S.S., Nizami, Q. and Salam, M. (2007) *Centella asiatica* (Linn.) Urban. A review. *Natural Product Radiance* 6, 158–170.

Jana, U., Sur, T.K., Maity, L.N., Debnath, P.K. and Bhattacharyya, D. (2010) A clinical study on the management of generalized anxiety disorder with *Centella asiatica. Nepal Medical College Journal (NMCJ)* 12, 8–11.

Jared, S.R. (2010) Enhancement of memory in rats with *Centella asiatica. Biomedical Research* 21, 429–432.

Jayasinha, P., Warnasuriya, D. and Dissanayake, H. (1999) *Centella asiatica – A Literature Survey.* Medicinal and Aromatic Plant Series No. 1, Information Service Centre, Industrial Technology Institute, Colombo.

Joshi, K. and Chaturvedi, P. (2013) Review article: therapeutic efficiency of *Centella asiatica* (L.) Urb. An underutilized green leafy vegetable: an overview. *International Journal of Pharmacy and Bio Sciences* 4, 135–149.

Kartnig, T. (1986) Clinical applications of *Centella asiatica* (L) Urb. In: Craker, L.E. and Simon, J.E. (eds) *Herbs, Spices, and Medicinal Plants: Recent Advances in Botany, Horticulture, and Pharmacology, Vol. 3.* Oryx Press, Phoenix, Arizona, pp. 145–173.

Khare, C.P. (2007) *Indian Medicinal Plants: An Illustrated Dictionary.* Springer, New York.

Kimura, Y., Sumiyoshi, M., Samukawa, K.I., Satake, N. and Sakanaka, M. (2008) Facilitating action of asiaticoside at low doses on burn wound repair and its mechanism. *European Journal of Pharmacology* 584, 415–423.

Kumar, P., Prasad, R., Singh, K.K. and Roy, B.K. (2009) Hepatoprotective effect of *Centella asiatica* against paracetamol induced liver damage in broiler chicken. *Indian Journal of Poultry Science* 44, 101–104.

Kuroda, M., Mimaki, Y., Harada, H., Sakagami, H. and Sashida, Y. (2001) Five new triterpene glycosides from *Centella asiatica. Natural Medicines* 55, 134–138.

Lee, J., Jung, E., Kim, Y., Park, J., Park, J., Hong, S., Kim, J., Hyun, C., Kim, Y.S. and Park, D. (2006) Asiaticoside induces human collagen I synthesis through TGF-beta receptor I kinase (TbetaRI kinase)-independent Smad signaling. *Planta Medica* 72, 324–328.

Lee, J.H., Kim, H.L., Lee, M.H., You, K.E., Kwon, B.J., Seo, H.J. and Park, J.C. (2012) Asiaticoside enhances normal human skin cell migration, attachment and growth *in vitro* wound healing model. *Phytomedicine* 19, 1223–1227.

Li, G.G., Bian, G.X., Ren, J.P., Wen, L.Q., Zhang, M. and Lü, Q.J. (2007) Protective effect of madecassoside against reperfusion injury after regional ischemia in rabbit heart *in vivo. Acta Pharmaceutica Sinica* 42, 475–480.

Liu, M., Dai, Y., Li Y, Luo, Y., Huang, F., Gong, Z. and Meng, Q. (2008a) Madecassoside isolated from *Centella asiatica* herbs facilitates burn wound healing in mice. *Planta Medica* 74, 809–815.

Liu, M., Dai, Y., Yao, X., Li, Y., Luo, Y., Xia, Y. and Gong, Z. (2008b) Anti-rheumatoid arthritic effect of madecassoside on type II collagen-induced arthritis in mice. *International Immunopharmacology* 8, 1561–1566.

Mamtha, B., Kavitha, K., Srinivasan, K.K. and Shivananda, P.G. (2004) An *in vitro* study of the effect of *Centella asiatica* (Indian pennywort) on enteric pathogens. *Indian Journal of Pharmacology* 36, 41–44.

Maquart, F.X., Chastang, F., Simeon, A., Birembaut, P., Gillery, P. and Wegrowski, Y. (1999) Triterpenes from *Centella asiatica* stimulate extracellular matrix accumulation in rat experimental wounds. *European Journal of Dermatology* 9, 289–296.

Matsuda, H., Morikawa, T., Ueda, H. and Yoshikawa, M. (2001) Medicinal foodstuffs. XXVII. Saponin constituents of gotu kola (2): structures of new ursane- and oleanane-type triterpene oligoglycosides, centellasaponins B, C, and D, from *Centella asiatica* cultivated in Sri Lanka. *Chemical and Pharmaceutical Bulletin* 49, 1368–1371.

Meena, H., Pandey, H.M., Pandey, P., Arya, M.C. and Ahmed, Z. (2012) Evaluation of antioxidant activity of two important memory enhancing medicinal plants *Baccopa monnieri* and *Centella asiatica*. *Indian Journal of Pharmacology* 44, 114–117.

Ming, Z., Liu, S. and Cao, L. (2004) Effect of total glucosides of *Centella asiatica* on antagonizing liver fibrosis induced by dimethylnitrosamine in rats. *Chinese Journal of Integrated Traditional and Western Medicine* 24, 731–734.

Mukherjee, P.K., Maity, N., Nema, N.K. and Sarkar, B.K. (2011) Bioactive compounds from natural resources against skin aging. *Phytomedicine* 19, 64–73.

Murray, M.T. (2012) *Centella asiatica* (gotu kola). In: Pizzorno, J.E. and Murray, M.T. (eds) *Text Book of Natural Medicine*, 4th edn. Elsevier Churchill Livingstone, St Louis, Missouri, pp. 649–654.

Nasir, M.N., Habsah, M., Zamzuri, I., Rammes, G., Hasnan, J. and Abdullah, J. (2011) Effects of asiatic acid on passive and active avoidance task in male Spraque–Dawley rats. *Journal of Ethnopharmacology* 134, 203–209.

Nhiem, N.X., Tai, B.H., Quang, T.H., Kiem, P.V., Minh, C.V., Nam, N.H., Kim, J.H., Im, L.R., Lee, Y.M. and Kim, Y.H. (2011) A new ursane-type triterpenoid glycoside from *Centella asiatica* leaves modulates the production of nitric oxide and secretion of TNF-α in activated RAW 264.7 cells. *Bioorganic and Medicinal Chemistry Letters* 21, 1777–1781.

Orhan, I.E. (2012) *Centella asiatica* (L.) Urban: from traditional medicine to modern medicine with neuroprotective potential. *Evidence-Based Complementary and Alternative Medicine* 2012, Article ID 946259. Available at: http://www.hindawi.com/journals/ecam/2012/946259/ (accessed 20 July 2015).

Park, B.C., Bosire, K.O., Lee, E.S., Lee, Y.S. and Kim, J.A. (2005) Asiatic acid induces apoptosis in SK-MEL-2 human melanoma cells. *Cancer Letters* 218, 81–90.

Pingale, S.S. (2008) Evaluation of effect of *Centella asiatica* on CCl_4 induced rat liver damage. *Pharmacologyonline* 3, 537–543.

Pittella, F., Dutra, R.C., Junior, D.D., Lopes, M.T. and Barbosa, N. (2009) Antioxidant and cytotoxic activities of *Centella asiatica* (L) Urb. *International Journal of Molecular Sciences* 10, 3713–3721.

Plohmann, B., Bader, G., Hiller, K. and Franz, G. (1997) Immunomodulatory and antitumoral effects of triterpenoid saponins. *Die Pharmazie* 52, 953–957.

Pragada, R.R., Veeravalli, K.K., Chowdary, K.P.R. and Routhu, K.V. (2004) Cardioprotective activity of *Hydrocotyle asiatica* L. in ischemia-reperfusion induced myocardial infarction in rats. *Journal of Ethnopharmacology* 93, 105–108.

Premila, M.S. (2006) *Ayurvedic Herbs: A Clinical Guide to the Healing Plants of Traditional Indian Medicine*. Haworth Press, Binghamton, New York.

Punturee, K., Wild, C.P., Kasinrerk, W. and Vinitketkumnuen, U. (2005) Immunomodulatory activities of *Centella asiatica* and *Rhinacanthus nasutus* extracts. *Asian Pacific Journal of Cancer Prevention* 6, 396–400.

Rahman, M., Sayeed, S.B., Haque, A., Hassan, M. and Islam, S.M.A. (2012) Phytochemical screening, antioxidant, anti-Alzheimer and anti-diabetic activities of *Centella asiatica*. *Journal of Natural Product and Plant Resources* 2, 504–511.

Rahman, P.M., Hossain, S., Rahaman, A., Fatima, N., Nahar, T., Uddin, B. and Basunia, M.A. (2013) Antioxidant activity of *Centella asiatica* (Linn.) Urban: impact of extraction solvent. *Journal of Pharmacognosy and Phytochemistry* 1, 27–32.

Rahmatullah, M., Ferdausi, D., Molli, A.H., Jahan, R., Chowdhury, M.H. and Haque, W.M. (2009) A survey of medicinal plants used by Kavirajes of Chalna Area, Khulna District, Bangladesh. *African Journal of Traditional, Complementary and Alternative Medicines* 7, 91–97.

Raju, G. and Shastry, M.S. (2002) *Guidelines for Harvesting, Storage, Drying and Grading & Structures Required for Value Addition and Storage.* AP-CF/Guidelines-GMCL & LAB/Final, prepared for Andhra Pradesh Forest Department, Hyderabad, India by Gram Mooligai Company Ltd. . Available at: http://www.apforests.gov.in/JFM%20CFM/CFM/Special%20Reports/FRLHT/Reports/11.%20 GMCL%20Guidelines%20Reports.pdf (accessed 1 June 2014).

Ramachandran, V. and Saravanan, R. (2013) Efficacy of asiatic acid, a pentacyclic triterpene on attenuating the key enzymes activities of carbohydrate metabolism in streptozotocin-induced diabetic rats. *Phytomedicine.* 20, 230–236.

Ramanathan, M., Sivakumar, S., Anandvijayakumar, P.R., Saravanababu, C. and Pandian, P.R. (2007) Neuroprotective evaluation of standardized extract of *C. asiatica* in mono sodium glutamate treated rats. *Indian Journal of Experimental Biology* 45, 425–431.

Randriamampionona, D., Diallo, B., Rakotoniriana, F., Rabemanantsoa, C., Cheuk, K., Corbisier, A., Mahillon, J., Ratsimamanga, S. and Jaziri, M.E.I. (2007) Comparative analysis of active constituents in *Centella asiatica* samples from Madagascar: application for *ex situ* conservation and clonal propagation. *Fitoterapia* 78, 482–489.

Rao, S.B., Chetana, M. and Uma Devi, P. (2005) *Centella asiatica* treatment during postnatal period enhances learning and memory in mice. *Physiology and Behaviour* 86, 449–457.

Rosalizan, M.S., Rohani, M.Y., Khatijah, I. and Shukri, M.A. (2008) Physical characteristics, nutrient contents and triterpene compounds of ratoon crops of *Centella asiatica* at three different stages of maturity. *Journal of Tropical Agriculture and Food Science* 36, 43–51.

Roy, D.C., Barman, S.K. and Shaik, M. (2013) Current updates on *Centella asiatica*: phytochemistry, pharmacology and traditional uses. *Medicinal Plant Research* 3, 20–36.

Sairam, K., Rao, C.V. and Goel, R.K. (2001) Effect of *Centella asiatica* Linn. on physical and chemical factors induced gastric ulceration and secretion in rats. *Indian Journal of Experimental Biology* 39, 137–142.

Saraf, S., Chhabra, S.K., Kaur, C.D. and Saraf, S. (2012) Development of photochemoprotective herbs containing cosmetic formulations for improving skin properties. *Journal of Cosmetic Science* 63, 119–131.

Satake, T., Kamiya, K., An, Y., Oishi, T. and Yamamoto, J. (2007) The anti-thrombotic active constituents from *Centella asiatica. Biological and Pharmaceutical Bulletin* 30, 935–940.

Schaneberg, B.T., Mikell, J.R., Bedir, E. and Khan, I.A. (2003) An improved HPLC method for quantitative determination of six triterpenes in *Centella asiatica* extracts and commercial products. *Pharmazie* 58, 381–384.

Shastry, M.S. (2002) *Value Addition Techniques for Commercially Important Medicinal Plants of Andhra Pradesh including Pharmacoepial Standards for 61 species.* AP_CF/LAB Report/Final prepared for Andhra Pradesh Forest Department, Hyderabad, India. Foundation for Revitalisation of Local Health Traditions, Bangalore, India. Available at: http://forest.ap.nic.in/JFM%20CFM/CFM/Special%20Reports/FRLHT/Reports/10.%20Lab%20Report.pdf (accessed 20 July 2015).

Shukla, A., Rasik, A.M. and Dhawan, B.N. (1999) Asiaticoside-induced elevation of antioxidant levels in healing wounds. *Phytotherapy Research* 13, 50–54.

Singh, S., Gautam, A., Sharma, A. and Batra, A. (2010) *Centella asiatica* (L.): a plant with immense medicinal potential but threatened. *International Journal of Pharmaceutical Sciences Review and Research* 4, 9–17.

Somchit, M.N., Sulaiman, M.R., Zuraini, A., Samsuddin, L., Somchit, N., Israf, D.A. and Moin, S. (2004) Antinociceptive and antiinflammatory effects of *Centella asiatica. Indian Journal of Pharmacology* 36, 377–380.

Song, J., Xu, H., Lu, Q., Xu, Z., Bian, D., Xia, Y., Wei, Z., Gong, Z. and Dai, Y. (2012) Madecassoside suppresses migration of fibroblasts from keloids: involvement of p38 kinase and PI3K signaling pathways. *Burns* 38, 677–684.

Stadtman, E.R. (2004) Role of oxidant species in aging. *Current Medicinal Chemistry* 11, 1105–1112.

Subban, R., Veerakumar, A., Manimaran, R., Hashim, K.M. and Balachandran, I. (2008) Two new flavonoids from *Centella asiatica* (Linn.) *Journal of Natural Medicines* 62, 369–373.

Suguna, L., Sivakumar, P. and Chandrakasan, G. (1996) Effects of *Centella asiatica* extract on dermal wound healing in rats. *Indian Journal of Experimental Biology* 34, 1208–1211.

Suntornsuk, L. and Anurukvorakun, O. (2005) Precision improvement for the analysis of flavonoids in selected Thai plants by capillary zone electrophoresis. *Electrophoresis* 26, 648–660.

Tee, E.S., Mohd Idris, N., Mohd Nasir, A. and Khatijah, I. (1997) *Nutrient Composition of Malaysian Foods*, 4th edn. Malaysian Food Composition Database Programme, Institute of Medical Research, Kuala Lumpur.

Tiwari, S., Gehlot, S. and Gambhir, I.S. (2011) Review: *Centella asiatica*: a concise drug review with probable clinical uses. *Journal of Stress Physiology and Biochemistry* 7, 38–44.

Upadhyaya, S. and Saikia, L.R. (2012) Evaluation of phytochemicals, antioxidant activity and nutrient content of *Centella asiatica* (L.) Urban leaves from different localities of Assam. *International Journal of Pharma and Bio sciences* 3, 656–663.

Wan, J., Gong, X., Jiang, R., Zhang, Z. and Zhang, L. (2013) Antipyretic and anti-inflammatory effects of asiaticoside in lipopolysaccharide-treated rat through up-regulation of heme oxygenase-1. *Phytotherapy Research* 27, 1136–1142.

Wanasuntronwong, A., Tantisira, M.H., Tantisira, B. and Watanabe, H. (2012) Anxiolytic effects of standardized extract of *Centella asiatica* (ECa 233) after chronic immobilization stress in mice. *Journal of Ethnopharmacology* 143, 579–585.

Wang, X.S., Dong, Q., Zuo, J.P. and Fang, J.N. (2003) Structure and potential immunological activity of a pectin from *Centella asiatica* (L.) Urban. *Carbohydrate Research* 338, 2393–2402.

WHO (1999) *WHO Monographs on Selected Medicinal Plants – Volume 1*. World Health Organization, Geneva, Switzerland. Available at: http://apps.who.int/medicinedocs/pdf/s2200e/s2200e.pdf (accessed 20 July 2015).

Wijeweera, P., Arnason, J.T., Koszycki, D. and Merali, Z. (2006) Evaluation of anxiolytic properties of gotukola (*Centella asiatica*) extracts and asiaticoside in rat behavioral models. *Phytomedicine* 13, 668–676.

Wunderlin, R.P. and Hansen, B.F. (2008) *Centella asiatica*. In: Atlas of Florida Vascular Plants. Institute for Systematic Botany, University of South Florida, Tampa, Florida. Available at: http://florida.plantatlas.usf.edu/Plant.aspx?id=709 (accessed 20 July 2015).

Xu, C.L., Wang, Q.Z., Sun, L.M., Li, X.M., Deng, J.M., Li, L.F., Zhang, J., Xu, R. and Ma, S.P. (2012a) Asiaticoside: attenuation of neurotoxicity induced by MPTP in a rat model of Parkinsonism via maintaining redox balance and up-regulating the ratio of Bcl-2/Bax. *Pharmacology, Biochemistry and Behaviour* 100, 413–418.

Xu, M., Xiong, Y., Liu, J., Qian, J., Zhu, L. and Gao, J. (2012b) Asiatic acid, a pentacyclic triterpene in *Centella asiatica*, attenuates glutamate-induced cognitive deficits in mice and apoptosis in SH-SY5Y cells. *Acta Pharmacologica Sinica* 33, 578–587.

Yoshida, M., Fuchigami, M., Nagao, T., Okabe, H., Matsunaga, K., Takata, J., Karube, Y., Tsuchihashi, R., Kinjo, J. and Mihashi, K. (2005) Antiproliferative constituents from Umbelliferae plants VII. Active triterpenes and rosmarinic acid from *Centella asiatica*. *Biological and Pharmaceutical Bulletin* 28, 173–175.

Yu, Q., Duan, H., Takaishi, Y. and Gao, W. (2006) A novel triterpene from *Centella asiatica*. *Molecules* 11, 661–665.

Zainol, M.K., Abd-Hamid, A., Yusof, S. and Muse, S. (2003) Antioxidant activity and total phenolic compounds of leaf, root and petiole of four accessions of *Centella asiatica* (L.) Urban. *Food Chemistry* 81, 575–581.

Zainol, N.A., Voo, S.C., Sarmidi, M.R. and Aziz, R.A. (2008) Profiling of *Centella asiatica* (L.) Urban extract. *The Malaysian Journal of Analytical Sciences* 12, 322–327.

Zhang, L., Li, H.Z., Gong, X., Luo, F.L., Wang, B., Hu, N., Wang, C.D., Zhang, Z. and Wan, J.Y. (2010) Protective effects of asiaticoside on acute liver injury induced by lipopolysaccharide/D-galactosamine in mice. *Phytomedicine* 17, 811–819.

Zheng, C.J. and Qin, L.P. (2007) Chemical components of *Centella asiatica* and their bioactives. *Journal of Chinese Integrative Medicine* 5, 348–351.

8 Chester

A.F. Alonge*
University of Uyo, Uyo, Nigeria

8.1 Botany

8.1.1 Introduction

The term 'vegetable' is not attributed to green leaves alone but also to the flowers and young seeds of some plants and even to the roots of herbaceous plants that are edible (i.e. not poisonous or toxic to the body). A vegetable is a plant whose fruits, shoots, stems, leaves and roots or other parts are used for food. They vary in function, i.e. in some cases the leaves serve as food or medicine while in others the stems, young roots and seeds may also serve similar or some other functions. Traditionally, the people of south-eastern Nigeria and other West African countries utilize chester plants (*Heinsia crinita*) for both food and therapeutic purposes (Fig. 8.1).

8.1.2 History/origin

Heinsia crinita (Afz.) G. Taylor (Rubiaceae), or chester is also known as bush apple, and in the local Efik dialect of Nigeria as *Atama*. It is a scrambling shrub with lasting and very conspicuous leafy calyx lobes, producing yellow or reddish fruits and sweet, acidic fruits which are edible.

8.1.3 Location

Chester is usually found sparsely distributed in tropical rain forest. The shrubs are most common in West Africa, mostly in Nigeria and might also be found in African countries such as Ghana, Cameroon, Cote d'Ivoire, Congo, Uganda and Upper Volta (Bassir and Umoh, 1975). In Nigeria, the plant is widely distributed in Ika, Itu, Etim Ekpo, Oron, Onna, Abak, Uyo, Ibesikpo Asutan and Etinan in the Local Government Area in Akwa Ibom State; it is also distributed in some Local Government Areas in Abia, Cross River, Imo and Enugu States of Nigeria. All of these areas are in the south-eastern region of Nigeria.

8.1.4 Morphology

H. crinita is a branching shrub. The flowers are whitish with yellowish throat hairs and

*Corresponding author, e-mail: akindelealonge@uniuyo.edu.ng

© CAB International 2016. *Leafy Medicinal Herbs: Botany, Chemistry, Postharvest Technology and Uses* (eds D.C.P. Ambrose *et al.*)

Fig. 8.1. Fresh chester leaves.

the fruits are yellowish with numerous seeds in each fruit and each seed is triangular in shape (Tindall, 1983). The branches are slightly pubescent, the leaves are elliptic–lanceolate, acutely acuminate, acute at the base and 5–10 cm long (Tindall, 1983). The white variety (*Afiatama*) has glossy, leathery light green leaves dense and elliptical in shape that appear fairly succulent, averaging about 8.0 cm long and 3.2 cm wide (Hutchinson, 1973). It is selectively cultivated by subsistence farmers because of its preferred taste.

The black variety (*Obubu-Atama*) has dull leathery, dark green leaves with scanty brown hairs along the ribs, 7.5 cm long and 3 cm wide on average. It is the black–green colour that gives it the name black variety as compared with the white variety. The black variety is harvested wild from forests for food but it is more popular for its supposed medicinal uses. It is used for enemas and as an abortifacient as well as a remedy for diarrhoea, peptic ulcers and other problems (Ebana *et al.*, 1995). It has a very bitter taste and must be thoroughly processed to remove the bitterness before it is used as food (Hutchinson, 1973).

The *Atama Idim* variety has a stem up to 3 m tall and is cylindrical and pubescent with short and reddish-brown hairs. The leaves are ovate, acute and rounded at the base, and pubescent with reddish-brown and short oppressed hairs on the veins, opposite and up to 10.3 cm long and 6 cm wide with stipules up to 2 mm long, and drop off early. The *Atama Ekpo* variety has a stem up to 2 m tall and is cylindrical, pubescent with short reddish-brown hairs. The young leaves are reddish-brown in colour and have short oppressed hairs on the veins, opposite and up to 1.3 cm long and 1 cm wide with stipules up to 1 mm long, and drop off early. This variety is not used as food because it is claimed to be poisonous (Dye, 1956).

8.2 Chemistry

Data on the proximate composition (moisture, ash, protein, crude fibre, fat and total carbohydrate contents) are given in Tables 8.1–8.3 for fresh, sun-dried and oven-dried leaves of chester.

8.2.1 Chemical composition

Nutritionally, vegetables are valuable because they serve as a cheap source of different types of nutrient. They are also a life-saving supplement of nutrients when there is a meagre diet at times of stress or during war (such as the civil war of Nigeria) (Eyo *et al.*, 1983).

The nutrient content of chester has been described in various accounts in the literature. Eyo *et al.* (1983) reported the contents of total dry matter (30.33%), crude protein (15.10%), crude fibre (13.90%), ether extract (fat content, 3.02%), N-free extract (59.30%), total ash (8.68%), calcium (11.07%) and phosphorus (0.17%).

Etuk *et al.* (1998) reported on the proximate, mineral, vitamin (A and C) and anti-nutrient composition of three varieties of chester namely, the white, black and Ekoi. The proximate composition in terms of % dry matter was ash (4.0–5.0), crude protein (9.45–14.7), ether extract (1.4–4.2), fibre (12.50–14.76), organic matter (95–96), carbohydrate (79–82.65) and calorific value (391–401 kcal/100 g). The moisture content in terms of wet weight was 33.0–45.2%. The vitamin C (ascorbic acid) content was also determined and found to range from 10 to 15 mg/100 g. The mineral composition was also determined and the results (as ppm) were: iron (13.24–19.86),

Table 8.1. Proximate composition of a sample of fresh chester leaves. From Alonge and Essien, 2014.

Component	Amount (%)
Ash	2.68
Crude fat	17.50
Crude fibre	17.5
Crude protein	1.75
Moisture	40.11
Total carbohydrate	81.54

Table 8.2. Proximate composition of a sun-dried sample of chester leaves. From Alonge and Essien, 2014.

Component	Amount (%)
Ash	3.13
Crude fat	9.50
Crude fibre	9.50
Crude protein	24.50
Moisture	15.45
Total carbohydrate	62.17

Table 8.3. Proximate composition of a sample of chester leaves oven dried at 105°C. From Alonge and Essien, 2014.

Component	Amount (%)
Ash	14.23
Crude fat	4.50
Crude fibre	28.00
Crude protein	3.50
Moisture	14.22
Total carbohydrate	64.45

magnesium (38.23–81.28), zinc (1.95–2.92) and calcium (622.90–1056.25). The results are summarized in Tables 8.4–8.7.

Etuk *et al.* (1998) also reported that, in terms of toxicants, the black variety of chester contains the highest contents of hydrocyanic acid and of tannins – as might be expected from its very dark colour and its bitterness. It also contains the highest alkaloid content of the three varieties.

Other research has demonstrated that chester contains all the (usual) amino acids found in protein, but has shown that there are variations in the levels of anti-nutritional substances reported by different workers. More research is needed on this aspect and studies should also be done on the contents of the B complex vitamins and vitamins D and K in chester.

8.2.2 Phytochemistry

This is the branch of chemistry dealing with the chemical processes associated with plant life and the chemical compounds produced by plants. The levels of toxicants or anti-nutritional substances reported as mg/100 g dry matter were oxalate (154.0), hydrocyanic acid (8.60) and phytic acid (110.0) (Eyo *et al.*, 1983). The amino acid contents of the leaf protein were found to be high in glutamic acid, aspartic acid and leucine, but low in methionine (a sulfur-containing amino acid) and histidine (Ifon, 1977; Ifon and Bassir, 1979; Umoh and Bassir, 1980; Eyo *et al.*, 1983).

8.3 Postharvest Technology

Food processing includes any unit operation that changes or converts raw material into a safe and edible form. Food processing also provides us with a means to extend the shelf life of otherwise perishable food; without food processing it would not be possible to sustain the needs of modern urban populations and the choice of food available would be very limited and largely seasonal. Changing lifestyles and family structure have resulted in a largely consumer-led demand for an ever-growing selection of food, particularly ready-prepared and partially prepared products. Virtually all food undergoes some form of processing before its consumption. At its most simple, processing can be peeling a piece of fruit or boiling potatoes. The oldest methods of food processing include sun drying, smoking, pickling and salting; these methods utilize the fact that water removal increases shelf life. Fermentation and freezing are also among the traditional methods of food processing.

Table 8.4. Proximate composition of three varieties of chester leaf. From Etuk *et al.*, 1998.

Variety	Organic compounds[a]	Moisture[b]	Ash[a]	Ether extract (fat)[a]	Crude protein[a]	Fibre[a]	Carbohydrate[a]	Calorific value[c]
Black	95.00	42.4	5.00	4.20	11.80	14.76	79.00	401.00
Ekoi	95.50	33.00	4.50	3.40	9.45	13.92	82.65	399.00
White	96.00	45.20	4.00	1.40	14.70	12.50	79.90	391.00

[a]% dry matter; [b]% wet weight; [c]kcal/100 g.

Table 8.5. Mineral content of three varieties of chester leaf (as ppm). From Etuk *et al.*, 1998.

Variety	Iron	Magnesium	Zinc	Calcium	Cadmium
Black	13.24	79.38	2.20	645.80	0.11
Ekoi	14.52	81.38	1.95	622.90	0.11
White	19.86	38.23	2.92	1056.25	0.21

Table 8.6. Vitamin A (retinal) and vitamin C (ascorbate) content of three varieties of chester leaf as mg/100 g dry weight. From Etuk *et al.*, 1998.

Variety	Vitamin A	Vitamin C
Black	18.89 ± 1.93	133.33 ± 3.72
Ekoi	ND[a]	ND
White	11.00 ± 1.73	121.51 ± 1.86

[a]ND = Not determined.

Table 8.7. Anti-nutrient (toxicant) composition of three varieties of chester as mg/100 g dry weight. From Etuk *et al.*, 1998.

Variety	Hydrocyanide	Total oxalate	Soluble oxalate	Tannin
Black	10.75	13.25	9.76	11.45
Ekoi	10.00	29.17	25.65	6.18
White	8.14	17.05	14.32	5.01

Canning, pasteurization and sterilization are techniques that have been used for many decades but are still important in the modern food processing industry. However, many new processes are now joining them and a number of others are already waiting in the wings (Jones *et al.*, 1979).

One of the problems with products is consumer interest in health and related issues, such as naturalness and added value. This interest has led to the development of products that have specific vitamins or minerals added to them. Fortified foods and drinks are used around the world as a cost-effective way of ensuring the nutritional quality of food supply. The addition of nutrients requires careful attention to food regulation and a suitable nutritional rationale so that the final product remains acceptable to consumers.

Food processing improves the nutritional value of certain foods. For example, severe heat treatment destroys trypsin inhibitors, which are anti-nutritional factors present in a range of foods. For instance, the harmful lectins present in red kidney beans are destroyed by extended boiling. Food processing can also increase the bioavailability of nutrients in foods and their organoleptic qualities. However, any form of food processing, even slicing, washing and cooking in the home, can result in the loss of heat-sensitive, oxygen-sensitive and otherwise highly sensitive nutrients, especially certain vitamins. The leaching of minerals into the cooking water can also occur with vegetables. The main commercial processes that cause nutrient loss are blanching, other application of heat, processing and drying.

In some cases, processed food actually retains more nutrients than the unprocessed form. The best example is frozen vegetables, which are harvested and frozen immediately after collection, whereas fresh vegetables may have been stored for several days before purchase or use. Even with unprocessed vegetables though, modern storage and transportation techniques can help to retain nutrients. An example of a potential health concern arising from food processing involves *trans* fatty acids (a group of unsaturated fatty acids). Vegetable oils are often hydrogenated to improve their oxidative stability and functional properties, as in the manufacture of margarine, and during this process *trans* fatty acids can be produced. These are metabolized in a similar manner to ordinary fatty acids after ingestion, and can cause an increase in 'bad' cholesterol, although the general consensus in the UK is that current intakes of *trans* fatty acid do not present a problem (Gooding, 1962).

Many new processing techniques have been developed in response to changing nutritional concepts and consumer demand for less processed food. The objective is to produce high-quality safe foods that are convenient, fresher and considered more natural. The nutritional implications of some of the techniques that are still in development have yet to be established and will almost inevitably determine whether or not these techniques achieve commercial success (Umoh and Bassir, 1977).

Functional foods are the last refinement in a continuum of products developed to provide added value. One commonly used definition of a functional food is a dietary ingredient that affects its host in a targeted manner so as to exert positive effects that may include or justify certain health claims. Within this context, there is increasing interest in probiotics and symbiotics. Probiotics are substances, e.g. oligosaccharides, that are not digested but which beneficially affect the host by selectively stimulating the growth of specific bacteria in the colon. It is now recognized that the composition of the bacterial population of the bowel is important for human health and can potentially be manipulated by the type of food eaten. The incorporation of live microorganisms (as probiotics) into food such as yoghurts influences the bacterial population in the gut. A third approach, the use of symbiotics, is a combination of the above two approaches (Umoh and Bassir, 1977).

The use of genetic modification in food production is a relatively new process and has many potential applications. For example, a plant can be modified to resist diseases, microbial attack or insect infestation, or to produce fruit with a better flavour and improved keeping qualities. Plants can also be developed to resist certain herbicides that are applied to kill weeds; other possibilities include development for drought resistance (very important in developing countries) and resistance to fruit damage. The use of genetic modification offers substantial potential benefits to the food industry and consumers, but it is recognized that some consumers may have reservations about this new and unfamiliar technology. To help in the recognition of foods that contain genetically modified material, regulations have recently been published that require all foods containing ingredients produced from genetically modified soya or maize to be labelled unless neither protein nor DNA resulting from the genetic modification is present in the food itself (van het Hof *et al.*, 1999).

8.3.1 Processing

The dehydration of vegetables is one of the oldest forms of food preservation known to man. It involves the removal of moisture by sun drying or mechanical drying. Apart from extending shelf life, drying aids in easier packaging and transportation due to the reduction in volume. The leaves of the chester plant are sun dried and used in African countries because of their otherwise limited shelf life; they are used in the preparation of soups. The demand for dehydrated food is rising due the focus on instant and convenience foods. (Gooding, 1962).

Dried vegetables can be produced by a variety of processes, which differ primarily in the type of drying method and depend on the type of food and quality required of the end products. Size reduction is an important pretreatment prior to drying (Ene-Obong and Obizoba, 1996). The pre-drying treatment involves preparation of the raw product, and includes product selection and sorting, and washing followed by size reduction.

Sun drying is a conventional method that has been practised for centuries. It is the cheapest method of drying for perishable food. In this method, the food is directly exposed to sunlight by placing it on the floor, in yards and also on roof tops. Cassava leaves, sweet potato, okra and other green edible leaves are also dried in the sun by farm families. In countries like Nigeria, blanching and salting is carried out before sun drying. Nutritive quality, storage and palatability are the key consideration during the drying of any foodstuff. However, sun drying and shade drying result in loss of micronutrients (Umoh and Bassir, 1980). Chester leaves are shade dried and mechanically dried after harvesting. They are then infused in water in different proportions. Inyang (2015) has reported the effect of treatments on the mineral and vitamin content of the leaves.

The removal of moisture from a typical food product will follow a series of drying rates, as illustrated in Figs 8.2 and 8.3. The initial removal of moisture (AB in Figs 8.2 and 8.3) occurs as the product and the water within the product experience a slight temperature increase. Following the initial stages of drying, significant reductions in moisture content will occur at a constant rate (BC) and at a constant product temperature. The constant-rate drying period occurs with the product at the wet bulb temperature of the air. In most situations, the constant-rate drying period will continue until the critical moisture content (WC in Fig. 8.2) is reached. Below the critical moisture content, the rate of moisture removal decreases with time. The falling-rate drying period (CE, see Fig. 8.1) will follow.

The choice of drying method is based on economic considerations and on the quality and characteristics of the raw material. Some of the drying methods adopted are sun or solar drying, stationary or batch drying processes (kiln, tower and cabinet driers) and continuous drying (tunnel drying, continuous belt trough drying, fluidized bed drying, explosion puffing, foam mat drying, spray drying, drum and microwave-heated drying and subatmospheric dehydration, viz. vacuum shelf, vacuum belt, vacuum drum and freeze drying.

Sun drying is susceptible to the vagaries of weather, though it is much used (see Fig. 8.4). Certain fruits, such as prunes, grapes, dates, figs, apricots and pears are sun dried. These crops are processed in substantial quantities without much in the way of technical aid. Sun-dried products do, however, have limited shelf life because the moisture is not completely removed from the product.

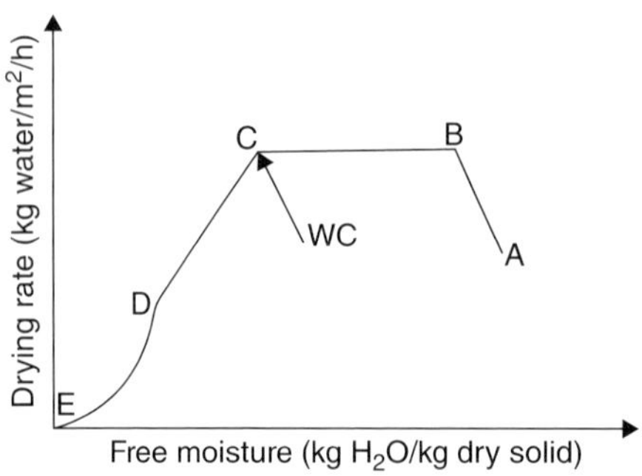

Fig. 8.2. Drying rate curve 1. A–B, initial removal of moisture; B–C, moisture content reduction at a constant rate; C–E, lower rate of drying towards end of drying process. WC, critical moisture content. From Singh and Heldman, 2001, p. 151.

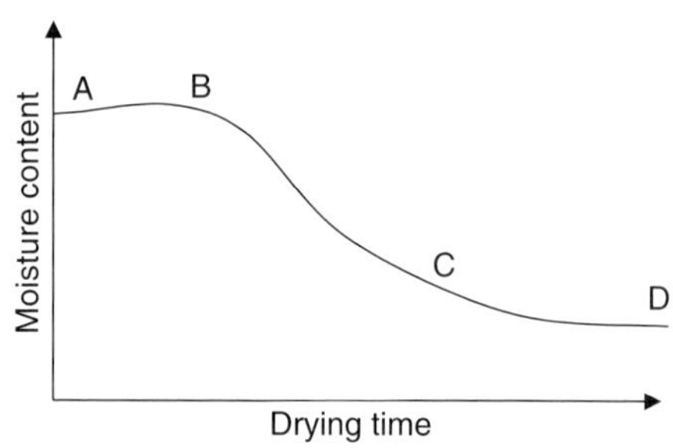

Fig. 8.3. Drying rate curve 2. A–B, initial removal of moisture; B–C, moisture content reduction at a constant rate; C–D, lower rate of drying towards end of drying process. From Singh and Heldman, 2001, p. 151.

Fig. 8.4. Solar drying of chester leaves in a trough.

Mechanical drying involves quicker and complete drying, though the capital investment is higher.

8.3.2 Value addition

Chester leaves are harvested and processed for preparation of soup. During the production season, they can be harvested, dried and stored (Fig. 8.5). Alonge and Essien (2014) noted that the extent of moisture removal depended on the temperature and humidity of the drying air. Both the sun drying method and the oven drying method were investigated and the products were compared with fresh samples to ascertain the changes that took place during drying of the leaves. It was concluded that the oven drying method was more effective, as the amount of nutrients increases with drying time, although the fat and carbohydrate content both decrease. Hence, the nutritive value and keeping quality of chester leaves can be increased by drying.

8.4 Uses

Generally, chester is consumed as cooked complement to major foodstuffs such as cassava, cocoyam, guinea corn, maize, millet, rice and plantain. In fact, most of the meals prepared using these staples are seen as incomplete if a good amount of cooked green leaves does not accompany them. As a result of the method of processing chester for preservation, the nutritive value of the vegetable can be affected. For example, the period of time that leaves spread in the sun are left to dry are likely to affect their nutritive value. However, there is little or no information on the effect of drying of the vegetable for preservation before it is being used for preparing soups. Ajibesin *et al.* (2008) reported that the Efiks in southern Nigeria use the leaves in vegetable soup and also to treat hypertension and abscesses.

8.4.1 General uses

Atama (chester in the Efik local dialect of Nigeria) has for several hundreds of years been exploited by traditional herbalists for the treatment of various ailments, including typhoid fever, diarrhoea and candidiasis (Andy *et al.*, 2008). Chester plants can be used in various forms to treat various diseases and this qualifies it to be called a medicinal plant (Andy *et al.*, 2008).

8.4.2 Pharmacological uses

Chester has a characteristic aroma which could possibly be attributed to the presence of

Fig. 8.5. Dried chester leaves.

terpenes and essential oils that convey its distinctive taste and aroma, and also contribute to its very high medicinal value as summarized below:

1. Oral enquiries of the indigenous people of Abak (in southern Nigeria) and herbalists revealed that chester is used to cure stomach pain such as that caused by ulcers, when it is eaten unprocessed. It is also reported that when burned, ground, dissolved in water and taken for few days, it is effective in the treatment of dysentery (Ebana *et al.*, 1995).
2. The dry leaves are pounded, steeped in water and applied to rheumatic joints to relieve pain (Ebana *et al.*, 1995).
3. The ground dry leaves are rubbed on to the forehead for the treatment of migraine headache (Ebana *et al.*, 1995).
4. A decoction of chester root is used for the treatment of general diseases and a poultice of the leaves is used to treat skin rashes and head lice (Ebana *et al.*, 1995).
5. The stems and branchlets are used as chewing sticks for dental hygiene by those who have mouth infections such as tooth and gum decay (Ebana *et al.*, 1995).
6. The leaves are used for treatment of craw craw and head lice in children (Abo *et al.*, 2011).

8.5 Summary

Chester, also known as bush apple, is a very useful leafy plant. It is a shrub that is mostly found in Africa. It contains important nutrients such as carbohydrates, proteins and minerals. It is consumed as a vegetable along with yams and other tubers in African countries. Chester leaves are used as an ingredient in soups by families in Nigeria. The leaves have a short shelf life and hence are sun dried for use. The effect of drying on the nutritive value of was studied by Alonge and Essien (2013). The nutritive value of the leaves increased on drying. The leaves have medicinal properties and are used in the treatment of various ailments.

References

Abo, K.A., Lawal, I.O. and Ogunkanmi, A. (2011) Evaluation of extracts of *Triclisia suboardata* Oliv and *Heinsia crinita* (Afz) G. Taylor for antimicrobial activity against some clinical bacterial isolates and fungi. *African Journal of Pharmacy and Pharmacology* 5, 125–131.

Ajibesin, K., Ekpo, B.A., Bala, D.N., Essien, E.E. and Adesanya, S.A. (2008) Ethnobotanical survey of AkwaIbom State of Nigeria. *Journal of Ethnopharmacology* 115, 387–408.

Alonge, A.F. and Essien, M.B. (2014) Effect of drying method on quality of chester (*Heinsia crinita*). In: *Proceedings of the 15th International Conference and 35th Annual General Meeting of the Nigerian Institution of Agricultural Engineers (NIAE) Held at the Federal University of Technology, Akure, Nigeria, 22nd–26th September, 2014.* Conference Proceedings Vol. 35, Nigerian Institution of Agricultural Engineers (A Division of the Nigerian Society of Engineers), pp. 285–292. Website: www.niae.net (accessed 13 January 2016).

Andy, I.E., Eja, M.E. and Mboto, C.I. (2008) An evaluation of the antimicrobial potency of *Lasianthera africana* (Beauv) and *Heinsia crinata* (G. Taylor) on *Escherichia coli, Salmonella typhi, Staphylococcus aureus* and *Candida albicans. Malaysian Journal of Microbiology* 4(1) 25–29.

Bassir, O. and Umoh, I.B. (1975) The nutritive adequacy of some Nigerian peasant diets. *Ecology of Food and Nutrition* 2, 297–306.

Dye, W.B. (1956) Studies on *Halogeton glomeratus. Weeds* 4, 55–57.

Ebana, R.U., Essien, A.T. and Ekpa, O.D. (1995) Nutritional and potential medicinal value of leaves of *Lasianthera africana* and *Heinsia crinata. Global Journal of Pure and Applied Sciences* 1, 1–8.

Ene-Obong, H.N. and Obizoba, I.C. (1996) Effect of domestic processing on the cooking time, nutrients, antinutrients and *in vitro* protein digestibility of the African yam bean (*Sphenostylis stenocarpa*). *Plant Foods for Human Nutrition* 49, 43–52.

Etuk, E.U., Bassey, M.N., Umoh, U.O. and Inyang, E.G. (1998) Comparative nutritional studies on three local varieties of *Heinsia crinita. Plant Varieties and Seeds* [later *International Journal of Plant Varieties and Seeds*] 11, 151–158.

Eyo, E.S., Mohme, E. and Abel, H.S. (1983) Chemical composition and amino acid content of *Gnetum africana, Heinsia crinata* and *Piper guineense. Nigerian Journal of Nutritional Sciences* 4, 57–62.

Gooding, E.G.B. (1962) The storage behaviour of dehydrated foods. In: Hawthorn, J. and Leitch, J.M. (eds) *Recent Advances in Food Science, Volume 2.* Butterworths, London, pp. 22–38.

Hutchinson, J. (1973) *The Families of Flowering Plants*, 3rd edn. Oxford University Press, Oxford, UK, pp. 184–194.

Ifon, E.T. (1977) The nutrient composition of some Nigerian leafy green vegetables and physiological availability of their iron contents. PhD thesis, Department of Biochemistry, University of Ibadan, Ibadan.

Ifon, E.T. and Bassir, O. (1979) The nutritive value of some Nigerian leafy vegetables. Part 1: vitamin and mineral content. *Food Chemistry* 5, 263–267.

Inyang, U.E. (2015) Effect of traditional pre-culinary hot water infusion time on the micronutrients and health protecting phytochemicals in *Heinsia crinita* leaf. *Journal of Food and Nutrition Sciences* 3, 191–195.

Jones, L.J., Juo, M.C.T., Kyle, P.E., Radding, S.B., Semrau, K.T. and Somogyi, L.P. (1979) *Overview of the Environmental Control Measures and Problems in the Food Processing Industries.* Document No. EPA-600/2-79-009, US Environmental Protection Agency, Industrial Environmental Research Laboratory Cincinnati, Ohio.

Singh, R.P. and Heldman, D.R. (2001) *Introduction to Food Engineering*, 3rd edn. Academic Press, New York.

Tindall, H.D. (1983) *Vegetables in the Tropics.* McMillan Press, London.

Umoh, I.B. and Bassir, O. (1977) Nutrient changes in some traditional peasant foods during cooking II. Proximate, mineral composition, metabolizable energy and protein calories percent. *West African Journal of Biological and Applied Chemistry* 20, 19–26.

Umoh, I.B. and Bassir, O. (1980) Nutrient changes in some Nigerian traditional peasant foods during cooking III. Essential amino-acid composition. *Nigerian Journal of Nutrition Science* 1, 48–53.

van het Hof, K.H., Brouwer, I.A., West, C.E., Haddeman, E., Steegers-Theunissen, R.P., van DusselDorp, M., West Strate, J.A., Eskes, T.K. and Haut Vast, J.G. (1999) Bioavailability of lutein from vegetable is 5 time higher than that of beta carotene. *The American Journal of Clinical Nutrition* 70, 261–268.

9 Coriander

Maripillai Munusamy Pragalyaashree* and Venkatachalam Thirupathi
Agricultural Engineering College and Research Institute (AEC & RI), Tamil Nadu Agricultural University, Tamil Nadu Agricultural University, Coimbatore, India

> Plants have been a rich source of medicines because they produce a host of bioactive molecules, most of which probably evolved as chemical defences against predation or infection.
>
> (Cox and Balick, 1994, p. 82)

9.1 Botany

9.1.1 Introduction

Coriander (*Coriandrum sativum*), is a delicate culinary and medicinal branched herb belonging to the Apiaceae family (Fig. 9.1). The name 'coriander' was derived from the French *coriandre*, which comes from the Latin *coriandrum*. It is thought to have been derived from the Greek word 'Koris', which means 'bug', which is believed to have been used because the seeds apparently smell like bed bugs. It is known by several other names, including cilantro (in North America), cilantrillo, Arab parsley, Chinese parsley, Mexican parsley, Dhanya and Yuen sai, and Pak Chee. *Cilantro* is the Spanish word for coriander leaves, and these are much used in Mexican cuisine in North America.

Owing to its flavouring properties, the herb is used in food preparations, perfumes and cosmetics, and it has a high economic value. It is also listed as a medicinal plant because of its ability to cure many diseases. The essential oils extracted from coriander have antibacterial, antioxidant, antidiabetic, anticancerous and antimutagenic activities due to the presence of various chemical compounds, viz. geranyl acetate, linalool, dihydrocarvone, anethole, camphor, α-pinene, hellandrene, linalyl acetate, limonene, *p*-cymene, decanal, etc.

9.1.2 History/origin

The medicinal importance of coriander was known to people over 3000 years ago. It is considered to have been one of the first cultivated spices in the world and the people of Greece have used and cultivated coriander since the second millennium BC. Chadwick (1976) reported that the coriander plant was used in two forms, the seeds as a spice and the leaves as food flavouring, and this was confirmed by archaeological evidence

*Corresponding author, e-mail: shreepragalyaa@gmail.com

© CAB International 2016. *Leafy Medicinal Herbs: Botany, Chemistry, Postharvest Technology and Uses* (eds D.C.P. Ambrose *et al.*)

Fig. 9.1. Fresh coriander leaves.

of the period. Other evidence from the early Bronze Age at Sitagroi in Macedonia also proved the cultivation of the plant at that time. Zahary and Hopf (2000) mentioned that the crop grew wild over an area of the Near East and in southern Europe. It was clear that coriander was cultivated by the ancient Egyptians from 1323 BC to add to wine in order to increase intoxication. The Romans are credited with bringing the herb to northern Europe, where it was used in combination with cumin and vinegar to rub on to meat as a preservative. Coriander is native to regions extending from southern Europe and North Africa to south-western Asia. Early settlers of the British colonies in North America first cultivated coriander in 1670.

9.1.3 Location

Coriander originated from the Mediterranean countries but nowadays it is mostly grown and used in Italy, India, Morocco, Eastern Europe, Latin America, Africa and South-east Asia.

9.1.4 Morphology

The herb is soft and delicate and grows to a height of 50 cm. The shape of leaf is broad at the base and slender and feathery at the top. The leaves are compound simple and are arranged in alternate pattern; there is one leaf per node along the stem. The edge of the leaf blade has either lobes or both teeth and lobes.

9.2 Chemistry

9.2.1 Chemical composition

The aroma of coriander leaves is mainly due to its content of aldehyde compounds with 6–10 carbon atoms, particularly decyl and nonyl aldehydes, and 2-decenoic acid, decanoic acid (also known as capric acid) and tetradecenoic acid. The quantity of volatile oil in the leaves is less than that of the fruit and it contains decyl and nonyl aldehydes and linalool (Potter, 1996). The leaves also contain 5% fat, 22% protein, vitamin C, sugars, coumarins and flavonoids including glycosides and chlorogenic and caffeic acids.

9.2.2 Nutritional composition

Coriander leaves have a good nutritional value. This herb has a high content of vitamins A, K and C, minerals such as iron, calcium and magnesium, and more antioxidants than most fruits and vegetables. Data on the nutritional value of the leaves are presented in Table 9.1.

9.2.3 Phytochemistry

Coriander has long been regarded as one of the world's important essential oil plants (Lawrence, 1993). It is suggested that the aroma of fresh herb oil is mainly due to aliphatic aldehydes (mainly C_{10}–C_{16} aldehydes). Many researchers have identified several volatile compounds (alkenals, alkanals, alkenols, and alkanols) from the vegetative

Table 9.1. Nutritional composition of coriander leaves. From various sources.

Constituent	Content/100 g	RDA[a]
Energy	23 kcal	1%
Carbohydrates	3.67 g	3%
Cholesterol	0 mg	0%
Dietary fibre	2.80 g	6.5%
Protein	2.13 g	4%
Total fat	0.52 g	2%
Vitamins		
Folates	62 μg	15.5%
Niacin	1.114 mg	7%
Pantothenic acid	0.570 mg	11%
Pyridoxine	0.149 mg	11%
Riboflavin	0.162 mg	12%
Thiamine	0.067 mg	5.5%
Vitamin A	6748 IU	225%
Vitamin C	27 mg	45%
Vitamin E	2.50 mg	17%
Vitamin K	310 mcg	258%
Electrolytes		
Sodium	46 mg	3%
Potassium	521 mg	11%
Minerals		
Calcium	67 mg	7%
Iron	1.77 mg	22%
Magnesium	26 mg	6.5%
Manganese	0.426 mg	18.5%
Phosphorus	48 mg	7%
Selenium	0.9 mg	2%
Zinc	0.50 mg	4.5%
Phytochemicals		
α-Carotene	36 μg	–
β Carotene	3930 μg	–
β-Cryptoxanthin	202 μg	–
Lutein/zeaxanthin	865 μg	–

[a]Recommended daily allowance.

organs of coriander and all of these are extracted by steam distillation or by the solvent extraction method. Among the many volatile compounds, (*E*)-2-decenal was identified as the most abundant in coriander leaves.

9.3 Postharvest Technology

Fresh coriander leaves are highly perishable produce with a very short shelf life under ambient conditions. As the need for fresh processed leaves has increased, growers and processors have continued to search for ways to enhance the shelf life of their produce so that it can withstand the conditions of the shipping process, which may be accompanied by problems such as mechanical failure of the shipping vehicles or storage facilities, and power failures.

The production of high-quality, convenient, processed coriander leaves poses unique challenges to food processors, which include the destruction of enzymes and substrates, loss of chlorophyll and formation of unwanted secondary metabolites. As a result, senescence and 'off' flavours may develop during respiration. Also, the exudates from the surface constitute a favourable medium for bacterial and fungal growth. When the material is handled, there are further opportunities for contamination and growth by microflora, and yet more health hazards may occur. Besides these problems, the leaves are highly perishable due to their high water content. Hence, it is essential to mitigate damage by proper handling and postharvest processing techniques. This section describes the various techniques used to improve the quality of the product and help to increase its shelf life, and also to add value by the extraction of essential oil.

9.3.1 Processing

Drying

Drying is an important unit operation used to increase the shelf life of food materials. It removes moisture that is responsible for microbial attack and poor shelf life. Coriander is one of the herbs that is much used for culinary purposes and has very little storage life. The shelf life of such herbs can be enhanced by different drying techniques, and the dried leaves can be powdered and incorporated into food and nutraceutical formulations.

The retention of essential nutrients after dehydration by several techniques, such as direct solar drying, shade drying, drying using an electric cabinet and pretreated shade drying, has been studied for different foods and varies from 25 to 90%. In general, the removal of moisture leads to substantial loss in nutrients and the heat that is used diminishes their aroma rapidly. Many drying methods are in vogue for the production of shelf-stable products, but none of the current methods are efficient in drying herbs such as coriander.

Modern drying methods such as microwave drying, refractance window and vacuum drying methods are used to maximize and retain the quality parameters such as texture, colour, flavour and nutrition (Sablani, 2006). Convective drying method is adopted for many of the heat sensitive products. Several research works had proven that air temperature is the most important factor that affects the drying rate and quality.

Air drying and its effects on the characteristics of coriander have been studied by various researchers (Pande *et al.*, 2000; Ahmed *et al.*, 2001; Kaur *et al.*, 2006). Pande *et al.* (2000) reported that the drying of coriander leaves revealed that there was no constant rate drying and exhibited only falling rate period. The suitable drying temperature to get quality product was between 40–50°C. Overall, the results obtained indicated that as aroma, colour and medicinal properties (all heat-sensitive properties) provide high market value to coriander leaves, to prevent the loss of these properties and to preserve them, drying should be done at a low temperature for a longer period of time.

Studies by Shaw et al. (2005, 2007) comparing microwave drying and convective thin layer drying indicated that microwave drying helps to retain the colour and nutritional qualities of coriander leaves better than convective drying.

Packaging

Coriander is highly perishable and has a very short shelf life for several reasons, including the following:

- It has a high water content, resulting in withering if water loss is excessive.
- It has high respiration rates compared with those of apples, oranges and pears.
- The leafy herbs are very delicate, which increases their susceptibility to damage and bruising.

Spoilage of coriander also occurs as a result of:

- mechanical damage during harvesting;
- handling damage;
- microbiological decay; and
- loss of cellular integrity with a subsequent spread of cellular damage during transport due to improper packing technologies.

The above-mentioned problems should be mitigated by proper handling and postharvest processing techniques. Modified atmosphere packaging is one of the techniques adopted to enhance the shelf life of coriander leaves.

MODIFIED ATMOSPHERE PACKAGING (MAP). MAP is the most commonly used technique for extending the shelf life of highly perishable commodities and effectively retards deterioration of fruits and vegetables. It utilizes polymeric packaging films with selective permeability for O_2, CO_2 and water vapour to create a modified atmosphere around the packaged product resulting from the respiration of the product and the permeability of the packaging material (Guevara *et al.*, 2003). MAP is mainly applied to horticultural products, and the reduced O_2 and increased CO_2 in the atmosphere surrounding the fresh produce has several positive effects: it decreases the rate of respiration, ethylene production and texture losses, but improves pigment retention, delays ripening and senescence, and reduces spoilage (Aguilera and Olivera, 2009). Various other research on MAP has reported that it is used to delay senescence and decay, and maintains very good visual quality of the product for 18–22 days at 0°C, 12–14 days at 5°C and 7–8 days at 7.5°C, but only 4–5 days at 10°C. This is due to low temperature and high humidity storage.

In a study on controlled atmosphere storage of coriander leaves, Loaiza and Cantwell (1997) evaluated the respiration rates and quality characteristics of freshly harvested leaves at different temperatures, for different times and under different atmospheric compositions (controlled atmospheres). The samples were stored under two sets of conditions. In the first experiment, samples were stored and analysed for up to 21 days at 0, 5 and 10°C inside glass containers. The humidified atmospheres flowing through the containers were air (21% O_2), air containing 10 or 20% CO_2, 3% O_2, and 3% O_2 containing 10 or 20% CO_2. The quality of the stored products was evaluated after 7, 14 and 21 days. In the second experiment, the leaf samples were kept at 0, 5 and 7.5°C for 22 days in air or in air containing 5 or 9% O_2. The stored samples were evaluated after 10, 14, 18 and 22 days. The respiration rates of freshly harvested leaves were moderately high and the ethylene production rates low at 5°C. Periodic evaluation of stored samples (kept in the dark at a range of temperatures in air or controlled atmospheres) noted visual quality, decay, aroma, off odour, colour and chlorophyll content. Leaves stored in air at 0°C had good visual quality for 18 and 22 days, while at 5 and 7.5°C, they maintained good quality up to 14 and 7 days, respectively. An atmosphere of air plus 5 or 10% CO_2 increased the shelf life of leaves stored at 7.5°C to about 14 days. Leaves stored in 3% O_2 plus CO_2 were similar to those stored in air plus CO_2. CO_2-enriched atmospheres (9–19%) caused dark lesions after 18 days and 20% CO_2 caused severe injury after 7 days. It was noted that visual quality could be maintained for up to 22 days, but the aroma decreased notably after 14 days of storage irrespective of storage conditions. It was concluded that controlled atmospheres with 5–10% CO_2 were beneficial in retaining colour and visual quality at 7.5°C, and that the best postharvest conditions for coriander leaves are low temperature and high humidity storage in air, when the samples can be stored up to 14 days.

Research was done by Luo *et al.* (2004) on the effect of MAPs with different oxygen transmission rates (OTRs) on package atmosphere and on the quality and microbial safety of coriander leaves. The different OTRs of package films chosen for the study were 3500, 6200 and 1700 ml/day/m^2. It was found that an OTR rate of 3500 ml/day/m^2 was equilibrated by the third day of storage and then started maintaining the same O_2 and CO_2 conditions (1.5–2.3 kPa and 3.6–4.1 kPa). High OTR (6200 ml/day/m^2) packages equilibrated at a higher O_2 pressure (3.6–5.6 kPa) and a lower CO_2 (2.7–3.3 kPa) until the 12th day of storage. Freshly cut leaves had the highest tissue integrity (lowest electrolyte leakage), with high visual scores, whereas samples kept at 1700 ml/day/m^2 OTR had atmospheres with less O_2 (0.02 kPa), as it was rapidly depleted, and higher CO_2 (7.7–9.0 kPa) from the 6th day of storage until the end of the storage period. It was observed that rapid depletion of oxygen resulted in the development of off odour and the loss of visual quality and aroma during storage, so that the product became unacceptable. There was also an increase in anaerobic microorganisms on coriander leaves stored in this package atmosphere. Control samples stored in perforated bags (without MAP) exhibited loss of moisture and wilting, and slow growth of aerobic microorganisms in the packages. It was concluded that OTRs of 3500 and 6200 ml/day/m^2 were effective in maintaining the freshness of coriander leaves until the 12th day of storage, and that an OTR of 1700 ml/day/m^2 and perforated bags failed to maintain the colour, moisture and microbial load of fresh cut leaves.

Pragalyaashree *et al.* (2013) conducted experiments to optimize the packaging materials and gas composition for enhancing the shelf life of coriander leaves by MAP. Three different thicknesses each of low-density polyethylene (LDPE) bags and polypropylene (PP) packaging materials were assessed for their permeability to O_2 and CO_2. The LDPE bag with a thickness of 152 μm recorded the lowest permeability to oxygen (1067 ml/m^2/day) and was selected for testing. The leaves were packaged in LDPE bags with product volume ratios of 1:11, 1:10 and 1:7 to assess the respiration rate under ambient and refrigerated conditions. The optimization of gas composition for MAP was done by calculating the respiration rate of the leaves using the Michaelis–Menten equation (Lakakul *et al.*, 1999). The bags were stored under ambient and refrigerated conditions for 30 days. A product volume ratio of 1:7 and gas composition of 5% O_2, 5% CO_2 and 90% N_2 were optimized based on the lowest respiration rate. No significance differences in quality were observed for characters such as physiological weight loss, chlorophyll, β-carotene content and microbial load for the optimized parameters. The keeping quality of leaves stored under ambient conditions had a shelf life of 4 days compared with 20 days for refrigerated conditions. The MAP-stored leaves kept under refrigerated conditions had a longer shelf life than those stored under ambient conditions.

Other methods of preservation

PRESERVATION OF LEAVES USING UV-C TREATMENT. Green leafy vegetables have been preserved by a number of methods, but thermal processing is a popular method that is being adopted by the food industries as it effectively inactivates microorganisms and enhances shelf life. However, the thermal approach commonly results in a loss of nutrients and results in poor quality products. This can be prevented by using non-thermal technologies that rely on techniques other than heating or cooling operations.

Ultraviolet (UV) light processing is one of the advanced emerging non-thermal processing methods and is currently undergoing intensive scientific and developmental evaluation. UV light irradiation can be used to inactivate several organisms and has been used for many years in the pharmaceutical, electronic and aquaculture industries as a disinfection medium. The UV light acts as a physical method for microbial disinfection. The effectiveness of UV-C light (short wave UV light of wavelength 200–400 nm) treatment depends on several factors, such as the dosage applied, intensity, exposure period and the distance between the source of radiation and the sample port.

Mahesh and Thirupathi (2007) studied the preservation of the quality of coriander leaves using UV-C radiation. Monochromatic UV light (254 nm) was supplied by low-pressure mercury vapour germicidal lamps. The UV treatment was done at four dosages (7.06, 8.00, 8.82 and 9.77 kJ/m^2) achieved by varying the exposure time at constant intensity. The treated samples were packed in polypropylene films and stored under ambient or refrigerated conditions. The effectiveness of the treatment on microorganisms was determined by recording the mean log population of bacteria and fungi in the treated samples at regular intervals during storage periods. The bacterial populations in the treated samples under ambient and refrigerated conditions were reduced from an initial value of 3.11 to 1.45 mean log population (cfu/g) at 8.82 kJ/m^2 after a period of 4 days, and to 0.75 mean log cfu/g after a period of 10 days. The fungal population was reduced from 1.87 to 1.05 and 0.83 cfu/g after 4 and 10 days under ambient and refrigerated conditions, respectively. Changes in the nutrient composition of the coriander leaves (Fe, vitamin C, protein and P content) were minimum under both ambient and refrigerated conditions at 8.82 kJ/m^2 UV dosage.

PRESERVATION OF LEAVES USING 1-METHYLCYCLOPROPENE (MCP). Demand for packaged fresh cut culinary herbs has increased. The preparation of fresh cut coriander leaves and maintenance of their shelf life remains challenged owing to their high perishability (Loaiza and Cantwell, 1997). The quality of fresh coriander leaves quality is best maintained under low temperature and high-humidity storage (Cantwell and Reid, 1993). Luo *et al.* (2004) observed that modified atmosphere packages with oxygen transmission rates of 6.5 and 29.2 pmol/s/m^2/Pa were effective in maintaining quality of fresh-cut coriander leaves but that the microbial populations were high. Hence, it is necessary to reduce or inactivate the microorganisms present by washing before packaging is done. Chlorine has been used on produce for this purpose for a long time (Wei *et al.*, 1985), but its efficacy in reducing the microbial population has been found to be less than was originally thought (Zhang and Farber, 1996), and there is a need to develop safer and more effective antimicrobial alternatives. Acidified sodium chlorite (ASC) has been approved as an alternative strong antimicrobial agent and was effective against various microorganisms inoculated on to cantaloupes and asparagus (Park and Beuchat, 1999).

Various studies on 1-methylcyclopropene (1-MCP) showed that it delays ethylene-induced senescence by delaying chlorophyll degradation but has no antimicrobial effect (Blankenship and Dole, 2003). The effects of 1-MCP and sanitizer (acidified sodium chlorite or sodium hypochlorite), either alone or in combination, indicated that 1-MCP delayed the decrease in O_2 and accumulation of CO_2, and acidified sodium chlorite effectively reduced the decay of samples due to coliforms (Kim *et al.*, 2007).

PRESERVATION THROUGH IRRADIATION. Fresh coriander leaves are much used in soups and salsas, and have resulted in outbreaks of food-borne illness (CDHS, 2000). The FDA (2001a,b) reported high rates of *Salmonella* and *Shigella* on coriander leaves and, hence, it has become essential to remove such food-borne pathogens. Various studies have demonstrated that irradiation effectively inactivated food-borne parasites, but the effect of the irradiation on the quality of the product was not reported. However, later findings confirmed that irradiation treatment resulted in a reduction of aroma value (Fan and Sokorai, 2002), a quality that usually decreases before there is an impact on visual quality. In their study, Fan and Sokorai (2002) investigated the effect of irradiation on volatile compounds of fresh coriander leaves by treating them with 0, 1, 2, or 3 kGy γ-radiation and then storing at 3°C up to 14 days. The volatile compounds present were identified using GC-MS (gas chromatography–mass spectrometry) 0, 3, 7 and 14 days after irradiation. The amount of volatile compounds in irradiated leaves was less than that in non-irradiated samples during storage, but the important volatile compounds (decanal and (*E*)-2-decenal) were not affected by irradiation in any consistent manner. Based on these results, it was suggested that coriander leaves can be irradiated with up to 3 kGy γ-radiation without much change in volatile compounds. Thus, irradiation can be used for pathogen inactivation without a major loss of volatile compounds.

9.3.2 Value addition

Extraction of essential oil

Essential oil is extracted from coriander leaves by two methods: hydrodistillation and solvent extraction.

HYDRODISTILLATION METHOD. This method is described as used by Bhuiyan *et al.* (2009). Well-grown fresh leaves were harvested and ground in a blender. They were then put into a modified Clevenger apparatus and subjected to hydrodistillation for 4 h. Moisture was removed from the collected oil by drying it over anhydrous sodium sulfate. The oil samples were then transferred to an airtight container and stored at 0°C. The samples were analysed using GC-MS (the electron impact ionization method). The results indicated that the oil contained 44 aromatic acids, with the major constituents decenoic acid (30.82%), (*E*)-11-tetradecenoic acid (13.4%), capric acid (12.7%), undecyl alcohol (6.4%) and tridecanoic acid (5.5%). The other constituents were present in small amounts only.

SOLVENT EXTRACTION METHOD. Solvents such as ethanol, petroleum ether and water can be used to extract essential oil from coriander leaves. In a study by Nirmala *et al.* (2013), the leaves were washed and dried in a drier with a good air circulation system. The dried samples were subjected to drying in a hot air oven at 60°C for 3 days and ground in a mechanical grinder to a coarse sample. Approximately 300 g of the sample was taken for extraction using the Soxhlet apparatus. Ethanol was used as a solvent and the extraction continued for 24 h. Siphoning of a colourless liquid from the apparatus indicated completion of the process. The liquid was then filtered using sterilized cotton and the solvent was removed by heating over a water bath to obtain a crude extract. A similar procedure was used to extract essential oil using petroleum ether and water. The process was carried out at room temperature and the oil was stored under refrigerated conditions.

9.4 Uses

9.4.1 Culinary uses

Fresh coriander leaves and seeds are used in all types of cuisine all over the world in many food preparations They are used to flavour curries, chutneys, breads, cakes, pickles, salsas, soups, salads and liqueurs. In Indian, Chinese and Thai dishes, and in Mexican cooking, chopped coriander leaves are used to garnish dishes.

Coriander leaves are often juiced with celery and apples to prepare a medicinal drink. They can also be added to smoothies, salsas, salads, guacamole, soups, pesto, tomatoes, beans and vegetable dishes.

9.4.2 Pharmacological uses

1. Coriander assists digestion by stimulating the secretion of gastric juices, which eases colic and flatulence.
2. The chemical substances derived from coriander leaves have antibacterial activity against *Salmonella choleraesuis*.
3. The extract from coriander leaves is used as folk medicine for various purposes, such as to improve the appetite, to relieve flatulence and indigestion, to treat arthritis, rheumatism, sore muscles and diabetes, and as an anxiolytic and anthelmintic, etc.
4. The essential oil has antimicrobial effects.
5. Coriander leaves alleviate irritations in the intestines, probably because of their plentiful magnesium, potassium and fibre content.
6. The regular consumption of coriander increases immunity and also purifies the blood owing to the presence of magnesium, potassium and fibre.
7. Coriander helps to stimulate insulin production and to prevent diabetes.
8. Coriander is used for digestive system problems, including stomach upset, loss of appetite, hernia, nausea, diarrhoea, bowel spasms and intestinal gas.
9. Coriander is also used to treat measles, haemorrhoids, toothache, worms and joint pain, as well as infections caused by bacteria and fungi.
10. Different parts of the herb contain various amounts of phenolic compounds and thus have different antioxidant activities. The phenolic content and antioxidant activity of leaf extracts is higher than those of stem extracts (Juhaimi and Ghafoor, 2011).

9.5 Summary

Coriander is an important culinary herb cultivated all over the world. It is economically important for its use as a flavouring agent in food products, perfumes and cosmetics. The herb is also credited with medicinal properties and the essential oil extracted from the leaves is used for curing many ailments. The delicate nature of coriander leaves does not allow it to be stored for a long period. Much research has been done on enhancing the shelf life of the leaves after harvesting, and a lot more work is needed to exploit the usage of the herb.

References

Aguilera, R. and Olivera, J.C. (2009) Review of design engineering methods and applications of active and modified atmosphere packaging systems. *Food Engineering Reviews* 1, 66–83.

Ahmed, J., Shivhare, U.S. and Singh, G. (2001) Drying characteristics and product quality of coriander leaves. *Food and Bioproducts Processing* 79, 103–106.

Bhuiyan, M.N.I., Begum, J. and Sultana, M. (2009) Chemical composition of leaf and seed essential oil of *Coriandrum sativum* L. from Bangladesh. *Bangladesh Journal of Pharmacology* 4, 150–153.

Blankenship, S.M. and Dole, J.M. (2003) 1-Methylycyclopropene: a review. *Postharvest Biology and Technology* 28, 1–25.

Cantwell, M. and Reid, M.S. (1993) Postharvest physiology and handling of fresh culinary herbs. *Journal of Herbs, Spices and Medicinal Plants* 1, 93–127.

Chadwick, J. (1976) *The Mycenaean World.* Cambridge University Press, Cambridge, UK.

CDHS (2000) *Foodborne Disease Outbreaks Reported in California (1998).* California Department of Health Services, Berkeley, California.

Cox, P.A. and Balick, M.J. (1994) The ethnobotanical approach to drug discovery. *Scientific American* 270(6), 82–87.

Fan, X. and Sokorai, K.J.B. (2002) Changes in volatile compounds of γ-irradiated fresh cilantro leaves during cold storage. *Journal of Agricultural and Food Chemistry* 50, 7622–7626.

FDA (2001a) FDA Survey of Imported Fresh Produce FY 1999, Field Assignment. US Food and Drug Administration, Washington, DC. Available at: www.fda.gov/Food/GuidanceRegulation/GuidanceDocumentsRegulatoryInformation/ProducePlantProducts/ucm118891.htm http://vm.cfsan.fda.gov/_dms/prodsurv.html (accessed 14 January 2016).

FDA (2001b) Survey of Domestic Fresh Produce: Interim Results, (June 30, 2001). US Food and Drug Administration, Washington, DC. Available at: http://www.fda.gov/Food/GuidanceRegulation/GuidanceDocumentsRegulatoryInformation/ProducePlantProducts/ucm118825.htm (accessed 14 January 2016).

Guevara, J.C., Yahia, E.M., Brito de la Fuente, E. and Biserka, S.P. (2003) Effects of elevated concentrations of CO_2 in modified atmosphere packaging on the quality of prickly pear cactus stems (*Opuntia* spp.). *Postharvest Biological Technology* 29, 167–176.

Juhaimi, F.A and Ghafoor, K. (2011) Total phenols and antioxidant activities of leaf and stem extract from coriander, mint and parsley grown in Saudi Arabia. *Pakistan Journal of Botany* 43, 2235–2237.

Kaur, P., Kumar, A., Arora, S. and Ghuman, B.S. (2006) Quality of dried coriander leaves as affected by pretreatments and method of drying. *European Food Research and Technology* 223, 189–194.

Lakakul, R., Beaudry, R.M. and Hernandez, R.J. (1999) Modeling respiration of apple slices in modified-atmosphere packages. *Journal of Food Science* 64, 105–110.

Lawrence, B.M. (1993) A planning scheme to evaluate new aromatic plants for the flavor and fragrance industries. In: Janick, J. and Simon, J.E. (eds) *New Crops*. Wiley, New York, pp. 620–627.

Loaiza, J. and Cantwell, M. (1997) Postharvest physiology and quality of cilantro (*Coriandrum sativum* L.). *HortScience* 32, 104–107.

Luo, Y., McEvoy, J.L., Wachtel, M.R., Kim, J.G. and Huang, Y. (2004) Package atmosphere affects post harvest biology and quality of fresh cut cilantro leaves. *HortScience* 39, 567–570.

Mahesh, N. and Thirupathi, V. (2007) Studies on preservation of green leafy vegetables by UV-C treatment. M. Tech thesis, Department of Food and Agricultural Process Engineering, Tamil Nadu Agricultural University, Coimbatore, India.

Nirmala, P., Sen, S.K., Karim, M.F.B., Hassan, N. and Faruque, A. (2013) Phytochemical and biological investigation of *Coriandrum sativum* (Cilantro) leaves. *International Journal of Innovative Pharmaceutical Sciences and Research* 1(1), 170–184.

Pande, V.K., Sonune, A.V. and Philip, S.K. (2000) Solar drying of coriander and methi leaves. *Journal of Food Science and Technology* 37, 592–595.

Park, C.M. and Beuchat, L.R. (1999) Evaluation of sanitizers for killing *Escherichia coli* O157:H7, *Salmonella* and naturally occurring microorganisms on cantaloupes, honeydew melons, and asparagus. *Dairy, Food and Environmental Sanitation* 19, 842–847.

Potter, T.L. (1996) Essential oil composition of cilantro. *Journal of Agricultural and Food Chemistry* 44, 1824–1826.

Pragalyaashree, M.M., Thirupathi, V., Kasthuri, R. and Raj Kumar, P. (2013) Enhancing the shelf life of culinary herbs by modified atmosphere packaging. *The Madras Agricultural Journal* 100, 612–618.

Sablani, S.S. (2006) Drying of fruits and vegetables: retention of nutritional/functional quality. *Drying Technology* 24, 123–135.

Shaw, M., Meda, V., Tabil, L. Jr and Leduc, P. (2005) *Development and Trends in Drying of Herbs and Specialty Crops in Western Canada.* Written for presentation at the CSAE/SCGR 2005 Meeting Winnipeg, Manitoba, June 26–29, 2005. CSAE/SCGR Paper No. 05-030, Canadian Society for Engineering in Agricultural, Food, and Biological Systems, Winnipeg, Manitoba, Canada. Available at: http://www.csbe-scgab.ca/docs/meetings/2005/05-030.pdf (accessed 14 January 2016).

Shaw, M., Meda, V., Tabil, L. Jr and Opoku, A. (2007) Drying and color characteristics of coriander foliage using convective thin layer and microwave drying. *Journal of Microwave Power and Electromagnetic Energy* 41(2), 56–65.

Wei, C.I., Cook, D.L. and Kirk, J.R. (1985) Use of chlorine compounds in the food industry. *Food Technology* 39, 107–115.

Zahary, D. and Hopf, M. (2000) *Domestication of Plants in the Old World*, 3rd edn. Oxford University Press, Oxford, UK.

Zhang, H. and Farber, J.M. (1996) The effects of various disinfectants against *Listeria monocytogenes* on fresh-cut vegetables. *Food Microbiology* 13, 311–321.

10 Curry Leaf Plant

Dawn C.P. Ambrose*
ICAR – Central Institute of Agricultural Engineering, Regional Centre, Coimbatore, India

10.1 Botany

10.1.1 Introduction

Indian cuisine is characterized by flavouring with many spices that also have medicinal properties. One of the leafy medicinal crops used as a spice is the curry leaf plant. Its botanical name is *Murraya koenigii* Spreng. and it belongs to the family Rutaceae. The leaves are widely used as a flavouring in the recipes of southern India, where the herb is grown as a homestead garden crop (Charles, 2013). Curry leaves are also widely used for food flavouring in Sri Lanka. They have a stimulatory effect on the tongue and a peculiar aroma. Curry leaves are used in a manner similar to bay leaf in culinary preparations. Apart from their use in food preparations, the leaves have many health benefits, and in India they are one of the ingredients of Ayurvedic formulations. The leaves taste pungent, bitter and lightly acidic, and they retain their flavour and other qualities even after drying (Sinha *et al.*, 2012). The various medicinal properties of the species that have been reported and the migration of users across the globe have created a demand for this medicinal crop. India produces curry leaves for both internal and external consumption, and is the leading exporter of the leaves in the world. The shelf life of the leaves can be extended by various processing techniques.

Derived from the Tamil word *kari*, meaning soup or spicy sauce, curry leaves are widely known in India as *karipatta* (Charles, 2013). They are also known by numerous vernacular names in India and in other countries. Some of these are listed in Table 10.1.

10.1.2 History

The curry leaf is mentioned in Tamil literature dating back as far as the 1st to 4th centuries AD as a flavouring agent for vegetables. It is also mentioned in the Kannada language a few centuries later.

10.1.3 Location

The curry plant is a native of India. It is a forest crop found in the Himalayas in the Indian subcontinent. In India, the plant is

*Corresponding author, e-mail: dawncp@yahoo.com

© CAB International 2016. *Leafy Medicinal Herbs: Botany, Chemistry, Postharvest Technology and Uses* (eds D.C.P. Ambrose *et al.*)

Table 10.1. Vernacular names of curry leaves in different languages. From the Spices Board, India.

Indian languages	
Assamese	Bisharhari, narsinghs
Bengali	Barsanga, kariphulli
Hindi	Bareanga, curry or kurry patta, gandhela, kathnim, mitha neem
Gujarati	Goranimb, kadhilimbdo
Kannada	Karibevue
Malayalam	Karriveppilei
Marathi	Gandla, jhirang, karhinimb, poospala
Oriya	Barsan, basango, bhuraunga
Punjabi	Curry patta
Sanskrit	Krishna nimba
Tamil	Karivempu, karuveppilei
Telugu	Karepaku
Other languages	
Chinese	Ga lei yihp
Dutch	Kerriebladeren
English	Curry leaves
French	Feuilles de cari
German	Curryblatter
Spanish	Hoja

found in West Bengal, Madhya Pradesh and in the southern and south-western states, namely Maharashtra, Tamil Nadu, Kerala and Andhra Pradesh. The plant originated in the tarai region of Uttar Pradesh, India, it is currently grown in Myanmar, Sri Lanka, China, Australia and the Pacific Islands. The curry plant reached Malaysia, South Africa and Réunion Island with south Indian immigrants. There are recent reports that it is being grown in Central African countries. The curry plant is an emerging commercial crop (Fig. 10.1) owing to its demand in the international market.

10.1.4 Morphology

The curry plant is a strong-smelling shrub or small tree and has leaves that are pinnate and about 15–30 cm long, with 9–27 alternate leaflets on a slender stalk; these are small, pubescent or glabrous, obliquely ovate, elliptic, lanceolate or rhomboid, and crenulate. On an average, a tree can yield 100 kg of leaves a year. The plant has an inflorescence of scented bisexual flowers,

Fig. 10.1. Curry leaf plant.

which are white in colour. The ripened fruits are of dark purple colour and ovoid, and give rise to dark green seeds.

Cultivars

Generally. the curry leaf plant is cultivated in the kitchen garden, but in south India, especially in Tamil Nadu and Karnataka, it is grown as a cash crop. 'Senkaambu' is a local cultivar grown in different parts of Tamil Nadu, especially in the Karamadai tract of Coimbatore District. The petiole is purplish red in colour. The leaves have a good aroma and flavour due to their high oil content. At the Department of Horticulture of the University of Agricultural Sciences at Dharwad in Karnataka, two genetically distinct cultivars, viz. 'DWD 1' and 'DWD 2' have been identified and being multiplied (Spice India, 2003). The seeds are used for propagation.

10.2 Chemistry

10.2.1 Chemical composition

Curry leaves contain volatile oils which are responsible for their aroma. They are a rich source of β-carotene and are known to promote hair growth and also to aid in good vision. The leaves also contain proteins, carbohydrates, fibre, minerals, etc. Table 10.2 presents data on the major nutrients found in fresh curry leaves.

Table 10.2. Composition of curry leaves. From the Spices Board, India.

Constituent	Content/100 g
Energy	108.0 kcal
Carbohydrates	18.7 g
Fat	1.0 g
Fibre	6.4 g
Minerals	4.0 g
Moisture	63.8 g
Protein	6.1 g
Vitamins	
Carotene	7560.0 µg
Thiamine	0.8 mg
Riboflavin	0.21 mg
Niacin	2.3 mg
Folic acid (free)	23.5 µg
Folic acid (total)	93.9 µg
Vitamin C	4.0 mg
Minerals/Elements	
Calcium	830.0 mg
Chlorine	198.0 mg
Chromium	0.006 mg
Copper	0.10 mg
Iron	0.93 mg
Magnesium	44.0 mg
Manganese	0.15 mg
Phosphorus	57.0 mg
Sulfur	81.0 mg
Zinc	0.20 mg
Phytochemicals	
Oxalic acid	132.0 mg
Phytin phosphorus	35.0 mg

The chemical composition of green (fresh) curry leaves at three stages of maturity, namely, tender, medium mature and mature as measured at the Central Food Technological Research Institute (CFTRI) of the Indian Council of Scientific and Industrial Research (CSIR) was as (expressed on a moisture free basis): protein, 5.44, 6.44 and 7.19%; fat, 3.3, 4.74 and 6.15%; sugars, 14.9, 17.9 and 18.9%; starch, 11.4, 14.2 and 14.6%; crude fibre, 6.0, 6.2 and 6.2%; volatile oil, 0.82, 0.55 and 0.48%; and total ash, 12.54, 12.7 and 13.1%) (Pruthi, 2001).

10.2.2 Phytochemistry

The screening of various curry leaf extracts for phytochemicals showed that they contain flavonoids, alkaloids, sterols, etc. The leaves have been found to have anthelmintic and anti-inflammatory properties, among others. The leaf extracts yield 2–5% volatile oil and are rich in carbazole alkaloids, including murrayastine, murrayaline, pyrayafoline and others. The major compounds responsible for the aroma and flavour of curry leaves are pinene, sabinene, caryophyllene, cadinol and cadinene. The leaves are also a rich source of calcium, but owing to the presence of oxalic acid in high concentration (total oxalates, 1.35%; soluble oxalates, 1.15%), its nutritional availability is affected. Tachibana *et al.* (2001) isolated 8,10′-{3,3′,11, 11′-tetrahydro-9,9′ dihydroxy-3,3′,5,8′-tetra methyl-3,3′-bis (4-methyl-3-pentenyl)}bis pyrano (3,2 a) carbazole from the leaf extract together with koenimbine, *O*-methyl-murrayamine, *O*-methyl-mahanine, isomahanine and bismahanine, and bispyrayafoline. Jain *et al.* (2012) have reported the isolation of glycozoline, 1-formyl-3-methoxy-6-methyl carbazole and 6,7- dimethoxy-1-hydroxy-3-methyl carbazole from dried leaves. The aerial parts of the plant are also reported to contain murrayanine and 8,8′ bis koenigine.

10.3 Postharvest Technology

10.3.1 Processing

Curry leaves can be processed in both fresh and dried form. Although the leaves are available throughout the year, they are highly perishable and cannot be kept fresh for more than a day. There is a need to increase research on postharvest handling in view of the growing global demand for the leaves. To enable cheaper shipment to other countries, which will play a pivotal role in global marketing, simpler techniques for extending the shelf life during transport are needed. There is a potential for increasing demand for the export of leafy spices from India if a suitable method is evolved for drying them while retaining their flavour (Omanakutty Amma *et al.*, 1984). The huge demand in India and abroad has made curry leaves a commodity of immense trade value, and both proper cultivation and postharvest management practices for this commercially important plant have to be developed to reach maximum returns (Lathan Kumar *et al.*, 2003).

Stripping

Stripping of leaves is a major operation that is involved in both the wet and dry methods of processing, and aids in the handling of large volumes of leaves; it also promotes quicker and uniform drying of the leaves. Presently, leaf stripping before further processing done by hand, which is laborious and time-consuming. It is also unhygienic and there is every possible risk of contamination during handling. Furthermore, as both the fresh leaves and the powdered form (masala) are exported, there is also a possibility of contamination being carried over to the processed product. A machine for stripping curry leaves has been developed and evaluated. The leaf stalks are fed into the machine via a nylon plate which is fitted with slots that are placed radially from the periphery of the centre hole; these enable the petiole to pass through the plate, and as a result of the to-and-fro motion of the plate, the leaves are stripped from the stalks. The stripped leaves are collected at the bottom of the machine and the stalks are removed through a separate outlet. The capacity of the machine is about 40–50 kg/h, and the stripping efficiency is about 90–95% (Naik *et al.*, 2013).

Pre-packaging

Food products undergo numerous physical, chemical and microbiological changes during storage. Their stability is a function of the changes occurring in various constituents, such as proteins, lipids, carbohydrates and water, as a result of environmental and processing factors (exposure to light, moisture, temperature, etc.). Any protective coating or barrier provided during processing, storage and handling not only retards deterioration of food, but may also enhance its quality. Suitable packaging can slow the deterioration rate and also may extend product shelf life. A variety of packages and approaches have been used to provide desirable effects. Examples of these include incorporating scavengers for removing moisture, and the use of gases and flavour-imparting or scavenging chemicals. These agents may be physically incorporated into the packaging material, or placed on or between the packaging and the food. Such approaches, designed to perform some desirable function other than providing an inert barrier, are called active packaging, interactive packaging and smart packaging. The use of plastics in the packaging of foods has been increasing at an accelerated rate. The reason for this is the reduction in the cost due to technological innovations and the inherent properties of plastic films, which make them very well suited to food packaging. Active packaging technology is a relatively novel concept designed to provide interactions between the food and packaging material, while sustaining the microenvironment within for extended shelf life of good quality and with microbial safety (King, 2006). The marketing of perishable commodities often requires some storage to balance day-to-day fluctuations between product harvesting and sale, and also long storage to extend marketing beyond the end of harvest season. Among horticultural products, leafy vegetables are ones that spoil rapidly, mainly because of leaf senescence due to physiological and biochemical changes. The most obvious symptoms of this are loss of fresh weight, drying, shrivelling, flavour changes and losses in chlorophyll, ascorbic acid and protein content (Halevy and Mayak, 1981).

A study on the pre-packaging of fresh curry leaves in different packaging materials, followed by storage under ambient (30 ± 2°C) and refrigerated (5 ± 1°C) conditions demonstrated that 20 μm thick polypropylene bags with a 0.1% vent area of 5 mm diameter could keep curry leaves fresh for 4 days under ambient storage; they were also suitable for sale in retail outlets. Under refrigerated conditions, samples kept well for 16 days in a 75 μm thick polyethylene bag (Ambrose *et al.*, 2014). Ambrose and Naik (2007) reported that self-help groups have been trained and encouraged to pre-pack curry leaves for retail outlets.

Drying

Curry leaves can be stored for a long time by removing their moisture, and several studies have been conducted on the drying of the leaves, which are then both used in the masala industries and exported. Studies on the drying of curry leaves in cross-flow, through-flow and vacuum dryers showed that the vacuum shelf-dried product had a better green colour than the other products (Prakash and Natarajan, 1974). Drying under controlled conditions also gave a superior quality product compared with natural drying. Organoleptic analysis of the powder showed that though the characteristic green colour was lacking in the naturally dried product, the products from cross-flow drying and oven drying were superior (Omanakutty Amma *et al.*, 1984). In another study, curry leaves were dried under sun and shade in the summer and in winter months. The drying time was longer in the winter than in the summer because of the difference in solar intensity. The results were compared with those from the herbal dryer of the CSIR's Central Research Institute for Dryland Agriculture (CRIDA), and it was found that the drying time was less than with sun and shade drying, and that leaf colour was retained (Pratiba *et al.*, 1998).

In a different study, the chemical composition of dehydrated curry leaves was analysed and they were incorporated into various products. The organoleptic and nutritive values of the prepared products were analysed and it was found that the calcium, iron and β-carotene content of the prepared products increased significantly as the amounts of incorporation increased. Hence, the nutrient content of different food products could be improved by incorporating dried curry leaves (Khatoon *et al.*, 2011). The microwave drying of curry leaves also yields a product with high lutein content.

The drying of curry leaves in a fluidized bed dryer at 45°C at an air velocity of 4 m/s under laboratory conditions resulted in a good-quality product with better retention of colour and flavour than drying at other temperatures and air velocities (Ambrose and Naik, 2014). Based on this laboratory study, a forced flow type dryer of 50 kg/batch capacity has been developed for curry leaf drying. The dryer consists of a drying chamber, a plenum chamber, a blower and a stirrer with stainless steel paddles operated by a hand wheel for intermittent stirring during drying (see Fig. 10.2). It takes 6 h to dry 50 kg curry leaves from an initial moisture content of 65% (wet basis, w.b.) to a final moisture content of 5% (w.b).

Radiation treatment

Radiation processing offers a very effective and safe alternative for disinfestation and microbial decontamination of spices and herbs. It is a cold process, sometimes also referred to as cold pasteurization, therefore it does not affect the delicate aroma and flavour compounds in spices. Radiation processing can be carried out on prepacked spices without running the risk of post-treatment contamination. The process is very effective compared with fumigation and does not leave any harmful residues on spices (Sharma, 2006). Radiation processing involves controlled application of the energy of ionizing radiations such as α-rays, X-rays and accelerated electrons to food commodities including spices. It can be used as a chemical-free technology and an alternative when conventional technologies are found to be inadequate.

Fig. 10.2. Curry leaf dryer.

The benefits of the technique are mainly because the process is heat free and the ionizing radiation has high penetrating power. Being a cold process, the technology is particularly appropriate for spices that are valued for their delicate aroma and flavour constituents. The irradiation of curry leaves prior to export is a possible technique for controlling pest infestation.

Extraction of essential oils

Curry leaves contain volatile oils which can be extracted using different extraction methods. There are several methods employed in the extraction of essential oils from herbs and spices, including steam distillation, hydrodistillation and solvent extraction.

Steam distillation is commonly used to extract essential oil from aromatic and medicinal crops. Heat-intolerant materials such as resins and oils are prone to be lost at higher boiling temperatures, and are not soluble in water. In steam distillation, the compounds are distilled at a temperature lower than the boiling point of the ingredients, which in essential oils is >200°C. These components are volatized at a temperature near to 100°C at atmospheric pressure when exposed to steam.

Solvent extraction is another method employed in the extraction of essential oils from plant sources. The process involves dissolving the extractable material (which is composed of aromatic compounds and waxes) in the solvent. The solvent can be reused by distilling the extract. The portion remaining after distillation is solid. This is further processed by warming and the addition of ethyl alcohol, which separates the aromatic molecules (essential oil) from the wax as they are more soluble in alcohol. The essential oil obtained by this extraction can be used to impart an aroma like that of fresh curry leaves and can be used when fresh curry leaf is not available. The chemical composition of the essential oils is tested by chromatographic methods.

Jain *et al.* (2012) review results from some of the studies that have been carried out on the composition of essential oils prepared by various methods from curry leaves originating from various agroclimatic conditions and geographical areas. Fresh leaves subjected to steam distillation have been reported to yield as much as 2.5% volatile oil which is light yellow in colour and aromatic with a spicy note (Supriya, 2004); lower contents have also been reported.

10.3.2 Value addition

Curry leaves can be processed to value-added products in the form of powder, essential oils, etc. which find use in food applications. Apart from the food uses, the essential oils are used in soap making, lotions, massage oils, diffusers, pot pourri, scent, air fresheners, body fragrances, perfume oils, aromatherapy products, bath oils, towel scenting products, incense, facial steams, hair treatments, and at spas, etc. Curry leaves can also be incorporated with wheat flour to prepare bakery products and into dough for making Indian bread. Experiments have shown that curry leaf can be incorporated into chicken-based products such as kebabs, nuggets, etc. Chicken nuggets with curry leaf powder incorporated were found to be highly acceptable to consumers (Kondaiah *et al.*, 2005). As curry leaves are a natural antioxidant, they can be used to prevent rancidity, and studies have been done on the use of curry leaves to control rancidity in such products as rice bran and palm olein oil.

10.4 Uses

10.4.1 General uses

Curry leaves have been used in south Indian cuisine since time immemorial for flavouring dishes. They aid in the emanation of a special aroma in the preparation of ghee by the melting of butter, a traditional practice in the rural areas of south India. Fresh curry leaves on steam distillation yield 2.5% volatile oil, which is used as a fixative in the soap industry (Lathan Kumar *et al.*, 2003).

10.4.2 Pharmacological uses

Curry leaves are normally discarded while eating as they are just regarded as having been added to flavour the food. However, more importantly, the leaves offer a number of health benefits without any side effects. Consumption of the leaves helps to arrest dysentery. The leaves promote wound healing when applied externally. After roasting, they can be made into a tea which helps to stop vomiting (Mhaskar *et al.*, 2000). Curry leaves are also a rich source of iron and can be used to prevent anaemia. In south India, fresh curry leaves, after grinding and drying, are boiled with coconut oil. The resulting oil seems to promote hair growth and also to prevent premature greying of the hair. Curry leaves have also been found to aid good digestion. They are used as a folk medicine to cure wounds and bruises, and in herbal formulations for the treatment of digestive disorders in the Ayurvedic system of medicine in India.

The pharmacological benefits of curry leaves have been described by various researchers. According to Vinuthan *et al.* (2004), curry leaves have been found to control blood sugar levels in the management of diabetes. They are also used as a nephroprotective agent in traditional medicine (Yankuzo *et al.*, 2010). Based on the clinical studies with diabetic mice, leaf extracts were effective in blood cholesterol and blood glucose levels, suggesting that they could be used to manage high cholesterol and type 2 diabetes diseases (Xie *et al.*, 2006).

10.5 Summary

The curry leaf plant is an age-old medicinal and aromatic plant widely used for the flavouring of dishes in Asian cuisine, particularly in South Asia. Curry leaves are demanded worldwide, and are imported to the US, European Union and West Asian markets, as the herb is grown in very few countries. The leaves contain antioxidants and are rich in iron and β-carotene. They can be processed in fresh and dried form. The dried leaves can be pulverized and the powder incorporated into food formulations. The essential oil extracted from the leaves is used in cosmetic applications. The leaves have many health benefits, such as antidiabetic and anti-anaemia activity, it lowers cholesterol and aids in digestion, along with many other health benefits. The crop has great promise for use in various biochemical and industrial applications in the future.

References

Ambrose, D.C.P. and Naik, R. (2007) Investigation on extending the shelf life of fresh curry leaf (curry patha). *FoodPack.Com* 1(9, March), 16–17. [ValueBase Publications, Mumbai, India.]. Copy available from the author.

Ambrose, D.C.P. and Naik, R. (2014) Studies on the mechanical drying of curry leaf. *International Journal of Processing and Post Harvest Technology* 5, 8–11.

Ambrose, D.C.P., Annamalai, S.J.K. and Naik, R. (2015) Effect of packaging in extending the shelf life of fresh curry leaves. *Journal of Applied Horticulture* 17, 165–168.

Charles, D.J. (2013) *Antioxidant Properties of Spices, Herbs and Other Sources.* Springer, New York, pp. 273–279.

Halevy, A.H. and Mayak, S.(1981) Senescence and postharvest physiology of cut flowers. Part II. *Horticultural Reviews* 3, 59-143.

Jain, V., Momin, M. and Laddha, K. (2012) *Murraya koenigii*: an updated review. *International Journal of Ayurvedic and Herbal Medicine* 2, 607–627.

Khatoon, J.A., Verma, N.C. and Sheikh, S. (2011) Utilization of dehydrated curry leaves in different food products. *Indian Journal of Natural Products and Resources* 2, 508–511.

King, K. (2006) Packaging and storage of herbs and spices: In: Peter, K.V. (ed.) *Handbook of Herbs and Spices, Volume* 3. Woodhead Publishing, Cambridge, UK, pp. 86–102.

Kondaiah, N., Anjaneyulu, A.S.R. and Mendiratta, S.K. (2005) Curry leaf powder – a promising spice ingredient for value added chicken products. Available at: http://www.poulvet.com/poultry/articles/tech_products/10.php (accessed 15 January 2016).

Lathan Kumar, K.J., Dassharma, K. and Mohandas, A. (2003) Curry leaf – an inevitable spice of Indian cuisine. *Spice India* 3(9), 8–9.

Mhaskar, K.S., Blatter, E. and Caius, J.F. (eds) (2000) *Kirtikar and Basu's Illustrated Indian Medicinal Plants, Volume I. XI*, 3rd edn. Indian Medical Science Series # 86–96. Sri Satguru Publications, Delhi.

Naik, R., Annamalai, S.J.K. and Ambrose, D.C.P. (2013) Development of power operated curry leaf (*Murraya koenigii*) stripper. *Agricultural Mechanization in Asia, Africa and Latin America* 44(1), 69–72.

Omanakutty Amma, M., Rajaraman, K., Sankarikutty, B. and Sumathikutty, M.A. (1984) Processing of curry leaves. *Indian Food Packer* 38(4), 32–36.

Prakash, V. and Natarajan, C.P. (1974) Studies on curry leaf (*Murraya koenigii* L.). *Journal of Food Science and Technology* 11, 284–286.

Pratibha, G., Srinivas, I., Korwar, G.R., Ramakrishna, Y.S. and Mayande, V.M. (1998) *Onfarm Value Addition with CRIDA Herbal Dryer.* Central Research Institute for Dryland Agriculture, Santoshnagar, Hyderabad, India. Available at: www.crida.in/Bulletins/Dryer.pdf (accessed 15 January 2016).

Pruthi, J.S. (2001) *Minor Spices of India: Crop Management and Post-Harvest Technology.* Indian Council of Agricultural Research, New Delhi.

Sharma, A. (2006) Irradiation to decontaminate herbs and spices. In: Peter, K.V. (ed.) *Handbook of Herbs and Spices*, Vol.3. Woodhead Publishing Limited, Cambridge, UK, pp. 60–73.

Sinha, P., Akhtar, J., Batra, N., Jain, H. and Bhardwaj, A. (2012) Curry leaves – a medicinal herb. *Asian Journal of Pharmaceutical Research* 2, 51–53.

Spice India (2003) Curry leaf. *Spice India* 3(9), 35–36.

Supriya, K.B. (2004) *Hand Book of Medicinal Plants.* Pointer Publishers, Jaipur, India.

Tachibana, Y., Kikuzaki, H., Lajis, N.H. and Nakatani, N. (2001) Antioxidative activity of carbazoles from *Murraya koenigii* leaves. *Journal of Agricultural and Food Chemistry* 49, 5589–5594.

Vinuthan, M.K., Girish, K.V., Ravindra, J.P. and Jayaprakash, N.K. (2004) Effect of extracts of *Murraya koengii* leaves on levels of blood glucose and plasma insulin in alloxan-induced diabetic rats. *Indian Journal of Physiology and Pharmacology* 48, 348–352.

Xie, J.-T., Chang, W.-T., Wang, C.-Z., Mehendale, S.R., Li, J., Ambihaipahar, R., Ambihaipahar, U., Fong, H.H. and Yuan, C.-S. (2006) Curry leaf (*Murraya koenigii* Spreng.) reduces blood cholesterol and glucose levels in *ob/ob* mice. *The American Journal of Chinese Medicine* 34, 279–284.

Yankuzo, H., Santosa, R.I., Ahmed, Q.U., Akter, S.F.U. and Talib, N.A. (2010) Nephroprotective effect of the leaves of *Murraya koenigii* L. Spreng. *in vivo.* In: *3rd International Conference on Advancement in Science and Technology (iCAST)* 2010, 27–29 November 2010, Vistana Hotel, Kuantan, Pahang, Malaysia. Abstract available at: http://irep.iium.edu.my/18885/ (accessed 15 January 2016).

11 Fenugreek

Gopal Amuthaselvi[1]* and Dawn C.P. Ambrose[2]
[1]*Krishi Vigyan Kendra, Sirugamani, Tiruchirappalli, Tamil Nadu, India;*
[2]*ICAR – Central Institute of Agricultural Engineering, Regional Centre, Coimbatore, India*

11.1 Botany

11.1.1 Introduction

Fenugreek belongs to the family Fabaceae. Its botanical name is *Trigonella foenum-graecum*. The leaves of the plant are used as a herb and the seeds as a spice. It is cultivated worldwide as a semi-arid crop. Fenugreek is an age-old medicinal herb that is used in many parts of the world (Srinivasan, 2006). The taxonomic position of fenugreek is as follows:

Kingdom: Plantae
Division: Magnoliophyta
Class: Magnoliopsida
Order: Fabales
Family: Fabaceae
Genus: *Trigonella*
Species: *foenum-graecum* Linn.

Consumption of the leaves of fenugreek has been found to control the sugar level in diabetic patients. In addition, fenugreek leaves are believed to aid in digestive problems, liver problems and anaemia (Deshmukh *et al.*, 2013). The vernacular name of the plant in many Indian languages is methi (or methe). Details of its various names in different languages are given in Table 11.1.

The leaves and seeds are consumed for culinary purposes across the globe. The dried leaves add flavour to dishes. The seeds, after drying, are powdered and used in the preparation of curry in Asian countries; they are also used in bakery products and in cheese, and as an insect repellent during the storage of grain. The fresh leaves, as well as sprouted seeds, are used in salads. The crop is of short duration and helps in fixing nitrogen in the soil.

Fenugreek leaves have been used in traditional Chinese medicine curing urinary related problems. They have also been used as a herbal remedy in African and Asian countries. Historical accounts indicate that fenugreek leaves were used for many medicinal purposes in countries such as India, Greece and China, and in Arab countries. These include the treatment of mouth ulcers and chapped lips, curing of baldness, reduction in abdominal pain and pain from abscesses, prevention of cardiovascular and hepatic disorders, treatment of arthritis, dropsy, spleen and liver enlargement and kidney ailments, and as an aid to lactation in breastfeeding women (Weiss, 2002; Tiran, 2003).

*Corresponding author. E-mail: g.amuthaselvi@gmail.com

© CAB International 2016. *Leafy Medicinal Herbs: Botany, Chemistry, Postharvest Technology and Uses* (eds D.C.P. Ambrose *et al.*)

Table 11.1. Vernacular names of fenugreek in different languages. From the Spices Board, India.

Arabic	Hulba
Chinese	K'u-Tou
Dutch	Fenegriek
French	Fenugrec
German	Bockshorklee
Italian	Fieno Greco
Japanese	Koroha
Portuguese	Alforva
Russian	Pazhitnik
Spanish	Alholva
Swedish	Bockshornklee

11.1.2 History

Fenugreek is mentioned in various historical texts, including one of the oldest medical documents, the Egyptian *Ebers Papyrus*, which dates to 1550 BC (Brier, 1998). It has been known in Asian countries for thousands of years, and in ancient Egyptian culture, it was used to scent embalmed mummies.

11.1.3 Location

Although fenugreek is a native of southeastern Europe and West Asia, it is now cultivated in countries across the globe. It is a cold season crop and is resistant to very low temperatures and to frost. It is best suited to tracts of moderate to low rainfall and is sown in all types of soil, but performs better in loam and clayey loam with proper drainage. Some of the countries where it is grown have a long history of use, while others only started cultivating the crop during the past two to three decades. Asia ranks first among the continents in terms of fenugreek production and acreage, followed by Africa. India is the leading country in the production of fenugreek seed, accounting for 90% of global production (Petropoulos, 2002). Among other Asian countries, Iran, Israel, China and Pakistan also have high levels of production. In Africa, fenugreek production is mostly concentrated in Egypt, Ethiopia, Kenya and Morocco. Spain, Turkey, Greece and Germany are notable for fenugreek cultivation in Europe, whereas Argentina is important for production in South America.

11.1.4 Morphology

Fenugreek is an annual plant that grows to around 60 cm tall. The leaves are similar to clover in shape (Fig. 11.1). The flowers are pea shaped and yellow or white, and appear in the leaf axils. Leaves from fenugreek are simple, distinctly petiolate and consist of three leaflets. The leaflets are slightly dented at the edge, oval to orbicular in shape and green in colour. They are arranged alternately throughout the plant (Slinkard *et al.*, 2006).

11.2 Chemistry

11.2.1 Chemical composition

The leaves are bitter in taste, and rich in minerals and vitamins. They are also rich in proteins, like other pulses, and can be used as a protein supplement. The dried leaves are known for their aroma. There are 19–25 mg β-carotene and 221–378 mg of ascorbic acid in 100 g of fresh fenugreek leaves (Yadav and Sehgal, 1997). Leaves contain 86.1% moisture, 4.4% protein, 0.9% fat, 1.5% minerals, 1.1% fibre, and 6% carbohydrates. The mineral and vitamin contents are calcium, iron, phosphorus, carotene, thiamine, riboflavin, niacin and vitamin C (Prajapati Ashish *et al.*, 2014).

11.2.2 Phytochemistry

Green leafy plants are good sources of carotenoids, tocopherols and flavonoids, which determine the antioxidant properties of a plant. Several phytochemical compounds have been extracted from the leaves of fenugreek. They are mostly flavonoids, tannins

Fig. 11.1. Fresh fenugreek leaves.

and other phenolic compounds, glycosides, alkaloids and steroids. Steroidal saponins are responsible for the bitter taste of the leaves. Graecunins are the major saponins present in fenugreek leaves. These compounds are glycosides of diosgenin, and are considered to be anti-nutritional factors.

11.3 Postharvest Technology

11.3.1 Processing

Drying

Fenugreek leaves are a highly perishable food commodity and cannot be stored for long. Drying is the usual preservation method for its long term-storage. Dried leaves of fenugreek are known as methi in India (Fig. 11.2). The effect of different pretreatments on the quality of dried fenugreek leaves has been studied by Bajaj *et al.* (1993), who demonstrated that their vitamin C content was highest in samples treated with 0.5% potassium metabisulfite, while blanching in water prior to dehydration resulted in good quality in terms of chlorophyll retention.

The sun drying of fenugreek leaves leads to loss of nutrients, especially vitamin C, and mechanical drying is a better alternative for retaining nutrient content. Various studies have shown that fenugreek leaves can be dried well in a solar dryer or by microwave drying, etc.

Fig. 11.2. Dried fenugreek leaves.

Karva *et al.* (2010) contrasted the drying of fenugreek leaves in a microwave oven (100% power, 2250 Hz), hot air oven (60°C), in the sun (38–42°C) and in the shade (24 ± 1°C) after washing and blanching of the leaves. Microwave drying resulted in samples with a better quality.

The drying behaviour of fresh leaves of fenugreek was investigated by Kalaskar *et al.* (2012). The leaves were dried in a cabinet dryer or sun dried after different pretreatments. Of these pretreatments and drying methods, leaves that had been blanched for 2 min in water containing 0.1% MgO + 0.5% KMS (potassium metabisulfite) + 0.1% $NaHCO_3$ and dried in cabinet dryer were superior in terms of maintaining minimum moisture and dehydration ratio, and maximum rehydration ratio, chlorophyll and vitamin C content throughout the storage period compared with sun-dried samples.

Fenugreek leaves can also be dried by washing and cutting into uniform pieces and blanching in water with known concentrations of MgO, KMS and $NaHCO_3$ added and drying in a mechanical dryer. The recovery of dried fenugreek leaves was 8.5–9.7% with 45.89 mg/100 g ascorbic acid. Sensorial studies indicated good acceptability of the samples by the consumers and the quality of the dried leaves was retained during storage for 5–6 months under ambient conditions (Singh *et al.*, 2012).

In a solar cabinet dryer, fenugreek leaves could be dried when the temperature was about 20–22°C more than the surrounding temperature during a clear sunny day. The average efficiency and the rate of energy utilization were high in a cabinet solar dryer, and the drying time was 43% less than for open sun drying. Nutritive analysis of the powdered samples showed an increase of all nutrients except for vitamin C (Navale *et al.*, 2014).

Packaging

The retention of shelf life is an important aspect of leafy crop products due to their perishable nature. The packaging of fenugreek leaves in polyethylene bags results in better shelf life and good nutritive value of the product (Negi and Roy, 2003). The primary processing of fenugreek plants can also be combined with packaging to produce a ready-to-use vegetable. This involves removing undesirable parts like the roots and yellowed leaves, and packing clean leaves in plastic baskets covered with cling film, thereby arresting wilting of the leaves (Gomez *et al.*, 2003).

The effect of different packaging materials on the physiological weight loss of eight different vegetables, including fenugreek leaves, was studied under household and laboratory refrigeration conditions in Karnataka, India. There was less weight loss under laboratory than household conditions. The weight loss was maximum in brown paper and minimum in polyethylene (PE) bags, and the percentage spoilage was much higher in low-density polyethylene (LDPE) bags than in ordinary PE bags (Korradi and Devendrappa, 2011).

Modified atmosphere packaging (MAP) could also be beneficial for storing fenugreek leaves for extended shelf life and quality.

11.3.2 Value addition

Fenugreek leaves are rich in iron and could be consumed as a vegetable to alleviate iron deficiency in women and children. Leaves could be dried, powdered and incorporated into flour for preparation of bakery products. Dried leaves of both methi (*T. foenum-graecum*) and kasuri methi (*T. corniculata*, a fenugreek that is cultivated in the Pakistani state of Kasur), could be used as a nutraceutical, as they are a rich source of iron, alkaloids, protein, etc. and have good antioxidant and antibacterial activities (Pasricha and Gupta, 2014).

11.4 Uses

11.4.1 General uses

The leaves of fenugreek are used as a vegetable in Indian cuisine, and also to enhance flavour when added to dishes as dried leaves. The seeds are dried, powdered and used as a spice. Kasuri methi is widely used in Indian and Pakistani cuisine. The fresh and tender leaves of fenugreek are also used in salads. Samudra methi is popular in Maharashtra state in India, where it is known as microgreens. The seeds of fenugreek are used as a herbal insecticide.

11.4.2 Pharmaceutical uses

Fenugreek was used for embalming mummies by the ancient Egyptians. In the *Ebers Papyrus*, fenugreek is listed as one of the ingredients for the treatment of burns. Hippocrates, the father of medicine, thought it a valuable soothing herb, while Dioscorides used it in treating swelling of the genitals. French researchers believe that it is a good aid to digestion. Basch *et al.* (2003) reported that fresh fenugreek leaves aid in digestive problems. Mouth ulcers could be cured by fenugreek tea. In India, paste prepared from fenugreek is used as a hair mask to control dandruff.

Other experiments indicate that fenugreek leaves can control diabetes. In a study by Devi *et al.* (2003), the diet of streptozotocin (STZ)-induced diabetic rats was supplemented with fenugreek leaves, and a number of antidiabetic effects demonstrated. Also in STZ-induced diabetic rats, Annida and Prince (2004) showed a reduction in the lipid profile by supplementation with fenugreek leaves (Annida and Prince, 2004).

In a review of the use of herbal medicines for the reduction of cholesterol, Thompson Coon and Ernst (2003) mentioned that the consumption of fenugreek leaves and seeds could reduce cholesterol in people with a high risk of cardiovascular disorders.

Fenugreek is rich in antioxidant and antitoxic potentials. In an *in vitro* study, Durga *et al.* (2014) demonstrated that an aqueous extract of fenugreek leaves protected red blood cells against the toxic effects of petrol exhaust nanoparticles and replenished their antioxidant levels.

11.5 Summary

Fenugreek leaves are consumed in every part of the world because of their various benefits. The chemical constituents of both the seeds and leaves have made them valuable as both food and medicine, in addition to being a rich source of nutrients. The leaves can be preserved in fresh form in ventilated and modified atmospheric packaging and also dehydrated to produce dried leaves. The fresh leaves are consumed as greens and the dried leaves are used as a flavouring agent in dishes owing to their distinct aroma. Supplementation of the diet with fenugreek aids in protecting the human body against ailments such as diabetes, high cholesterol, etc.

References

Annida, B. and Prince, P.S.M. (2004) Supplementation of fenugreek leaves lower lipid profile in streptozotocin-induced diabetic rats. *Journal of Medicinal Food* 7, 153–156.

Bajaj, M., Aggarwal, P., Minhas, K.S. and Sidhu, J.S. (1993) Effect of blanching treatments on the quality characteristics of dehydrated fenugreek leaves. *Journal of Food Science and Technology* 30, 196–198.

Basch, E., Ulbricht, C., Kuo, G., Szapary, P. and Smith, M. (2003) Therapeutic applications of fenugreek. *Alternative Medicine Review* 8, 20–27.

Brier, B. (1998) *Ancient Egyptian Magic*. William Morrow, New York.

Deshmukh, D., Pawade, C., Nalawade, V., Dongare, K. and Padwal, K. (2013) Anti-inflammatory activity of ethanolic extract of *Trigonella foenum graecum* and *Ziziphus jujube*. *International Research Journal for Inventions in Pharmaceutical Sciences* 1(1), 30–33.

Devi, B.A., Kamalakkannan, N. and Prince, P.S.M. (2003) Supplementation of fenugreek leaves to diabetic rats. Effect on carbohydrate metabolic enzymes in diabetic liver and kidney. *Phytotherapy Research* 17, 1231–1233.

Durga, M., Nathiya, S. and Devasena, T. (2014) Protective role of fenugreek leaf extract and quercetin against petrol exhaust nanoparticle induced lipid peroxidation and oxidative stress in rat erythrocytes *in vitro*. *Asian Journal of Pharmaceutical and Clinical Research* 8, 237–241.

Gomez, S., Roy, S.K. and Pal, R.K. (2003) Primary processing of fenugreek (*Trigonella foenum graecum* L.) – an eco-friendly approach for convenience and quality. *Plant Foods for Human Nutrition* 58(3), 10 pp. Available at: http://link.springer.com/article/10.1023/B%3AQUAL.0000040366.42996.1f (accessed 18 January 2016).

Kalaskar, A.B., Sonkamble, A.M. and Patil, P.S. (2012) Studies on drying and dehydration of fenugreek leaves. *International Journal of Processing and Post Harvest Technology* 3(1), 15–17.

Karva, S., Bharati, P. and Chimmad, B. (2010) Postharvest processing of green leafy vegetables for iron security. *Karnataka Journal of Agricultural Sciences* 23, 306–310.

Korradi, V.V. and Devendrappa, S. (2011) Analysis of physiological weight loss of vegetables under refrigerated conditions. *Journal of Farm Sciences* 1(1), 61–68.

Navale, S.R., Upasni Supriya, Harpale, V.M. and Mohite, K.C. (2014) Effect of solar drying on the nutritive value of fenugreek leaves. *International Journal of Engineering and Advanced Technology* 4, 133–136.

Negi, P.S. and Roy, S.K. (2003) Changes in beta carotene and ascorbic acid content of fresh amaranth and fenugreek leaves during storage by low cost technique. *Plant Foods for Human Nutrition* 58, 225–230.

Pasricha, V. and Gupta, R.K. (2014) Nutraceutical potential of methi (*Trigonella foenum graecum* L.) and kasuri methi (*Trigonella corniculata* L.). *Journal of Pharmacognosy and Phytochemistry* 3(4), 47–57.
Petropoulos, G.A. (2002) *Fenugreek – the Genus Trigonella*. Taylor and Francis, London and New York.
Prajapati Ashish, D., Sancheti, V.P. and Shinde Puja, M. (2014) Review article on fenugreek plant with it's [sic] medicinal uses. *International Journal of Phytotherapy Research* 4(4), 39–55.
Singh, S., Singh, K.P. and Koley, T.K. (2012) Drying of fenugreek leaves. In: *Annual Report 2011–12*. Indian Institute of Vegetable Research, Varanasi, India, pp. 63–64.
Slinkard, A.E., McVicar, R., Brenzil, C., Pearse, P., Panchuk, K. and Hartley, S. (2006) Fenugreek in Saskatchewan. Fact Sheet, Saskatchewan Ministry of Agriculture, Regina, Canada. Available at: http://www.agriculture.gov.sk.ca/Default.aspx?DN=c6428c37-cab6-4e93-b862-e20a55af3586
Srinivasan, K. (2006) Fenugreek (*Trigonella foenum-graecum*): a review of health beneficial physiological effects. *Food Reviews International* 22, 203–224.
Thompson Coon, J.S. and Ernst, E. (2003) Herbs for serum cholesterol reduction: a systematic review. *Journal of Family Practice* 52, 468–478.
Tiran, D. (2003) The use of fenugreek for breast feeding woman. *Complementary Therapies in Nursing and Midwifery* 9(3), 155–156.
Weiss, E.A. (2002) Fenugreek. In: Weiss, E.A. (ed.) *Spice Crops*. CAB International, Wallingford, UK, pp. 77–85.
Yadav, S.K. and Sehgal, S. (1997) Effect of home processing and storage on ascorbic acid and, β-carotene content of bathua (*Chenopodium album*) and fenugreek (*Trigonella foenum graecum*) leaves. *Plant Foods for Human Nutrition* 50, 239–247.

12 Lemongrass

Salome Amarachi Chime* and Ikechukwu V. Onyishi
University of Nigeria, Nsukka, Nigeria

12.1 Botany

12.1.1 Introduction

Lemongrass is a herb belonging to the grass family Poaceae. It is grown in South-east Asia and Sri Lanka. The herb is used in Asian cuisine and has a lemony flavour. In India, it is known as 'choomana poolu' and is used in Ayurvedic medicine and as an ingredient in the perfumery industry. *Cymbopogon flexuosus* Stapf originated from Kerala, is very hardy and grows in humid climatic condition requiring plenty of sunlight. The species *C. citratus* is believed to be a native of Malaysia. In English, the herb is commonly known as lemongrass, citronella grass or fever grass. In other languages and designations, it is called Herba Andropogonis (a pharmaceutical designation), hashisha al-limun (Arabic), limonova treva (Bulgarian), sabalin (Burmese), chou geung (Cantonese), vlaska (Croatian), citroengras (Dutch), citronelo (Esperanto), verveine des Indes (French), herba de limón (Galician), zitronengras (German), essef limon (Hebrew), sera (Hindi), cimbopogone (Italian), remongurasu (Korean), erva-principe (Portuguese), limonnoe sorgo (Russian), zacate de limon (Spanish), limon out (Turkish), serai (Indonesian), sereh (Dutch) and takrai (Thai).

Cymbopogon is a very large genus, consisting of over 500 species, of which the species *C. citratus* and *C. flexuosus* are generally called lemongrass.

12.1.2 History/origin

Lemongrass is an aromatic plant that grows in tropical and subtropical South-east Asia and Africa. In India, it is cultivated in the southern states and the north-eastern hills. The oldest known record of lemongrass is from the Philippines in 17th century. The herb is claimed to have been introduced to Jamaica in 1799 and to Haiti and the USA in 1917. The cultivation of lemongrass as a commercial crop was reported in the USA and Haiti in 1947. It is not known when or by whom it was introduced to Sri Lanka; according to the reports available at the national herbarium in 1905, a Sri Lankan researcher called J.F Jovit had acquired several plants of 'kochin sera' (*C. citratus*)

*Corresponding author, e-mail: salome.chime@unn.edu.ng

© CAB International 2016. *Leafy Medicinal Herbs: Botany, Chemistry, Postharvest Technology and Uses* (eds D.C.P. Ambrose *et al.*)

from south India and had planted them at Bandarawela Farm for research purposes (Bor, 1960). Several large-scale cultivated areas of lemongrass existed in Sri Lanka for some decades, but currently there are only a few small ones.

The use of lemongrass for medicinal uses has been known since antiquity, and it has been claimed to cure a long list of diseases.

12.1.3 Location

Lemongrass is grown in Brazil, Guatemala, Argentina, Mexico, West Indies, Vietnam, Thailand, Sri Lanka, Bangladesh and China, among others. Wild species of lemongrass are reported to have been found in Latin America and Asia (Lonkar *et al.*, 2013). In India, it is grown in the states of Kerala, Karnataka, Tamil Nadu, Sikkim, Bengal, Madhya Pradesh, Arunachal Pradesh and Maharashtra. It is also grown throughout Africa, in the Democratic Republic of the Congo (DRC), Central African Republic, Nigeria, Ghana, Angola, Gabon, Chad, Madagascar and the Comoros Islands.

The ideal climate for lemongrass cultivation is humid with plenty of sunshine. Seeds are produced from healthy plants, and are threshed manually from the dried inflorescence and used for propagation.

12.1.4 Morphology

Lemongrass is a large perennial and fast-growing aromatic grass with clumped, slightly branched cylindrical culms that arise from a short rhizome and often reach up to 2 m (Fig. 12.1). The stems (culms) are composed of sheaths, which are parallel veined and are narrow at the base and sharp at the top. The leaves are up to 1 m long and 2 cm wide. The plants flower at the mature stage. New plants emerge from the tillers produced from the rhizomes. The plant produces a network of roots and rootlets that rapidly exhaust the soil.

Fig. 12.1. Lemongrass plant.

12.2 Chemistry

12.2.1 Chemical composition

Lemongrass contains a large amount of essential oils (1–2% on a dry weight basis), of which the major component is the terpenoid aldehyde citral, which imparts its aromatic odour (Carlson *et al.*, 2001; Schaneberg and Khan, 2002). Research reports have indicated that overall essential oil content is greater during the early growth stage in *C. flexuosus* (Singh *et al.*, 1989). In general, the yield of essential oil is correlated with biomass yield. The production of higher quality oil with high citral content (75%) is determined by the proportion of young leaves to older leaves when the leaves are harvested at any given point.

The biological effects often ascribed to lemongrass are attributed to the primary bioactive constituents of extracts from its leaves, stems and roots. The three major methodologies used to extract bioactive compounds lead to the production of decoctions, infusions and essential oils. Oven drying, boiling and filtration produces a decoction, while grinding and boiling produces infusions. Steam distillation and drying with anhydrous sodium sulfate produces essential oils from the leaves.

12.2.2 Phytochemistry

Various research studies have indicated the presence of bioactive compounds including myrcene, citral (α-citral or geranial, and β-citral or neral), heptenone, racemic limonene (dipentene), linalool, borneol, geraniol, β-myrcene, 6-methyl-5-hepten-2-one, undecan-2-one and citronellol (references include: Schaneberg and Khan, 2002; Pengelly, 2004; Mirghani *et al.*, 2012).

A large number of studies have reported data on the chemical composition and phytochemistry of lemongrass (including: Alves *et al.*, 1960; Chisowa *et al.*, 1988; Aftab et al., 2011; Zheljazkov *et al.*, 2011; Ewansiha *et al.*, 2012; Idrees *et al.*, 2012; Tajidin *et al.*, 2012). These studies have indicated that the chemical composition of lemongrass extracts depends on the location where the plant is growing, genetic differences between plants, part of the plant used, method of extraction, age/stage of maturity, season of harvesting and the species. After the extraction of citral from lemongrass, the leftover oil consists of limonene and myrcene, besides methyl heptenone, linalool, geranyl acetate, nerol and geraniol. Clearly, these could produce minor fractions in high purity that could be valuable.

12.3 Postharvest Technology

12.3.1 Processing

After harvesting, the first treatment involves washing the plants in order to remove sand and other extraneous materials. Any further treatment depends on the final end product desired. In the preparation of lemongrass powder, for example, the following postharvest processing steps may be involved as shown in Fig. 12.2.

- The leaves are first washed.
- Cutting: the whole leaves may be cut into 5 cm pieces. This helps the plant to dry faster as well as to pack well in various pieces of equipment, e.g. a milling machine for further processing.
- Blanching: blanching is done in hot water at 80°C for 1 min or in 1% Na_2CO_3 at 80°C for 1 min.

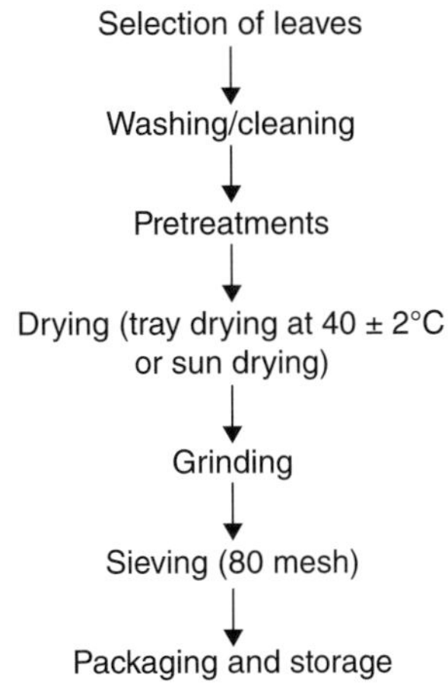

Fig. 12.2. Processing of lemongrass powder. From Lonkar *et al.*, 2013.

- Drying: this may be done by sun drying or shade drying, or by oven drying at a temperature not above 40–45°C.
- Milling: lemongrass powder is prepared by milling the dried leaves. Milling can be carried out using a hammer mill. Lemongrass has a very wide demand in nutritional, medicinal and flavouring industries, but it cannot be stored fresh for a long time under ambient conditions because it rots. Hence, lemongrass powder is preferred and is in huge demand in the world market (Lonkar *et al.*, 2013).
- Sieving: this can be carried out using a 0.25 mm sieve in order to remove oversized pieces and produce very fine powder. The oversized pieces can then be milled again in order to produce more fine powder.
- Packaging: powder for use as tea is often packaged in moisture and light resistant containers for storing under the appropriate conditions (Lonkar *et al.*, 2013).

Lemongrass powder may also be presented in various dosage forms which may include:

- granulated lemongrass powder;
- tablets; and
- capsules (Chime *et al.*, 2012a,b).

Granules may be prepared by a wet or dry granulation method and encapsulated in a hard gelatin capsule shell or tabletted using a tablet press (Chime *et al.*, 2012a,b).

Extraction

In one approach, the powdered leaves can be extracted by maceration in 90% methanol for 48 h. The extract is filtered and concentrated using a rotary evaporator attached to a vacuum pump. The crude powdered extract that is obtained can be stored and used in the preparation of various dosage forms, e.g. tablets and granules (Chime *et al.*, 2012a,b).

Figure 12.3 summarizes other methods that are used to extract bioactive compounds from lemongrass.

Distillation

Distillation is the process of extracting essential oil from lemongrass. Lemongrass oil is obtained mainly by steam distillation. After shade drying to a reduced moisture content, the leaves (or stems) are reduced in size and subjected to steam distillation. The oil is extracted after distillation at a certain temperature and after a certain extraction time to yield essential oil. Dichloromethane can be used to separate the essential oil from the condensate. The oil obtained is further purified by using anhydrous sodium sulfate. The insoluble particles present in the oil are filtered off. Using leaves infected with rust may lead to change of colour of the extracted oil and this can be corrected by a steam rectification process (Mirghani *et al.*, 2012; Srivastava *et al.*, 2013). The oil comprises 75–85% citral and has a lemon-like odour and is yellowish in colour; the recovery is 0.5–0.8%. The whole process can be completed within 4 h (Srivastava *et al.*, 2013).

Newer techniques viz. the use of supercritical fluids, ultrasound and microwave extraction can also be used to extract oil from lemongrass. Microwave-assisted extraction has been found to be successful under laboratory conditions. In terms of high recovery of oil and simplicity, microwave-assisted hydrodistillation (MAHD) and solvent-free microwave extraction (SFME) have been found to

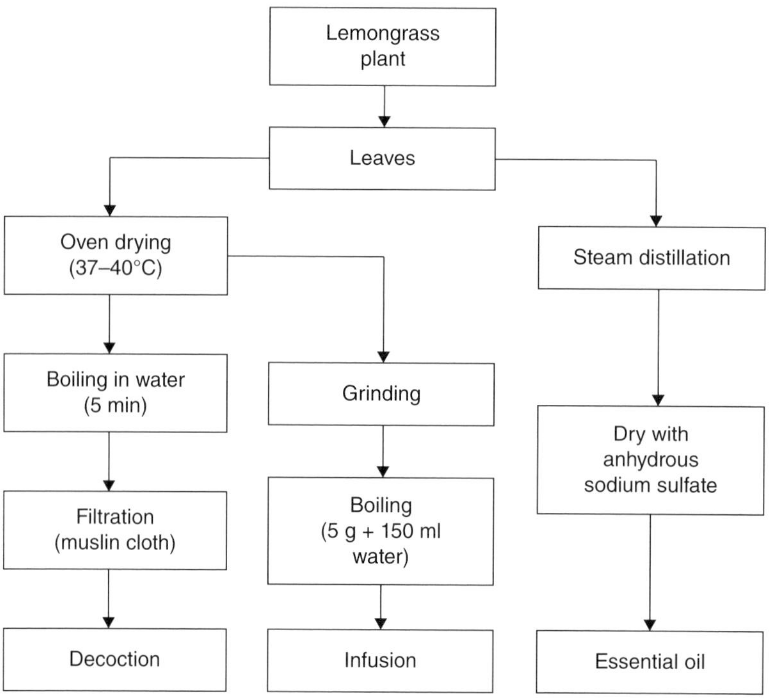

Fig. 12.3. Processes for extracting bioactive compounds from lemongrass. From Olorunnisola *et al.*, 2014.

be suitable at the industrial level. The use of a modified household microwave oven with a microwave frequency of 2450 MHz at atmospheric pressure could facilitate the hydrodistillation of essential oil from lemongrass.

When cleaned and dried stalks (Fig. 12.4) are reduced in size and subjected to MAHD, the resulting oil is a light yellow (Chen and Spiro, 1994; Singh *et al.*, 2014). Solvent free microwave extraction is similar to MAHD, however without using solvents which is faster but with less oil yield (Singh *et al.*, 2014).

Grading

The quality of lemongrass oil is determined by the content of citral, which is mainly used for the manufacturing of vitamin A (Directorate Plant Production, 2009).

Packaging and storage

As essential oils are volatile, lemongrass leaves need to be handled with care before they are subjected to steam distillation. They also need to be packed firmly to prevent the formation of steam channels. Stainless steel/ aluminium or galvanized iron containers and glass bottles can be used for the storage of the essential oil, which, because it is acidic, would destroy plastic and rubber. The oil should not be exposed to heat or light and it should be refrigerated if the containers are opened. Spoilage has begun if the oil is much darker or more viscous than normal (Directorate Plant Production, 2009).

Fig. 12.4. Dried lemongrass stalks.

12.3.2 Value addition

Essential oil

The citral-rich oil from lemongrass has germicidal, medicinal and flavouring properties. Due to the predominance of citral, the oil has a strong lemon-like odour, and because of this characteristic smell, the oil is used in soap industries and also as an ingredient in insect repellent preparations. Citral is used as a major source in perfumery, cosmetics, beverages and the synthetic preparation of vitamin A. India held a monopoly both in the production of lemongrass essential oil and its world trade during the early 1950s, but the crop is now widely cultivated in a number of countries, so this situation has changed. The citral content of lemongrass essential oil has been reported in varying ranges, such as 44.3–91.4% and 79–91.5% (Lonkar *et al.*, 2013).

Lemongrass has better antioxidant capacity than coriander leaves, ginger, tomato and garlic (Nambiar and Matela, 2012). Lemongrass may also serve as a good source for biofuel production.

Lemongrass powder

Traditionally, lemongrass powder is prepared by grinding the dried leaves (or stems). Per 100 g, it contains approximately 22.2 calories along with 4.59 g protein, 0.96 g sugar and 1.80 mg ascorbic acid (Lonkar *et al.*, 2013). Lemongrass powder is used for culinary purposes and for treating various ailments (Husain, 1993). As already noted, it can be encapsulated in capsule shells and or tabletted for use. In various countries, the powder is used as lemongrass tea to relieve congestion, coughing, bladder disorders, headaches, fever, stomach ache, digestive problems, diarrhoea, gas, bowel spasms, vomiting and flu symptoms; it is also used to promote perspiration and as a possible cholesterol lowering agent (Lonkar *et al.*, 2013).

Powdered lemongrass is marketed under the name of ‘Sereh powder’. It can be used in recipes such as those for lemongrass rice, sweet lemongrass blend and lemongrass beef.

12.4 Uses

12.4.1 General uses

Food ingredient

The stem and fresh leaves of lemongrass are used in culinary preparations in the Eastern world and other regions because of their distinct lemon flavour. The dried grass is blended with teas and used as herbal drink (Directorate Plant Production, 2009). It is also often used as an ingredient in curries, marinades and seafood dishes and soup, salads, etc.

Cosmetics and industrial uses

Lemongrass is used in products including soaps, perfumes, candles, and mosquito and other insect repellents (Directorate Plant Production, 2009; Nambiar and Matela, 2012).

Folk medicine

Lemongrass is used in traditional medicine for the treatment of diseases such as digestive disorders, fevers, menstrual disorders, and rheumatism and other joint pains.

12.4.2 Pharmacological uses

Manvitha and Bidya (2014) and Olorunnisola *et al.* (2014) have presented reviews on the biological and pharmacological activities of lemongrass. Some details of the individual pharmacological uses of lemongrass are briefly discussed below.

Antibacterial properties

There are a number of reports that indicate that lemongrass has good antibacterial properties (Syed *et al.*, 1990; Wannissorn *et al.*, 2005; Oleyede, 2009). The major bioactive compounds in the oil include α-citral and β-citral (Onawunmia *et al.*, 1984).

Antifungal activity

The essential oil inhibits the growth of pathogens and the development of mycotoxins. It is among the most active agents against human dermatophytes and is active against dermatophytic species such as *Trichophyton mentagrophytes*, *T. rubrum*, *Epidermophyton floccosum* and *Microsporum gypseum*. Other studies have shown that the oil from lemongrass inhibits keratinophilic fungi, ringworm and food storage fungi, plant pathogenic fungi, etc. (Kishore *et al.*, 1993; Mishra and Dubey, 1994; Wannissorn *et al.*, 1996; Abe *et al.*, 2003; Anaruma *et al.*, 2010; Shah *et al.*, 2011).

Antiprotozoal activity

Lemongrass essential oil exerts antiprotozoal activity against *Crithidia deanei* (Pedroso *et al.*, 2006) and has amoebicidal activity (Blasi *et al.*, 1990).

Antidiarrhoeal activity

A lemongrass infusion has been found to cure diarrhoea (Tangpu and Yadav, 2006), and the consumption of lemongrass tea has been found to be a good remedy for digestive disorders.

Anthelmintic activity

A leaf extract of lemongrass oil and its formulation as an emulsion has been reported to have anthelmintic (antihelmintolytic) potential (Gore *et al.*, 2010; Dama *et al.*, 2011).

Antimalarial activity

Based on *in vivo* studies using mice infected with *Plasmodium berghei*, lemongrass has antimalarial activity (Tchoumbougnang *et al.*, 2005). In another study of mice infected with *P. yoelii nigeriensis*, aqueous extracts of *C. citratus* in the drinking water cleared the parasites (Onabanjo *et al.*, 1993).

Hypolipidaemic, hypoglycaemic and antihypertensive properties

Investigations have shown that lemongrass extract has hypoglycaemic and hypolipidaemic properties (Adejuwon and Esther, 2007) and can reduce obesity and hypertension problems (Olorunnisola *et al.*, 2014).

Antioxidant activity

The antioxidant and free radical scavenging properties of lemongrass are due the phenolic compounds such as flavonoids present in the leaves. Cheel *et al.* (2005) studied the antioxidant and free radical scavenging activity of extracts, infusions and decoctions of the aerial parts, as well as their activity in inhibiting lipoperoxidation. The authors also identified the compounds responsible for these activities and compared their results with those of other published studies.

Antimutagenic and cytotoxic properties

Several studies have indicated that extracts of lemongrass have antimutagenic properties (Vinitketkumnuen *et al.*, 1994; Suaeyun *et al.*, 1997; Puatanachokchai *et al.*, 2002; Shah *et al.*, 2011).

Anti-inflammatory and anti-nociceptive properties

Various researchers have reported that lemongrass extracts and other preparations exhibit anti-inflammatory activities (Oleyade, 2009; Gore *et al.*, 2010; Figueirinha *et al.*, 2010; Tiwari *et al.*, 2010; Cruz and Batista, 2011; Dama *et al.*, 2011).

Anti-hepatotoxic activity

The aqueous leaf extracts of lemongrass showed anti-hepatotoxic action against cisplatin-induced hepatic toxicity in rats (Arhoghro *et al.*, 2012).

Anxiolytic properties

Lemongrass tea consumption has shown anxiolytic effects and other neurobehavioural effects (Blanco *et al.*, 2007; Olorunnisola *et al.*, 2014).

Anti-filarial activity

Fresh lemongrass was found to be effective in the treatment of filariasis (Shah *et al.*, 2011).

12.5 Summary

Lemongrass can be taken as a decoction or essential oil, used as a tea, taken with meals as a salad ingredient, and used in the treatment of various ailments in patients of all ages. The essential oil resulting from steam distillation of lemongrass is used as a fragrance and flavouring ingredient, while the powdered leaves and stalk can be used as teas. The powders can also be tabletted or encapsulated and utilized in the treatment of diseases. Lemongrass has a wide application both industrially and pharmacologically. Efforts should be made to propagate, cultivate and process lemongrass in order to produce the dosage forms and edible products and make them easily available for use and sale internationally.

References

Abe, S., Sato, Y., Inoue, S., Ishibashi, H., Maruyama, N. and Takizawa, T. (2003) Anti-*Candida albicans* activity of essential oils including lemongrass (*Cymbopogon citratus*) oil and its component, citral. *Nippon Ishinkin Gakkai Zasshi* 44, 285–291.

Adejuwon, A.A. and Esther, O.A. (2007) Hypoglycemic and hypolipidemic effects of fresh leaf aqueous extract of *Cymbopogon citratus* Stapf in rats. *Journal of Ethnopharmacology* 112, 440–444.

Aftab, K., Ali, M.D., Aijaz, P., Beena, N., Gulzar, H.J. and Sheikh, K. (2011) Determination of different trace and essential element[s] in lemon grass samples by X-ray fluorescence spectroscopy technique. *International Food Research Journal* 18, 265–270.

Alves, A.C., Prista, L.N. and Souza, A.F. (1960) A preliminary note on the phytochemical study of *Cymbopogon citratus*. *Garcia de Orta* 8, 629–638.

Anaruma, N.D., Schmidt, F.L., Duarte, M.C.T., Figueira, G.M., Delarmelina, C., Benato, E.A. and Sartoratto, A. (2010) Control of *Colletotrichum gloeosporioides* (Penz.) Sacc. in yellow passion fruit using *Cymbopogon citratus* essential oil. *Brazilian Journal of Microbiology* 41, 66–73.

Arhoghro, E.M., Kpomah, D.E. and Uwakwe, A.A. (2012) Curative potential of aqueous extract of lemon grass (*Cymbopogon citratus*) on cisplatin induced hepatotoxicity in albino Wistar rats. *Journal of Physiology and Pharmacology Advances* 2, 282–294.
Blanco, M.M., Costa, C.A., Freire, A.O., Santos, J.G. and Costa, I.M. (2007) Neurobehavioral effect of essential oil of *Cymbopogon citratus* in mice. *Phytomedicine* 16, 265–270.
Blasi, D.V., Debrot, S., Menound, P.A., Gendre, L. and Schowing, J. (1990) Amoebicidal effect of essential oils *in vitro*. *Journal Toxicologie Clinique Experimentale* 10, 361–373.
Bor, N.L. (1960) *The Grasses of Burma, Ceylon, India and Pakistan*. Pergamon Press, Oxford, London, New York.
Carlson, L.H.C., Machado, R.A.F., Spricigo, C.B., Pereira, L.K. and Bolzan, A. (2001) Extraction of lemongrass essential oil with dense carbon dioxide. *The Journal of Supercritical Fluids* 21, 33–39.
Cheel, J., Theoduloz, C., Rodriguez, J. and Schmeda, H.G. (2005) Free radical scavengers and antioxidants from lemongrass (*Cymbopogon citratus* (DC.) Stapf). *Journal of Agricultural and Food Chemistry* 53, 2511–2517.
Chen, S.S. and Spiro, M. (1994) Study of microwave extraction of essential oil constituents from plant materials. *Journal of Microwave Power and Electromagnetic Energy* 4, 231–241.
Chime, S.A., Ugwuoke, C.E.C., Onyishi, I.V., Brown, S.A., Ugwu, C.E. and Onunkwo, G.C. (2012a) Formulation and evaluation of *Cymbopogon citratus* dried leaf-powder tablets. *African Journal of Pharmacy and Pharmacology* 6, 3274–3279.
Chime, S.A., Brown, S.A., Ugwoke, C.E.C., Agubata, C.O., Ubah, J.O. and Onunkwo, G.C. (2012b) Formulation of methanolic extract of *Cymbopogon citratus* tablets: *in vitro* evaluation. *Drug Invention Today* 4, 397–400.
Chisowa, E.H., Hall, D.R. and Farman, D.I. (1998) Volatile constituents of the essential oils of *Cymbopogon citratus* Stapf grown in Zambia. *Flavour and Fragrance Journal* 13, 29–30.
Cruz, M.T. and Batista, M.T. (2011) *Cymbopogon citratus* as source of new and safe anti-inflammatory drugs: bio-guided assay using lipopolysaccharide-stimulated macrophages. *Journal of Ethnopharmacology* 133, 818–827.
Dama, G.Y., Tare, H.L., Gore, M.S., Deore, S.R. and Bidkar, J.S. (2011) Comparative hemintholytic potential of extracts obtained from *Cymbopogon citratus* (DC) Stapf. *Journal of Ethnopharmacology* 12, 279–286.
Directorate Plant Production (2009) *Essential oil crops: Production Guidelines for Lemongrass: Lemongrass Production*. Produced in collaboration with SAEOPA (Southern African Essential Oil Producers Association) and the KARWIL Consultancy. Department of Agriculture, Forestry and Fisheries, Directorate Communication Services, Pretoria. Available at: http://www.nda.agric.za/docs/Brochures/ProGuiLemonGrass.pdf (accessed 19 January 2016).
Ewansiha, J.U., Garba, S.A., Mawak, J.D. and Oyewole, O.A. (2012) Antimicrobial activity of *Cymbopogon citratus* (lemon grass) and its phytochemical properties. *Frontiers in Science* 2, 214–220.
Figueirinha, A., Cruz, M.T., Francisco, V., Lopes, M.C. and Batista, M.T. (2010) Anti-inflammatory activity of *Cymbopogon citratus* leaf infusion in lipopolysaccharide-stimulated dendritic cells: contribution of the polyphenols. *Journal of Medicinal Food* 13, 681–690.
Gore, M.S., Tare, H.L., Deore, S.R., Bidkar, J.S. and Dama, G.Y. (2010) Hemintholytic potential of *Cymbopogon citratus* leaves extract and its formulation as an emulsion. *International Journal of Pharmaceutical Sciences and Research* 1(10), 174–177.
Husain, A. (1993) *Essential Oil Plants and Their Cultivation*. Central Institute of Medicinal and Aromatic Plant (CIMAP), Lucknow, India.
Idrees, M., Naeem, M., Khan, M.N., Aftab, T., Khan, M.M.A. and Moinuddin (2012) Alleviation of salt stress in lemongrass by salicylic acid. *Protoplasma* 249, 709–720.
Kishore, N., Mishra, A.K. and Chansouria, J.P. (1993) Fungitoxicty of essential oils against dermatophytes. *Mycoses* 36, 211–215.
Lonkar, P.B., Chavan, U.D., Pawar, V.D., Bansode, V.V. and Amarowicz, R. (2013) Studies on preparation and preservation of lemongrass (*Cymbopogon flexuosus* (Steud) Wats) powder for tea. *Emirates Journal of Food and Agriculture* 25, 585–592.
Manvitha, K. and Bidya, B. (2014) Review on pharmacological activity of *Cymbopogon citratus*. *International Journal of Herbal Medicine* 1(6), 5–7.
Mirghani, M.E.S., Liyana, Y. and Parveen, J. (2012) Bioactivity analysis of lemongrass (*Cymbopogon citratus*) essential oil. *International Food Research Journal* 19, 569–575.

Mishra, A.K. and Dubey, N.K. (1994) Evaluation of some essential oils for their toxicity against fungi causing deterioration of stored food commodities. *Applied and Environmental Microbiology* 60, 1101–1105.

Nambiar, S.V. and Matela, H. (2012) Potential functions of lemon grass (*Cymbopogon citratus*) in health and disease. *International Journal of Pharmaceutical and Biological Archive* 3, 1035–1043.

Oleyede, O.I. (2009) Chemical profile and antimicrobial activity of *Cymbopogon citratus* leaves. *Journal of Natural Products* 2, 98–103.

Olorunnisola, S.K., Asiyanbi, H.T., Hammed, A.M. and Simsek, S. (2014) Biological properties of lemongrass: an overview. *International Food Research Journal* 21, 455–462.

Onabanjo, A.O., Agbaje, E.O. and Odusote, O.O. (1993) Effects of aqueous extracts of *Cymbopogon citratus* in malaria. *Journal of Protozoology Research* 3, 40–45.

Onawunmia, G.O., Yisak, W.A. and Ogunlana, E.O. (1984) Antibacterial constituents in the essential oil of *Cymbopogon citratus* (DC.) Stapf. *Journal of Ethnopharmacology* 12, 279–286.

Pedroso, R.B., Nakamura, T.U., Filho, B.P.D., Cortez, D.A.G., Cortez, L.E.R., Morgado-Diaz, J.A. and Nakamura, C.V. (2006) Biological activities of essential oil obtained from *Cymbopogon citratus* on *Crithidia deanei. Acta Protozoologica* 45, 231–240.

Pengelly, A. (2004) Chapter 7: Essential oils and resins. In: *The Constituents of Medicinal Plants: An Introduction to the Chemistry and Therapeutics of Herbal Medicine*, 2nd edn. CAB International, Wallingford, UK, pp. 85–109.

Puatonachokchai, R., Kishida, H., Denda, A., Murata, N., Konishi, Y. and Vinitketkumnuen, U. (2002) Inhibitory effects of lemon grass (*Cymbopogon citratus* Stapf) extract on the early phase of hepatocarcinogenesis after initiation with ethyinitrosamine in male Fischer 344 rats. *Cancer Letters* 183, 9–15.

Schaneberg, B.T. and Khan, I.A. (2002) Comparison of extraction methods for marker compounds in the essential oil of lemongrass by GC. *Journal of Agricultural and Food Chemistry* 50, 1345–1349.

Shah, G., Richa, S., Vivek, P., Narender, S., Bharpur, S. and Mann, A.S. (2011) Scientific basis for the therapeutic use of *Cymbopogon citratus* Stapf (lemon grass). *Journal of Advanced Pharmaceutical Technology and Research* 2, 3–8.

Singh, N., Luthra, R., Sangwan, R.S. and Thakur, R.S. (1989) Metabolism of monoterpenoids in aromatic plants. *Current Research on Medicinal and Aromatic Plants* 11, 174–197.

Singh, N., Shrivastava, P. and Shah, M. (2014) Microwave-assisted extraction of lemongrass essential oil: study of the influence of extraction method and process parameters on extraction process. *Journal of Chemical and Pharmaceutical Research* 6, 385–389.

Srivastava, V., Dubey, S. and Mishra, A. (2013) A review on lemongrass: agricultural and medicinal aspect[s]. *International Research Journal of Pharmacy* 4(8), 42–44.

Suaeyun, R., Kinouchi, T., Arimochi, H., Vinitketkumnuen, U. and Ohnishi, Y. (1997) Inhibitory effects of lemon grass (*Cymbopogon citratus* Stapf) on formation of azoxymethane-induced DNA adducts and aberrant crypt foci in the rat colon. *Carcinogenesis* 18, 949–955.

Syed, M., Khalid, M.R. and Chaudhary, F.M. (1990) Essential oils of Gramineae family having antibacterial activity Part 1 (*Cymbopogon citratus, C. martinii* and *C. jawarancusa* oils). *Pakistan Journal of Scientific and Industrial Research* 33, 529–531.

Tajidin, N.E., Ahmad, S.H., Rosenani, A.B., Azimah, H. and Munirah, M. (2012) Chemical composition and citral content in lemongrass (*Cymbopogon citratus*) essential oil at three maturity stages. *African Journal of Biotechnology* 11, 2685–2693.

Tangpu, V. and Yadav, A.K. (2006) Antidiarrhoeal activity of *Cymbopogon citratus* and its main constituent, citral. *Pharmacologyonline* 2, 290–298.

Tchoumbougnang, F., Zollo, P.H., Dagne, E. and Mekonnen, Y. (2005) *In vivo* anti malaria activity of essential oils from *Cymbopogon citratus* and *Ocimum gratissimum* on mice injected with *Plasmodium berghei. Planta Medica* 71, 20–23.

Tiwari, M., Dwivedi, U.N. and Kakkar, P. (2010) Suppression of oxidative stress and pro-inflammatory mediators by *Cymbopogon citratus* D. Stapf extract in lipopolysaccharide stimulated murine alveolar macrophages. *Food and Chemical Toxicology* 48, 2913–2919.

Vinitketkumnuen, U., Puatanachokchai, R., Kongtawelert, P., Lertprasertsuke, N. and Matsushima, T. (1994) Antimutagenicity of lemon grass (*Cymbopogon citratus*, Stapf) to various known mutagens in *Salmonella* mutation assay. *Mutation Research* 341, 71–75.

Wannissorn, B., Jarikasem, S. and Soontorntanasart, T. (1996) Antifungal activity of lemon grass and lemon grass oil cream. *Phytotherapy Research* 10, 551–554.
Wannissorn, B., Jarikasem, S., Siriwangchai, T. and Thubthimthed, S. (2005) Anti-bacterial properties of essential oils from Thai medicinal plants. *Fitoterapia* 76, 233–236.
Zheljazkov, V.D., Cantrell, C.L., Astatkie, T. and Cannon, J.B. (2011) Lemongrass productivity, oil content, and composition as a function of nitrogen, sulfur, and harvest time. *Agronomy Journal* 103, 805–812.

13 Mint

Maria do Carmo Ferreira* and Aline de Holanda Rosanova
Federal University of São Carlos, São Carlos, São Paulo, Brazil

13.1 Botany

13.1.1 Introduction

The Lamiaceae or mint family is one of the most diverse and widespread dicotyledonous plant families. Several species in this family have external glandular structures that produce volatile oil and are highly aromatic (Giuliani and Maleci Bini, 2008). The family includes about 236 genera and 6900–7200 species, including many culinary herbs, such as mint, basil, rosemary, marjoram and thyme (Venkateshappa and Sreenath, 2013). The genus *Mentha* is considered to be the most important in this family because its essential oil has a high economic value and is used in several different industrial sectors, such as food, flavouring, fragrances, cosmetics and pharmaceuticals. Mint species have been traditionally used in natural (or complementary) medicine and ethnomedicine against a wide variety of diseases. They are also used for culinary purposes, owing to the pleasant and aromatic flavour of their leaves (Andrews, 1996; Bhat *et al.*, 2002; Abbaszadeh *et al.*, 2009; Kunnumakkara *et al.*, 2009; Chawla and Thakur, 2013; Sujana *et al.*, 2013). As well as being known as mint in English, the genus *Mentha* is popularly known as minzen in Germany, menta in Spain, menthe in France, menta in Italy, munt in the Netherlands and hortelã in Brazil and Portugal.

Mentha species are perennial herbs, meaning they will either remain green all winter or go dormant over the winter season and revive in the spring. The herbs grow fast and spread quickly in open areas, and for these reasons, they are considered to be invasive (Abbaszadeh *et al.*, 2009). According to Harley and Brighton (1977), the genus *Mentha* includes five sections, 25–30 species and hundreds of varieties. The precise number of taxonomically valid species in the genus is unknown, as species cross freely, producing many intermediate forms (Bhat *et al.*, 2002). The different species include peppermint (*M. piperita*), spearmint (*M. spicata*), corn or menthol mint (*M. arvensis*), apple mint or pineapple mint (*M. suaveolens*), pennyroyal (*M. pulegium*), chocolate mint (*M.* × *piperita*), basil mint (*M.* × *piperita*), water mint (*M. aquatica*), ginger mint (*M. gracilis*) and horsemint (*M. longifolia*). All but the pennyroyal (*M. pulegium*) varieties can be eaten safely (Barrett, 2009; Buckland and Drost, 2009; Kunnumakkara *et al.*, 2009; Duea and Murphy, 2011).

*Corresponding author, e-mail: mariaf@ufscar.br

© CAB International 2016. *Leafy Medicinal Herbs: Botany, Chemistry, Postharvest Technology and Uses* (eds D.C.P. Ambrose *et al.*)

According to Chen *et al.* (2012), only four species, namely spearmint, American wild mint (*M. canadensis*), ginger mint and peppermint, are cultivated worldwide for commercial oil production. The most common and cultivated mint types are the peppermint and spearmint varieties (Abbaszadeh *et al.*, 2009). Spearmint (named *M. spicata* because of the plant's spiky shape) is native mainly from part of Europe and south-west Asia, but its exact original location is uncertain. Some references point out that the name comes from one of its first known locations – the monastery at St Pierre, in France. This particular variety has many aromatic and medicinal properties and is widely used in manufactured products, medicine and cooking (Abbaszadeh *et al.*, 2009; Lyle, 2010; Sulieman *et al.*, 2011). Peppermint (also known as 'peppery mint') is originally from England and was discovered in the 17th century as a wild plant. Nowadays it is found in various countries, in both cultivated and wild forms (Raghavan, 2006). This variety is a sterile hybrid derived from a cross between water mint and spearmint. Its leaves have a strong scent and flavour, and for these features, peppermint is considered the 'king of all mints' (Abbaszadeh *et al.*, 2009; Lyle, 2010; Duea and Murphy, 2011; Chawla and Thakur, 2013; Sujana *et al.*, 2013).

Other commercially less important cultivars include water mint, a variety that grows in wet places; horsemint, a wild mint introduced into America from Europe; and apple mint, which is often used as an ornamental plant. Other varieties are bergamot mint (*M. citrata*), which has a savoury and fruity aroma, regular, small-leaved or creeping mint (*M.* × *villosa*), a very common variety in Brazil, and pennyroyal, whose essential oil is used in aromatherapy, and is also high in pulegone, an extremely toxic volatile organic compound (Barrett, 2009; Duea and Murphy, 2011).

13.1.2 History/origin

The name 'mint' originates from the Latin word 'menthe', which is rooted in the Greek word 'minthe'. According to Greek mythology, Minthe was a beautiful nymph and Pluto, the god of the underworld, fell in love with her. Pluto's wife, Proserpine, became furious when she got to know about her husband's love for the nymph and punished Minthe by turning her into an ordinary plant. As compensation for being unable to restore Minthe to her original state, Pluto created a characteristic and pleasant scent that she would give off whenever she was crushed. The name Minthe was further changed to *Mentha* and used as the genus of aromatic plants that includes all the mint varieties (Barrett, 2009; Kunnumakkara *et al.*, 2009; Gerritsen, 2010; Lyle, 2010).

According to the literature, the first mint species seem to have appeared in European and Mediterranean countries, and they have been important herbs from the beginning of civilization. As they grow very easily, their medicinal, aromatic and flavouring properties have been enjoyed for centuries (Barrett, 2009). There are reports on mint usage dating back to as long ago as Biblical times, when mint was accepted as payment to the temple, and it is even known that the Pharisees paid tax with mint leaves. Dried mint leaves have also been found in Egyptian tombs (Barrett, 2009; Gerritsen, 2010). Mint has been traditionally known as the herb of fun and hospitality. The Romans and Greeks commonly crowned themselves with it and decorated tables using it for their parties, and the ancient Romans used to spread mint in their homes for freshness (Barrett, 2009; Palmer, 2012). Like the ancient Romans, during many years before the advent of finished floors, people used to spread herbs on the dirty floors of houses. These gave off a slight scent where they were scattered, and when they were trampled on, a strong and pleasant aroma scented the place. Although many types of aromatic plants served this purpose, mint was preferred over alternatives owing to its stronger effect (Andrews, 1996).

The usage of mint for medicinal purposes is also traditional, though scientific knowledge on its antimicrobial, antiviral, antioxidant, antitumor and other potentially relevant medical properties is quite recent (Palmer, 2012; Chawla and Thakur, 2013).

13.1.3 Location

Most mint varieties grow best in rich and moist soils that are slightly acidic. Cultivation is benefited by a wet environment and exposure to full sun to partial shade, while some species require full light to develop, and at least 12 h of daylight to bloom. The spread occurs by dividing rhizomes, which must be planted at a space of 20 × 30 cm. The plant adapts well to sandy soils rich in organic matter, with pH values between 6.0 and 7.0 (Radünz, 2004). Cultivation may take up to 4 years, and produce two or three crops a year. Cultivation procedures may also affect the contents of bioactive constituents and aromatic compounds (Radünz, 2004; Maia, 2007). However, in general, mint species tolerate a wide range of conditions, and grow well in places of varying climatic conditions and soil types (Abbaszadeh *et al.*, 2009; Buckland and Drost, 2009; Chawla and Thakur, 2013). As a result, even though mint is cultivated mostly in the temperate and subtemperate regions, it is widely distributed throughout the world (Bhat *et al.*, 2002; Abbaszadeh *et al.*, 2009; Barrett, 2009).

The leading countries in mint production are India, the USA, China and Brazil. India accounts for nearly 80% of the world production of mint oil, with an estimated production of 15,000 t/year of essential oil and a mint cultivation area of over 150,000 ha in 2005/6 (Tzanetakis *et al.*, 2010), mainly of menthol mint. According to Chen *et al.* (2012), China, despite being a large producer, presently faces a variety degeneration problem that is causing a decrease in mint oil quality and yield.

In the USA, the cultivated mint area covers around 50,000 ha, with peppermint and spearmint corresponding, respectively, to roughly 80% and 20% of the cultivation area and crop value. The states of Oregon and Washington are the largest producers, followed by Idaho, Indiana and Wisconsin (Tzanetakis *et al.*, 2010). Along with the USA, Japan and the UK are the biggest world producers and exporters of peppermint essential oil and raw material. Other countries, such as Germany, Russia, Italy, Bulgaria, Greece and Norway, also cultivate peppermint, but their production volumes are smaller compared to the three former countries (Bhat *et al.*, 2002; Sústriková and Salamon, 2004). The cultivated spearmint areas in the USA are located mainly in Indiana and Michigan. The production of spearmint also takes place in France, the UK, Italy, the former Yugoslavia, Hungary, Bulgaria, Russia, South Africa, Thailand and Vietnam (Bhat *et al.*, 2002).

Other countries, such as China, Taiwan and India, cultivate bergamot mint commercially. These countries, along with Thailand, Japan and Brazil, are producers of Japanese mint (*M. arvensis*), also known as corn or menthol mint, wild mint, field mint and pudina (Bhat *et al.*, 2002; Kunnumakkara *et al.*, 2009).

13.1.4 Morphology

Although mint species differ somewhat in form and growth habit, they share many common morphological characteristics. Most mints are aromatic and perennial herbs that grow quickly up to 10–120 cm tall. Moreover, they have horizontal, wide-spreading underground rhizomes, overground stolons and vertical, square and branched stems (Abbaszadeh *et al.*, 2009; Chawla and Thakur, 2013). The leaves are arranged in opposite pairs, with profiles varying from simple oblong to lance shaped (see an example in Fig. 13.1). They often have a downy surface and serrated margins, and their colours range from dark green and grey–green to blue or purple (Abbaszadeh *et al.*, 2009; Chawla and Thakur, 2013). The flowers are morphologically described as tubular, with clusters ('verticils') produced on an erect spike, and colours varying from white to purple. The flower corolla is two lipped with four subequal lobes, and the upper lobe is usually the largest. The fruit is a small, dry capsule containing from one to four seeds (Abbaszadeh *et al.*, 2009; Chawla and Thakur, 2013).

Given the large number of varieties, only the two most popular types, spearmint and

Fig. 13.1. Fresh mint shoot with leaves.

peppermint, are described here. Blamey and Grey-Wilson (1989) described spearmint as a herbaceous rhizomatous perennial plant growing up to 30–100 cm tall (average about 60 cm), with variably hairless to hairy stems and foliage, and a wide-spreading fleshy underground rhizome. The leaves are opposite, 5–9 cm long and 1.5–3 cm wide, with a serrated margin. The flowers are produced in slender spikes, each flower pink or white, 2.5–3 mm long and broad. Peppermint may grow up to 90 cm in height. The stems are usually reddish purple and smooth. The leaves are fragrant, toothed and hairy on the underside. The flowers are bisexual and zygomorphic, and of a pinkish or purple colour (Abbaszadeh *et al.*, 2009).

Despite the similarities among the varieties, the different mint types can be distinguished by morphological analysis. A comparison of spearmint and peppermint shows that peppermint has wider, shorter and slightly darker green leaves, and that the stalks have a purplish colour. In comparison with some other mint types, they are both most obvious due to their sharp scent (Andrews, 1996; Lyle, 2010).

13.2 Chemistry

13.2.1 Chemical composition

Mints contain a large amount of fibre and of vitamins A, B_6, C and K, folate, folic acid, thiamine and riboflavin, as well as minerals such as calcium, potassium, sodium, magnesium, manganese, phosphorus and iron (Raghavan, 2006). So eating fresh mint (Fig. 13.1) is healthy, though the ingestion of dried mint (Fig. 13.2) is also recommended, as it will have significantly higher nutritional values than a fresh serving (Buckland and Drost, 2009; Palmer, 2012; Chawla and Thakur, 2013).

The Agricultural Research Service of the US Department of Agriculture (USDA) reports that 100 g of fresh peppermint leaves contain 78.65 g water, 3.75 g protein, 0.94 g total lipid, 14.89 g carbohydrates and 8.0 g fibre. A serving of 100 g of leaves contains the minerals calcium (243 mg), iron (5.08 mg), magnesium (80 mg), phosphorus (73 mg), potassium (569 mg), sodium (31 mg) and zinc (1.11 mg), and the vitamins C (31.8 mg), B_6 (0.129 mg) and A (212 μg), thiamine (0.082 mg), riboflavin (0.266 mg), niacin (1.706 mg) and folate (114 μg).

Even though the direct consumption of fresh or dried mint is substantial, a significant part of cultivation is to raise mint for the extraction of essential oils, which can be done by various processes. There is a strong demand for mint essential oils worldwide, as they contain a large variety of volatiles and aromatic chemical compounds, such as menthol, carvone, menthone, pulegone and menthyl acetate. These substances give the mints their characteristic aromas and flavours. The oil composition varies depending on the species and plant variety, and also on cultivation and handling. Factors such as the part of the plant used for extraction (leaves, stems, etc.), growth stage, harvest season, geographic origin and environmental conditions (climate, soil, topography) influence both oil yield and composition. The largest amount of mint oils produced over the world is extracted from leaves of peppermint, spearmint and corn mint. The leaves of these species contain, respectively, around

Fig. 13.2. Dried mint shoot and leaves.

1–3, 0.5 and 1–2% of essential oils of complex composition (Oksman-Caldentey and Barz, 2002).

Over 200 constituents have been found in peppermint oil (Oksman-Caldentey and Barz, 2002; Palmer, 2012; Chawla and Thakur, 2013), and while the different chemical compounds found in distinct mint plants may vary greatly, menthol is found in nearly all of them (Bhat *et al.*, 2002; Barrett, 2009; Palmer, 2012; Pirbalouti *et al.*, 2013). The average values of the main compounds of peppermint oil from two literature sources are given in Table 13.1.

Data on the chemical composition of spearmint oil reported by Bhat *et al.* (2002) are given in Table 13.2. Chawla and Thakur (2013) reported that spearmint essential oil contains 50–70% of carvone and dihydrocarvone, but also includes dihydrocuminyl acetate, limonene, menthone, menthol and 1,8-cineol in small quantities.

According to Bhat *et al.* (2002), corn mint oil contains 70–80% menthol, 10% menthyl acetate, 8% menthone, and limonene, pinene and caryophyllene (in traces). Chawla and Thakur (2013) report contents of 28–34% menthol, 16–31% menthone, 6–13% isomenthone, 5–10% limonene and a higher content of α- and β-pinenes.

Table 13.1. Data on the chemical composition of peppermint (*Mentha piperita*) leaf essential oil from two sources.

Constituent	Composition (%) (Bhat *et al.*, 2002)	Composition (%) (Chawla and Thakur, 2013)
Isomenthone	NA[a]	2–8
Limonene	Traces	2.5
Menthofurane	NA	2–8
Menthol	50–55	26–46
Menthone	10	16–36
Menthyl acetate	20	3.8–7
β-Pinene	NA	1.5–2
Pulegnone	NA	1.4–4

[a]Not available.

Table 13.2. Data on the chemical composition of spearmint (*Mentha spicata*) leaf essential oil. From Bhat *et al.*, 2002.

Constituent	Composition (%)
Carvone	58
Dihydocarveol	7
Dihydrocarveol acetate	12
Dipentene	10
Limonene	8

Martins *et al.* (2007) characterized the essential oil from leaves of regular mint and obtained a total volatile content of 99.5%. From 28 compounds detected in the original essential oil, 13 were monoterpenes and ten were sesquiterpenes. Piperitenone oxide was the major component (35.4%).

It is clear from those results that the oil composition may vary widely, even for plants of the same species and variety. Nowadays, the analysis and identification of essential oils composition are performed using gas chromatographic (GC) techniques, as these are suited for detecting volatile compounds. In most cases, GC is coupled to flame ionization (FID) or mass spectrometry (MS) detection modes, which are suitable for terpenoid determinations. Analysis is limited to a relatively low number of compounds that are stable upon vaporization and separation (Turek and Stintzing, 2013).

High performance liquid chromatography (HPLC) is also a very popular method for the analysis of essential oils. The analytical sensitivity of this method can be enhanced by its association with other detectors, such UV or photodiode array (PDA) (Gurib-Fakim, 2006). HPLC coupled with mass spectrometry (HPLC/MS) is another technique used for compound identification.

13.2.2 Phytochemistry

A wide range of bioactive phytochemical constituents of medicinal plants produces defined physiological actions on the human body and may play a protective role against some pathogens. Some of the most important phytochemicals of medicinal plants are tannins, terpenes, alkaloids, flavonoids, steroids, saponins, anthraquinones, coumarins and sterols (Gershenzon *et al.*, 2000; Ullah *et al.*, 2011).

The extensive use of mint species for medicinal purposes occurs mostly because their essential oil contains two classes of secondary metabolites, and different structural types of phenolic compounds. The secondary metabolites act as antioxidants, anti-inflammatory compounds, antispasmodics, antiemetics, diaphoretics or antiviral agents (Mimica-Dukic and Bozin, 2008). Menthol is the main monoterpene in mint essential oils, followed by menthone, and the derivatives of both these compounds (e.g. acetyl menthol, isomenthone, pulegone). Other secondary metabolites include alkaloids, tannins and steroids. The phenolic compounds include the polyphenol rosmarinic acid and flavonoids (Mimica-Dukic and Bozin, 2008, Ullah *et al.*, 2011; Palmer, 2012; Sujana *et al.*, 2013).

Menthol is mostly found in older leaves and has antioxidant, anti-inflammatory, antifungal, antibacterial, antiseptic, antispasmodic and stimulant properties (Barrett, 2009; Chawla and Thakur, 2013; Sujana *et al.*, 2013). The phenolic compounds have a wide range of pharmacological actions, including antioxidant, hepatoprotective, anti-inflammatory, antidiabetogenic and anti-ulcer activities (Mimica-Dukic and Bozin, 2008).

13.3 Postharvest Technology

The proper timing for harvesting herbs depends on their intended use. To preserve their flavour and aroma, mint plants should be harvested when their essential oil content is at its peak. As the oils of such plants are concentrated in their leaves, it has been recommended that the leaves be harvested before flowering and at dawn, although according to Rohloff *et al.* (2005), mint plants are normally harvested in full bloom. The same authors report that peppermint oil yield increased from early to full bloom and that the flavour compounds, menthol and menthone, reached their optimum at full bloom.

13.3.1 Processing

The water content of fresh mint leaves varies, on average, from 75 to 85% (wet basis, w.b.) (Rosanova and Ferreira, 2014), and like the majority of plants, they suffer rapid deterioration under ambient conditions. One practical approach to ensure optimized yield and quality of essential oils is direct distillation immediately after harvesting (Tarhan *et al.*, 2010). However, when fresh plant processing is not feasible, dehydration and thermal drying are the most popular preservation techniques. Additionally, drying reduces shipping weights and minimizes packing requirements, as after dehydration, mint plants may lose up to 85% of their weight. The mint leaves shrink significantly when moisture content is reduced, as observed by Costa *et al.* (2014) and Rosanova and Ferreira (2014) (see also Fig. 13.2).

Drying and storage

There is no agreement in the literature about a proper moisture level for storing herbs in general. Likewise, a particular moisture range considered adequate for mint storage has not been found in the references that have been consulted. It is well known though that the essential oil chemical composition is highly dependent on the plant processing and storage conditions (Turek and Stintzing, 2012). The terpenoids contained in essential oils are likely volatile and thermolabile and, depending on their respective structure, may be easily oxidized or hydrolysed (Scott, 2005). According to Martins *et al.* (2002), leaf and flower moisture must be reduced to around 5–10% (w.b.) for storage, while Farias (2003) establishes values around 8–14%. The choice of drying technique will depend on the quantity to be processed and on the specified quality attributes for dry plant and essential oils.

When processing small quantities, bunches of mint stems can be dehydrated by hanging them outside from lines so that they are exposed to a fresh breeze until all the moisture is evaporated. Another common procedure is to use trays upon which thin layers of leaves are spread and exposed to warm air either in an oven or under moderate sunlight. The material must be revolved periodically to avoid moisture gradients that may lead to non-uniform moisture distribution in the final product. After dehydration, the stalks and other hard parts can be rejected, and the remaining crisp parts crumbled to produce a powder that may be conveniently stored in glass or metal containers.

As the plant contains many volatile and thermolabile compounds, thermal drying has to be performed with care to avoid the loss of active ingredients caused by too high a temperature. Batch convective dryers are extensively used for drying aromatic herbs and medicinal plants in large quantities. Many studies have been reported in the literature focusing on investigating how the drying conditions affect the extraction yield and composition of mint essential oils. However, given the great number of mint varieties and the lack of standardized procedures, the results obtained are often restricted to a narrow range of conditions or even discrepant among themselves.

Radünz (2004) investigated the thermal drying of leaves of regular mint leaves at temperatures ranging from 40 to 80°C and compared the oil extraction yield obtained from dry plants with that from fresh material. There was an increase in the oil extraction yield from mint dried at temperatures up to 50°C (0.87 g/g of dry matter) in comparison with the yield obtained from fresh material (0.57 g/g of dry matter). At higher temperatures, the extraction yield decreased, and at 80°C it was greatly reduced (to 0.37 g/g of dry matter). The composition of the oil obtained from dried and fresh plants was also compared, and it was found that while the main constituents of mint essential oil were preserved after drying, their content was significantly reduced by the drying temperature, with 50°C being the temperature that best preserved them. The best drying temperature will therefore depend on the intended usage for the oil, as its chemical constituents are not equally affected by drying.

The necessity of constantly revolving the material to avoid moisture gradients in the final product is a difficulty that may limit the use of a static bed configuration for drying plants on a commercial scale. The use of moving beds, such as the rotary drum (Tarhan *et al.*, 2010), or the innovative rotating drum proposed by Rosanova and Ferreira (2014), may offer alternatives for overcoming this drawback.

Tarhan *et al.* (2010) dried peppermint in a rotary drum dryer with a capacity for processing up to 15 kg of leaves and programmed to operate under two different drying schemes: the air could be heated either at a constant temperature or to give a rectangular wave-shaped temperature profile. Drying was performed over 15–18 h at the constant temperature profile or over 12–15 h for the rectangular wave-shaped temperature profile. Hot air drying caused considerable darkening of the dried peppermint leaves. The essential oil contents of the samples dried according to these two procedures were relatively unaffected by the drying scheme (2.08–2.7 ml per 100 g dry matter). The menthol content of the leaves increased from 32.5% (fresh) up to 44.5% (dried), while their menthone content decreased from 24.9% (fresh) down to 9.1% (dried). The variations within replications were relatively high in terms of the menthol (25.4–44.5%) and menthone (9.1–18.5%) contents in the dried leaves, and such variations were attributed to the uncontrolled morphological and physiological changes in the fresh peppermint leaves that were associated with their maturity. According to the authors, these variations are the main obstacle to the reproducibility of results and hence to obtaining dried peppermint products with consistent quality.

Rosanova and Ferreira (2014) described the development of a rotating drum dryer for drying the leaves and stems of regular mint. The dryer consists of a cylindrical drum with perforated walls into which air is blown, resulting in a radial air flux through the samples. The drum is rotated at a low velocity (2 rpm), just enough to keep the material mixing throughout the drying process. The equipment proved to be effective, and at an air temperature of 80°C, it took about 9 h to reduce the moisture of the leaves to 9.4% (w.b.) and the stems to 6.2% (w.b.). There was strong shrinkage of both leaves and stems as the moisture was removed.

Pääkkönen (1999) reported on the use of an infrared convective static bed dryer to dry peppermint. The dryer had three chambers distributed over steel mesh trays, and a capacity of 13.5 kg for batch samples. The thermal energy was provided by eleven infrared lamps, with a spectrum ranging from a wavelength <2 μm to 4–5 μm, and with nominal powers of 13 kW for chamber 1, and 9 kW for chambers 2 and 3. Intermittent radiation periods were tempered with airflow. The drying was performed at an average temperature of 45°C, and the total oil content and composition were compared with those obtained from oven-dried samples. The author reported on higher drying rates using infrared radiation than oven drying at 40°C for 2 or 3 days. The total oil content did not differ significantly when dried with oven heat or infrared radiation, but the menthol content was slightly higher for the sample dried with infrared radiation. This combined infrared/convection process may be an appealing alternative for drying herbs, though it does require further investigation.

Thin-layer drying is a classical approach for assessing the drying behaviour of herbs and aromatic leaves under different conditions. This technique allows the effects of process variables such as the air temperature and velocity on the moisture reduction and drying rates to be identified. This may be done by simple experiments based on measurements of small samples weight as a function of time. Because the internal gradients are small enough to be neglected in such samples, a lumped analysis may be applied for modelling. Using this approach, simple equations can be fitted to predict drying rates. Although the operating conditions in commercial-scale equipment are often very different from those in thin-layer drying, many researchers rely on this approach as being effective for providing fundamental information on drying processes and for obtaining suitable equations to estimate drying rates.

Akpinar (2010) investigated thin-layer drying of mint leaves using an indirect forced convection solar dryer and drying under open sun with natural convection. Several empirical equations were tested for predicting the drying rates, and the Wang and Singh model was found to be the best for predicting the kinetics of both forced solar drying and natural sun drying. A comparison between the two drying techniques indicated no difference in the quality of the dried mint leaves. Sallam *et al.* (2015) also evaluated prototype direct and indirect solar dryers operating under natural and forced convection modes for drying whole mint leaves. They tested ten empirical equations for predicting thin-layer drying rates in mint solar drying and verified that all of them represented drying behaviour to an acceptable degree of accuracy, the only exception being the Wang and Singh equation, which did not fit their experimental data well.

Costa *et al.* (2014) investigated the drying of small samples (18–42 g) of leaves of regular mint in a horizontal flow convective dryer under different air temperatures (36–64°C) and air velocities (1.0–2.0 m/s). Drying was mostly affected by the air temperature, with a much weaker influence of air velocity, although the latter might be relevant in some operational conditions. The samples essentially offered an internal resistance to mass transfer, which was reduced by an increase in air temperature. Even though the leaf samples consisted of very heterogeneous particulate media, an empirical model based on a neural network technique performed well to estimate moisture content within the range of conditions tested. The classical thin-layer drying methods were also tested and best fit of the kinetic data was obtained with Page and Henderson–Pabis equations.

Lima-Corrêa *et al.* (2014) analysed the use of thin-layer drying technique to describe the drying of basil leaves (which have a similar structure to mint leaves) and demonstrated that is not always possible to apply this approach to the convective drying of leaves, particularly when these leaves have a high initial moisture content and shrink considerably during drying, as so mint leaves. They affirmed that, considering the intrinsic variability in the composition and physical properties of leaves, the random packing structure of samples exposed to air flow and the deformation caused by shrinking, it is very difficult to reproduce similar samples and identical drying conditions. Therefore, discrepancies among results from different authors are very common, even for plants of a same species and variety.

Extraction of essential oils

The use of the phytochemicals from a plant depends upon an appropriate extraction method. Some traditional techniques are Soxhlet extraction, maceration and hydrodistillation (Azmir *et al.*, 2013). The Soxhlet extractor is the most common and widely employed extractor for obtaining essential oils from plants and natural products, despite its use being very time-consuming. It requires previously dehydrated plants and the choice of solvent is a key factor for an efficient extraction. The method may destroy some plant components and consumes relatively large amounts of polluting solvents, which are important drawbacks for its use. Ethanol and methanol are some of the solvents used to extract the terpenoids and tannins present in mint essential oils (Azmir *et al.*, 2013).

The use of maceration enables extraction to be done cheaply and is adequate for small-scale operation. In this method, the plant materials are ground into smaller pieces and then extracted by contact with an appropriate solvent in a closed vessel. Three types of hydrodistillation can also be used: water distillation, water and steam distillation, and direct steam distillation. Hydrodistillation is a method that does not use organic compounds and may be applied to fresh material. However, the use of a high extraction temperature may cause the loss of volatile compounds and it may not be suitable for thermolabile compounds (Vankar, 2004).

Some novel techniques have recently been introduced to overcome the limitations of conventional extraction methods, and these include supercritical fluid extraction, ultrasound-assisted extraction, microwave–assisted extraction and pressurized solvent extraction (Kaufmann and Christen, 2002; Azmir *et al.*, 2013). These techniques are recognized as being fast and efficient, and some of them are even categorized as 'green techniques', as they comply with standards set by the US Environmental Protection Agency. Such non-conventional techniques are described and discussed in detail elsewhere (Azmir *et al.*, 2013).

13.3.2 Value addition

Drying is the best known alternative for adding value to fresh herbs, and dehydrated mint leaves and their essential oils have potential applications in several commercial sectors, including the pharmaceutical, food and chemical industries (Oksman-Caldentey and Barz, 2002). Additionally, because it is highly aromatic, dry mint may be used directly, e.g. for flavouring dishes or in tea infusions, and so may be commercialized in supermarkets and food shops, and also sold to restaurants. Beyond these common uses for culinary purposes and in industry, many other applications are now being explored and may open opportunities for the development of new value-added products based on mint plants. An example is the microencapsulation of essential oils, in which the sensitive ingredients and volatile compounds may be entrapped in solid carriers to increase their protection against oxidation, promote easier handling, reduce evaporation and control their release (Pegg and Shahidi, 2007). Baranauskiene *et al.* (2006) investigated the microencapsulation of peppermint essential oil in several commercial food starch-based matrices by spray-drying. All of the modified starch matrices tested retained oil very efficiently, and it was possible to retain up to 97.4% of peppermint oil volatiles after spray-drying, although the matrices varied in their release of volatiles under different conditions.

Kanatt *et al.* (2007, 2008) investigated the antioxidant activity of mint extracts for use in the preservation of meat and meat products. The antioxidant activity of the extract is comparable to that of the synthetic antioxidant BHT (butylated hydroxytoluene), owing to its content of total phenolic and flavonoid compounds. They demonstrated that, if blended with chitosan, a substance with excellent antimicrobial activity, mint extract produced a potent mixture that was effective in extending the shelf life of meat products.

13.4 Uses

13.4.1 General uses

Mint leaves and stems, as well as their essential oils, are used for various medicinal purposes, as food flavouring and as a fragrance in cosmetics (Oksman-Caldentey and Barz, 2002; Abbaszadeh *et al.*, 2009). The use of mint as a flavouring in cooking dates from a long time ago and has been recorded in different parts of the world (Thailand, North Africa, England and Greece). The Greeks and Romans used mint for flavouring sauces, food and wines. Its use for culinary purposes remains widespread (Andrews, 1996; Barrett, 2009), and its usage as a (natural) flavouring agent is reported as coming third worldwide after vanilla and citrus flavours (Arslan *et al.*, 2010; Sallam *et al.*, 2013). Mint is in the category of tender and delicate herbs used for culinary purposes as a garnish. Such herbs are added preferably in the final cooking stages, or added raw to salads or relishes, to maintain their pungency and retain their green colour.

Some mint varieties are particularly suited for cosmetic applications. Bergamot mint and its oil, for instance, have high cosmetic value. The distinctive lavender and floral scent of bergamot mint make it a highly valuable scent for soaps, perfumes and colognes, and for use as bath teas and in masks for facial care; it is also used for exfoliation purposes. The leaves and the petals, either dried or fresh, may be added to scent hot bathing water.

13.4.2 Pharmacological uses

For centuries, folk healers have used mint herbs to treat ailments such as colic and digestive disorders. Mint is still widely employed for medicinal purposes, but nowadays with a scientific foundation that provides better understanding of its pharmacological effects and of the best uses of this plant (Palmer, 2012; Chawla and Thakur, 2013). From the many species and hundreds of varieties and cultivars of the genus *Mentha*, two stand out for their medicinal applications: spearmint and peppermint (Abbaszadeh *et al.*, 2009; Barrett, 2009). However, in an ethnobotanical survey recently conducted by Cartaxo *et al.* (2010) to evaluate promising medicinal plants for bioprospecting studies in a rural community in the state of Ceará, Brazil, regular mint (*M.* × *villosa*) was classified as the second most versatile plant with potential for medicinal use out of a total of 119 species.

According to information published on the website of the Medical Center of the University of Maryland (https://umm.edu/health/medical/altmed/herb/peppermint), peppermint is used to soothe an upset stomach or to aid digestion. It has an anaesthetizing effect, and is therefore used to treat headaches, skin irritations, anxiety associated with depression, nausea, diarrhoea, menstrual cramps and flatulence. Peppermint is also an ingredient in topical ointments used as chest rubs in the treatment of cold symptoms. Evidence of its pharmacological activities has been reported by several researchers. Significant antimicrobial and antiviral activities, strong antioxidant and antitumour activities, and some antiallergenic potential of peppermint have been detected in tests *in vitro* (McKay and Blumberg, 2006). Several studies support the use of peppermint for indigestion and irritable bowel syndrome (Kline *et al.*, 2001; Cappello *et al.*, 2007; Magge and Lembo, 2011). A medical study showed the benefits of an enteric-coated peppermint oil formulation to reduce abdominal distention and flatulence. Nearly 80% of the patients who took peppermint also showed alleviation of abdominal pain (Liu *et al.*, 2006). There is also evidence that the topical application of peppermint is effective in soothing skin irritations caused by hives, poison ivy or poison oak (Herro and Jacob, 2010). In animal models, peppermint has demonstrated relaxant effects on gastrointestinal tissue, analgesic and anaesthetic effects in the central and peripheral nervous systems, immunomodulatory activity and chemopreventive potential (McKay and Blumberg, 2006). The prescription of menthol has also been reported as a medication for gastrointestinal disorders, common cold and musculoskeletal pain (Patel *et al.*, 2007).

13.5 Summary

In this chapter, a review of published information on the biological and chemical characteristics of mint leaves and their essential oils has been presented. The most common processing techniques and uses of these herbs and their derived products have also been addressed. The large number of species in the genus *Mentha* and their accessibility certainly contribute to the great versatility of these plants in their possible commercial applications and potential manufactured products, while at the same time, they introduce important issues concerning the identification of constituents and the standardization of analysis methodologies and processing techniques. Exploring the full potential of mints for use in different industrial sectors is a challenge for multidisciplinary fields of knowledge, including the basic sciences (botany, chemistry, pharmacology) and technical areas (agriculture, biotechnology, food engineering and chemical engineering, among others). The development of therapeutics based on mint phytochemicals is probably the most promising application to be explored, but further research is required in order to provide safe products on a commercial scale.

References

Abbaszadeh, B., Valadabadi, S.A., Farahani, H.A. and Darvishi, H.H. (2009) Studying of essential oil variations in leaves of *Mentha* species. *African Journal of Plant Science* 3, 217–221.

Akpinar, E.K. (2010) Drying of mint leaves in a solar dryer and under open sun: modeling, performance analyses. *Energy Conversion and Management* 51, 2407–2418.

Andrews, G. (1996) *Growing and Cooking with Mint: Storey Country Wisdom Bulletim A-145.* Storey Publishing, North Adams, Massachusetts.

Arslan, D., Özcan, M.M. and Menges, H.O. (2010) Evaluation of drying methods with respect to drying parameters, some nutritional and colour characteristics of peppermint (*Mentha* x *piperita* L.). *Energy Conversion and Management* 51, 2769–2775.

Azmir, J., Zaidul, I.S.M., Rahman, M.M., Sharif, K.M., Mohamed, A., Sahena, F., Jahurul, M.H.A., Ghafoor, K., Norulaini, N.A.N. and Omar, A.K.M. (2013) Techniques for extraction of bioactive compounds from plant materials: a review. *Journal of Food Engineering* 117, 426–436.

Baranauskiene, R., Zukauskaite, J., Bylaite, E. and Venskutonis, P.R. (2006) Aroma retention and flavour release of peppermint essential oil encapsulated by spray-drying into food starch based matrices. In: Wandrey, C. and Poncelet, D. (eds) *Proceedings of XIVth International Workshop on Bioencapsulation & COST 865 Meeting, Lausanne, Switzerland, Oct. 5–7, 2006.* EPFL (École Polytechnique Fédérale de Lausanne), Lausanne, Switzerland, Contribution P-06. Available at: http://impascience.eu/bioencapsulation/340_contribution_texts/2006-10-05_P0-6.pdf (accessed 20 January 2016).

Barrett, J. (2009) *What Can I Do with My Herbs? How to Grow, Use, and Enjoy These Versatile Plants*, 1st edn. Texas A&M University Press, College Station, Texas.
Bhat, S., Maheshwari, P., Kumar, S. and Kumar, A. (2002) *Mentha* species: *in vitro* regeneration and genetic transformation. *Molecular Biology Today* 3, 11–23.
Blamey, M. and Grey-Wilson, C. (1989) *The Illustrated Flora of Britain and Northern Europe*, 1st edn. Hodder & Stoughton, London.
Buckland, K. and Drost, D. (2009) *Mint in the Garden*. Publication No. Horticulture/Garden/2009-05pr, Utah State University Cooperative Extension, Logan Utah. Available at: https://extension.usu.edu/files/publications/publication/Horticulture_Garden_2009-05pr.pdf (accessed 19 January 2016).
Cappello, G., Spezzaferro, M., Grossi, L., Manzoli, L. and Marzio, L. (2007) Peppermint oil (mint oil) in the treatment of irritable bowel syndrome: a prospective double blind placebo-controlled randomized trial. *Digestive and Liver Disease* 39, 530–536.
Cartaxo, S.L., Souza, M.M.A. and Albuquerque, U.P. (2010) Medicinal plants with bioprospecting potential used in semi-arid northeastern Brazil. *Journal of Ethnopharmacology* 131, 326–342.
Chawla, S. and Thakur, M. (2013) Overview of mint (*Mentha* L.) as a promising health-promoting herb. *International Journal of Pharmaceutical Research and Development* 5, 73–80.
Chen, X.H, Zhang, F. and Yao, L. (2012) Chloroplast DNA molecular characterization and leaf volatiles analysis of mint (*Mentha; Lamiaceae*) populations in China. *Industrial Crops and Products* 37, 270–274.
Costa, A.B.S., Freire, F.B., Ferreira, M.C. and Freire, J.T. (2014) Convective drying of regular mint leaves: analysis based on fitting empirical correlations, response surface methodology and neural networks. *Acta Scientiarum* 36, 271–278.
Duea, A.W. and Murphy D. (2011) *The Complete Guide to Growing Windowsill Plants: Everything You Need to Know Explained Simply*. Atlantic Publishing Group, Ocala, Florida.
Farias, M.R. (2003) Assessment of quality of raw vegetables. In: Simões, C.M.O., Schenkel, E.P., Gosmann, G., Mello, J.C.P., Mentz, L.A. and Petrovick, P.R. (eds) *Farmacognosia: From Plant to Medicament*, 5th edn. Federal University of Rio Grande do Sul and Federal University of Santa Catarina Publishers, Porto Alegre, Rio Grande do Sul and Florianópolis, Santa Catarina, Brazil. [In Portuguese.]
Gerritsen, V.B. (2010) Mint condition. *Protein Spotlight* No. 113, 1–2. Available at: http://web.expasy.org/spotlight/back_issues/113/ (accessed 20 January 2016).
Gershenzon, J., McConkey, M.E. and Croteau, R.B. (2000) Regulation of monoterpene accumulation in leaves of peppermint. *Plant Physiology* 122, 205–213.
Giuliani, C. and Maleci Bini, L. (2008) Insight into the structure and chemistry of glandular trichomes of *Labiatae*, with emphasis on subfamily *Lamioideae*. *Plant Systematics and Evolution* 276, 199–208.
Gurib-Fakim, A. (2006) Medicinal plants: traditions of yesterday and drugs of tomorrow. *Molecular Aspects of Medicine* 27, 1–93.
Harley, R.M. and Brighton, C.A. (1977) Chromosome numbers in the genus *Mentha* L. *Botanical Journal of the Linnean Society* 74, 71–96.
Herro, E. and Jacob, S.E. (2010) *Mentha piperita* (peppermint). *Dermatitis* 21, 327–329.
Kanatt, S.R., Chander, R. and Sharma, A. (2007) Antioxidant potential of mint (*Mentha spicata* L.) in radiation-processed lamb meat. *Food Chemistry* 100, 451–458.
Kanatt, S.R., Chander, R. and Sharma, A. (2008) Chitosan and mint mixture: a new preservative for meat and meat products. *Food Chemistry* 107, 845–852.
Kaufmann, B. and Christen, P. (2002) Recent extraction techniques for natural products: microwave assisted extraction and pressurized solvent extraction. *Phytochemical Analysis* 12, 105–112.
Kline, R.M., Kline, J.J., Di Palma, J. and Barbero, G.J. (2001) Enteric-coated, pH-dependent peppermint oil capsules for the treatment of irritable bowel syndrome in children. *Journal of Pediatrics* 138, 125–128.
Kunnumakkara, A.B., Chung, J.-G., Koca, C. and Dey, S. (2009) Mint and its constituents. In: Aggarwal, B.B. and Kunnumakkara, A.B. (eds) *Molecular Targets and Therapeutic Uses of Spices: Modern Uses for Ancient Medicine*. World Scientific Publishing Co., Singapore, pp. 373–402.
Lima-Corrêa, R.A.B., Ribeiro, K.C., Freire, J.T. and Ferreira, M.C. (2014) A critical analysis about using of a thin layer drying concept to model the drying of leaves. In: Andrieu, J., Peczalski, R. and Vessot-Crastes, S. (eds) *Proceedings of the 19th International Drying Symposium (IDS 2014)*, Lyon, France, August 24–27. CD-ROM, EDP Sciences, Les Ulis, France, pp. 1–10.
Liu, J.P., Yang, M., Liu, Y.X., *et al.* (2006) Herbal medicines for treatment of irritable bowel syndrome. *Cochrane Database of Systematic Reviews* 1, CD004116.

Lyle, K.L. (2010) *The Complete Guide to Edible Wild Plants, Mushrooms, Fruits and Nuts: How to Find, Identify and Cook Them.* FalconGuides, Globe Pequot Press, Guilford, Connecticut.

Magge, S. and Lembo, A. (2011) Complementary and alternative medicine for the irritable bowel syndrome. *Gastroenterology Clinics* 40, 245–254.

Maia, J.T.L.S. (2007) Cultivation of aromatic and medicinal herbs intercropped with vegetables. M.Sc. dissertation, Graduating Program in Agricultural Sciences, Federal University of Minas Gerais, Montes Claros, Minas Gerais, Brazil. [In Portuguese.]

Martins, E.R., Castro, D.M., Castellani, D.C. and Dias, J.E. (2002) *Medicinal Plants*, 4th edn. Federal University of Viçosa, Minas Gerais, Brazil. [In Portuguese.]

Martins, A.P., Craveiro, A.A., Machado, M.I.L., Raffin, D.N., Moura, T.F., Novák, Cs. and Éhen, A. (2007) Preparation and characterization of *Mentha* × *villosa* Hudson oil–β-cyclodextrin complex. *Journal of Thermal Analysis and Calorimetry* 88, 363–371.

McKay, D.L. and Blumberg, J.B. (2006) A review of the bioactivity and potential health benefits of peppermint tea (*Mentha piperita* L.). *Phytotherapy Research* 20, 619–633.

Mimica-Dukic, N. and Bozin, B. (2008) *Mentha* L. species (Lamiaceae) as promising sources of bioactive secondary metabolites. *Current Pharmaceutical Design* 14, 3141–3150.

Oksman-Caldentey, K.-M. and Barz, W.H. (2002) *Plant Biotechnology and Transgenic Plants.* Marcel Dekker, New York.

Pääkkönen, K. (1999) Infrared drying of herbs. *Agricultural and Food Science in Finland* 8, 19–27.

Palmer, S. (2012) *The Plant-Powered Diet: The Lifelong Eating Plan for Achieving Optimal Health, Beginning Today*, 1st edn. Experiment LLC, New York.

Patel, T., Ishiuji, Y. and Yosipovitch, G. (2007) Menthol: a refreshing look at this ancient compound. *Journal of the American Academy of Dermatology* 53, 873–878.

Pegg, R.B. and Shahidi, F. (2007) Encapsulation, stabilization and controlled release of food ingredients. In: Shafiur Rahman, M. (ed.) *Handbook of Food Preservation*, 2nd edn. CRC Press, Boca Raton, Florida.

Pirbalouti, A.G., Amirkhosravi, A., Bordbar, F. and Hamedi, B. (2013) Diversity in the chemical composition of essential oils of *Ziziphora tenuior* as a potential source of pulegone. *Chemija* 24, 234–239.

Radünz, L.L. (2004) Effect of air temperature in content and composition of guaco (*Mikania glomerata* Sprengel) and regular mint (*Mentha* × *villosa* Huds) essential oils. MSc Dissertation, Graduating Program in Agricultural Engineering, Federal University of Viçosa, Minas Gerais, Brazil. [In Portuguese.]

Raghavan, S. (2006) *Handbook of Spices, Seasoning and Flavourings*, 2nd edn. CRC Press and Taylor & Francis Group, Boca Raton, Florida.

Rohloff, J., Dragland, S., Mordal, R. and Iversen, T.-H. (2005) Effect of harvest time and drying method on biomass production, essential oil yield, and quality of peppermint (*Mentha* × *piperita* L.). *Journal of Agricultural and Food Chemistry* 53, 4143–4148.

Rosanova, A.H. and Ferreira, M.C. (2014) Secagem de Hortelã em Secador de Cesto Rotativo [Drying of mint in rotating basket dryer]. In: XX Congresso Brasileiro de Engenharia Química (COBEQ 2014), CentroSul – *Florianópolis/SC, 19 a 22 de Outubro* [*Proceedings of XX Brazilian Conference on Chemical Engineering, October 19–22, Florianópolis, Santa Catarina, Brazil*]. *Blucher Proceedings* 6, Volume 1(2) February 2015. Available on CD-ROM [In Portuguese.] Full text available at: http://www.proceedings.blucher.com.br/article-details/secagem-de-hortel-em-secador-de-cesto-rotativo-17330 (accessed 20 January 2016).

Sallam, Y.I., Aly, M.H., Nassar, A.F. and Mohamed, E.A. (2015) Solar drying of whole mint plant under natural and forced convection. *Journal of Advanced Research* 6, 171–178.

Scott, R.P.W. (2005) Essential oils. In: Worsfold, P., Townshend, A. and Poole, C. (eds) *Encyclopedia of Analytical Science*, 2nd edn. Elsevier, Oxford, UK, pp. 554–561.

Sujana, P., Sridhar, T.M., Josthna, P. and Naidu, C.V. (2013) Antibacterial activity and phytochemical analysis of *Mentha piperita* L. (peppermint) – an important multipurpose medicinal plant. *American Journal of Plant Sciences* 4, 77–83.

Sulieman, A.M.E., Abdelrahman, S.E. and Rahim A.M.A. (2011) Phytochemical analysis of local spearmint (*Mentha spicata*) leaves and detection of the antimicrobial activity of its oil. *Journal of Microbiology Research* 1, 1–4.

Sústriková, A. and Salamon, I. (2004) Essential oil of peppermint (*Mentha* × *piperita* L.) from fields in eastern Slovakia. *Horticultural Science (Prague)* 31, 31–36.

Tarhan, S., Telci, I., Tuncay, M.T. and Polatci, H. (2010) Product quality and energy consumption when drying peppermint by rotary drum dryer. *Industrial Crops and Products* 32, 420–427.

Turek, C. and Stintzing, F.C. (2012) Impact of different storage conditions on the quality of selected essential oils. *Food Research International* 46, 341–353.

Turek, C. and Stintzing, F.C. (2013) Stability of essential oils: a review. *Comprehensive Reviews in Food Science and Food Safety* 12, 40–53.

Tzanetakis, I.E., Postman, J.D., Samad, A. and Martin, R.R. (2010) Mint viruses: beauty, stealth, and disease. *Plant Disease* 94, 4–12.

Ullah, N., Khurram, M., Amin, M.U., Afridi, H.H., Khan, F.A., Khayam, S.M.U., Ullah, S., Najeeb, U., Hussain, J. and Khan, M.A. (2011) Comparison of phytochemical constituents and antimicrobial activities of *Mentha spicata* from four northern districts of Khyber Pakhtunkhwa. *Journal of Applied Pharmaceutical Science* 7, 72–76.

Vankar, P.S. (2004) Essential oils and fragrances from natural sources. *Resonance – Journal of Science Education* 9, 30–41.

Venkateshappa, S.M. and Sreenath, K.P. (2013) Potential medicinal plants of Lamiaceae. *American International Journal of Research in Formal, Applied and Natural Sciences* 3, 82–87.

14 Moringa

Anthonia O. Oluduro,[1]* Dawn C.P. Ambrose,[2] Aregbesola Oladipupo Abiodun[1] and Alice L. Daunty[3]
[1]*Obafemi Awolowo University, Ile-Ife, Nigeria;* [2]*ICAR – Central Institute of Agricultural Engineering, Regional Centre, Coimbatore, India;* [3]*Mcrennet Foods, Chennai, Tamil Nadu, India*

14.1 Botany

14.1.1 Introduction

Moringa oleifera, also commonly called moringa or drumstick, is a cultivated tree crop belonging to the family, Moringaceae. It is known as the 'miracle tree' due to its various medicinal benefits. The name of the genus comes from 'murunggi' or 'muringa' in the Tamil and Malayalam languages. Other common names by which the tree is known are the horseradish tree and the ben oil tree. The tree is widely found in Asian countries such as India, Sri Lanka and the Philippines, and also in South America and African continent. In India, it is cultivated widely in the southern states and is drought resistant and rapid growing.

As well as the leaves, various other parts of the tree (the pods, seeds and bark) are used in various medicinal and industrial applications. The leaf yield of the tree is dependent on the season, with the highest yield obtained in the summer. In countries such as India, the leaves are harvested once in every 3 months if the tree is exclusively used for its foliage. The leaves of *M. oleifera* can be eaten as vegetable and also used as a nutritional supplement in the form of dried leaves and powder, which have greater shelf life (Fahey, 2005; Arabshahi *et al.*, 2007). After stripping from their stalks, the leaves are used as a vegetable in various recipes in India and Africa; they can also be used as a pot herb in the preparation of sauces (Price, 1985, rev. 2007). As they are a rich source of minerals, in many African countries, the leaves are used in combating malnutrition problems, and moringa leaf powder is used as a food supplement to combat malnutrition.

14.1.2 History

Historical writings reveal that ancient kings and queens used moringa leaves and fruit in their diets to maintain mental alertness and a healthy skin (Mahmood *et al.*, 2010). The moringa plant is native to northern India, where it was first described around 2000 BC as a plant with many medicinal values. The health benefits of moringa were discovered by the Greeks, who then introduced it to the Romans. As well as having been used by ancient civilizations, moringa has gained popularity across the globe in recent times.

*Corresponding author, e-mail: aoluduro2003@yahoo.co.uk

© CAB International 2016. *Leafy Medicinal Herbs: Botany, Chemistry, Postharvest Technology and Uses* (eds D.C.P. Ambrose *et al.*)

14.1.3 Location

Moringa is believed to be native to sub-Himalayan tracts of northern India but is now found worldwide in the tropics and subtropics. It is grown in the southern states of India as a commercial crop (see Fig. 14.1), and is also cultivated in Africa, South America, the Philippines and other regions/countries owing to its nutritional benefits. The trees require good sunlight for their growth, and they grow well at altitudes under 500 m and in any type of soil, although well-drained sandy or loamy soil with a pH of 6.3–7.0 is the ideal soil type. The presence of a long taproot makes the tree resistant to periods of drought. The trees can be easily grown from seed or from cuttings.

14.1.4 Morphology

Moringa is a fast-growing tree which can reach a maximum height of 7–12 m and a diameter of 20–40 cm at chest height (Foidl *et al.*, 2001). Moringa leaves are green and double or triple pinnate, 20–70 cm long, with the leaflets elliptical or obovate and 1–2 cm long. The leaves grow mostly at the branch tips and are greyish and downy when young. They have a long petiole with 8–10 pairs of pinnae each bearing two pairs of opposite leaflets. There are glands at the bases of the petioles and pinnae (Morton, 1991, in Foidl *et al.*, 2001). The leaves are arranged alternatively as compound leaves on the twigs. The twigs are green when fresh and turn brown when dried.

Fig. 14.1. A plantation of *Moringa oleifera*.

14.2 Chemistry

14.2.1 Chemical composition

The Trees for Life organization has reported that, ounce-for-ounce, moringa leaves contain more vitamin A than carrots, more calcium than milk, more iron than spinach, more vitamin C than oranges, and more potassium than bananas, and that their protein quality rivals that of milk and eggs. The nutrient comparison for fresh and dehydrated moringa leaves, with other common foods is presented in Table 14.1. The protein content of the leaves is high (20–35% on a dry weight basis), and its amino acid balance is unusual in plant foods. Foidl and Paul (2008) present somewhat different data on the amounts of nutrients present in 100 g of fresh moringa leaves (vitamin A, 7564 IU; vitamin C, 51.7 mg; calcium, 185 mg; and potassium, 337 mg), and according to Yameogo *et al.* (2011), moringa leaves contain 27.2% protein, 5.9% moisture, 17.1% fat, and 38.6% carbohydrates on a dry matter basis.

14.2.2 Phytochemistry

For the preparation of health drugs, knowledge is necessary of the phytochemicals present in medicinal plants. Moringa leaf extract is reported to contain phytosterols,

Table 14.1. Nutrient comparison of fresh and dried moringa leaves per 100 g compared with that of some common foods. From Fuglie, 1999, rev. 2001.

Nutrient	Fresh moringa leaves	Dried moringa leaves	Common foods
Calcium	440 mg	2003 mg	Milk 120 mg
Potassium	259 mg	1324 mg	Bananas 88 mg
Protein	6.7 g	27.1 g	Yoghurt 3.1 g
Vitamin A	6.8 mg	18.9 mg	Carrots 1.8 mg
Vitamin C	220 mg	17 mg	Oranges 30 mg

triterpenoids, flavonoids and saponins. Terpenoids and saponins are compounds that have been found suitable for treating diabetes, and alkaloids and saponins for controlling cardiovascular anomalies, while flavonoids have antioxidant properties. Glycosides have also been reported in the ethanolic extract of moringa leaves, and the leaves have also been found to contain sugars such as rhamnose in abundance.

14.3 Postharvest Technology

14.3.1 Processing

Moringa leaves can be used fresh as a minimally processed vegetable, but they can also be used as the dried leaves. The various operations involved in the production of the dried leaves are described below.

Stripping of leaves

Stripping is the primary processing of moringa leaves and is done manually. In African countries, the leaflets are removed from the tender stalks by human labour before the drying operation. Unwanted and yellowed leaves are discarded. However, in India, the harvested branches are field dried and the leaves are shed naturally (Fig. 14.2). Stripping before drying aids quicker and more uniform drying of the leaves with resulting better quality; it also reduces the volume of material that needs to be handled. A prototype moringa leaf stripper has been developed at the Central Institute of Agricultural Engineering (CIAE), Regional Centre, Coimbatore, India, for mechanically stripping the leaflets. The machine is operated by a 2 hp single phase electric motor and is capable of stripping 100 kg leaves in an hour (CIAE News, 2014).

Fig. 14.2. Field drying of *Moringa oleifera*.

Washing

For the preparation of minimally processed produce, washing is an important step. Water can reduce potential contamination by pathogenic microorganisms, but it can also transfer them, so the use of sanitizers is important for removing soil, debris and any pesticide residues, but especially in avoiding cross-contamination between clean and contaminated product. In a study of washing of moringa leaves, Mishra *et al.* (2012) washed the leaves in running water to remove dirt, and then soaked them in 1% saline solution for 5 min to remove microbes. Further washing was done with 70% ethanol and distilled water to remove the dust and pathogens present on the leaf surface. Washing studies were also conducted by Daunty (2014) who soaked and rinsed the fresh and mechanically stripped leaves in plain water and chlorinated water. The microbial load before and after washing was estimated. In the fresh leaves, the initial microbial load was too numerous to count at 10^{-5} dilution but at 10^{-6} dilution it was 212 $\times 10^7$ cfu. The total viable count for leaves washed with plain water was 165×10^7 at 10^{-5} dilution. Leaves washed with chlorinated water (100 ppm sodium hypochlorite) had the least microbial count compared with all other treatments, with 4×10^7 cfu at 10^{-5} dilution and no colonies at 10^{-6} dilution.

Packaging

Packaging minimizes postharvest losses by protecting against mechanical damage, microbes, pests, dust, air pollution and moisture loss by acting as a barrier between the product and the environment. The influence of different packaging materials on minimally processed drumstick leaves was studied during storage. Pretreated (cleaned) drumstick leaves were packaged in low density polyethylene (LDPE) and polypropylene materials of different thicknesses (150, 250 and 350) and stored under ambient (25 ± 2°C), and refrigerated (5 ± 2°C) conditions. The quality of the leaves was analysed during storage. The results showed that 350 gauge LDPE maintained leaf quality in terms of colour, β-carotene content, etc. and also with a smaller microbial load (Arun Kumar *et al.*, 2013).

Other packaging treatments have also been tested for extending the shelf life of fresh moringa leaves. Leaves packed in 40 μm thick polypropylene bags with a 1% vent remained fresh under ambient conditions for two days and could be stored for up to 14 days when packed in 80 μm LDPE bags with a 1% vent under refrigerated conditions, based on the colour as measured by a colour meter. Quality analysis of the packaged samples revealed that not much difference was observed between them except in ascorbic acid (vitamin C) content. Sensory evaluation studies also showed little difference in taste, colour, texture and overall acceptability between the stored samples and the fresh sample when cooked for consumption. The market results based on consumer acceptance were promising (Daunty, 2014).

Drying

Drying is one of the oldest methods of food preservation and represents a very important aspect of food processing. Traditionally in India, farmers dry harvest moringa stems in the field under the sun, which results in more microbes and dust on the leaves and is therefore an unhygienic practice. Sun drying also results in a loss of nutrients from the leaves and a deterioration in quality. Several studies on the drying of moringa leaves have been conducted.

Leaves were dried in a convective type dryer at a temperature range 50–80°C and a constant air velocity of 0.5 m/s. The sample dried at 60°C had a better colour than those dried at 50, 70 and 80°C. The chroma (colour difference from the control [fresh] leaves) of the samples dried at 50–80°C ranged from 4.9238 to 9.5258, and was the least for the sample dried at 60°C, which validated the sensory results that were obtained (Premi *et al.*, 2010). A study of the nutritive value of moringa leaves dried by different methods showed that nutrient retention was higher in shade-dried than sun and oven dried leaves. (Joshi and Mehta, 2010).

Arun Prabhu *et al.* (2011) reported that dehydrating moringa leaves by oven, wind and sun drying yielded more or less similar results when used in combination with germicide, salt and turmeric treatments. The process of moisture removal prevents a favourable environment for the growth of microbes, and so the dried product can be stored for a longer period of time. Preservation techniques involving dehydration could be effectively used to extend the shelf life of moringa leaves without altering their nutritional value.

Gyamfi *et al.* (2011) dried moringa leaves by air, freeze and oven drying. The mineral content was more in the powdered samples obtained by freeze drying, followed by air drying and oven drying. However, air drying could be the preferred drying method as it is better economically and the air-dried powder also recorded higher concentrations of elements.

In another study, moringa leaf powder was prepared by sun, shade, cabinet and oven drying. The air temperature for cabinet and oven drying was fixed at 60°C. The study showed that cabinet drying allowed superior retention of nutrients compared with the other drying methods (Satwase *et al.*, 2012).

Ambrose *et al.* (2013) attempted to upgrade the drying method to produce a better quality end product, by studying the effect of different drying methods on the colour of the product. Harvested moringa leaves (PKM–I variety) were stripped, washed and dried in a polyhouse dryer or a forced flow mechanical dryer (CIAE-IEP model), and under sun and shade for comparison. It took 6 h to dry from an initial moisture content of 85% (wet basis, w.b) to 5% (w.b) in the polyhouse dryer, 3 h in the mechanical dryer, a day under sun drying and 2 days under shade drying. The colour scores measured using a HunterLab colour spectrophotometer indicated that mechanical drying, followed by polyhouse drying, were superior to shade and sun drying.

In another study by Ambrose *et al.* (2015), fresh moringa leaves (PKM-1) were mechanically stripped and dried by sun drying and by solar, vacuum and cabinet drying (at 45, 50, 55 and 60°C). The drying characteristics of the leaves showed that most drying took place in the falling rate period, and that drying rate increased with drying time. Various thin layer drying models were fitted to the experimental data for stripped and dried leaves. Among them, the Wang and Singh model gave the best fit with the highest R^2 and least χ^2 values. Cabinet drying at 50°C resulted in better quality in terms of colour and nutrient content.

14.3.2 Value addition

Dried moringa leaves can be further processed as a value added product in the form of powder. When the leaves are plentiful, drying and pulverizing them as moringa leaf powder provides an easy storage method. The powder can be consumed by adding to food preparations as a supplement. In India, it has a great export potential, and moringa leaf processing is carried out in Coimbatore and Erode districts of Tamil Nadu by small-scale processors. The leaves are exported in dried form as powder, capsules, tablets, moringa tea leaf etc., and are used as health supplements (Ambrose *et al.*, 2014).

In African countries, moringa leaves are used as a flavorant, and can be added to meat preparations (Teye *et al.*, 2013).

Salem *et al.* (2013) investigated the nutritional benefits of dried moringa leaves added to Labneh cheese, and found that this led to an increase in the vitamin and mineral contents. Sensory studies revealed that the product was acceptable to consumers. Dried moringa leaves can also be added to buttermilk to improve its nutritional value. Amala, a product made from yam flour, can be fortified by the addition of 2.5% moringa leaf powder to the flour in order to improve its nutritional value (Karim *et al.*, 2013).

14.4 Uses

14.4.1 General uses

Both fresh and dried moringa leaves are used in food preparations in many countries. Young leaves of the plant are consumed as cooked vegetable, and the leaflets can be used in any vegetable recipe. The fresh leaves can be used in the preparation of soups and salads. The dried leaves can be used as tea and also as a pot herb in the preparation of soup and porridge. After grinding, dried leaves store well for a long time and can be used as a flavorant or as a health supplement, for which use there are also capsule and tablet preparations. Moringa leaves are also considered to be a panacea for malnutrition (Ambrose and Daunty, 2015) because of their rich nutrient content. Apart from being used as a food, the leaves are also fed as mulch to animals to increase their milk yield. They also assist nitrogen fixation in the soil when used as a manure.

14.4.2 Pharmacological uses

The moringa tree is a valuable crop due to its various health benefits. The leaves, being rich in iron, are used in childhood malnutrition in children, and also as a healthy iron substitute in lactating and pregnant women. The tree is known as 'mother's milk' in the African continent, as it increases milk production in lactating women.

Traditional health practitioners have suggested that tumours could be cured by moringa leaves. Studies on the effect of moringa leaf extracts and fractions therefrom on a cancerous cell line and mouse cancer model showed that they had anticancer properties towards the cells and increased the mouse survival rate. The anticancer fraction was characterized (Krishnamurthy *et al.*, 2015).

The bioactive phytochemical β-sitosterol found in aqueous extracts of *Moringa* leaves has been found to have anti-ulcer properties, and Marcu (2005) indicated that moringa has the following health benefits: it reduces cholesterol levels and triglycerides ('bad' fats); it controls blood sugar and helps normal sugar and energy balance; it provides vitamins and minerals that are necessary for maintaining normal physiology; and it offers powerful anti-ageing and anti-inflammatory substances, many with anticancer properties. It has also been reported that the antioxidants present in moringa leaves aid in the treatment of oxidative stress, liver ailments and stomach disorders. In addition, moringa is effective against nervous disorders, including headaches, migraines, hysteria, and epilepsy (Richardson, 2009, cited in Bey (2010).

14.5 Summary

The overall aspects of moringa in terms of its nutritional benefits, processing and uses have been reported elsewhere, and it is regarded as the 'miracle tree' known for its nutritional benefits and grown at a commercial scale in many countries. It is also grown in the kitchen garden for household purposes in countries such as India. The leaves, which are high in nutrient content can be processed either as a fresh product or as dehydrated/dried leaves and powder. Mechanical processing of the leaves results in better quality and more hygienic products than does conventional manual processing. Moringa is considered as a crop for combating malnutrition in many African countries.

References

Ambrose, D.C.P. and Daunty, A.L. (2015) *Moringa oleifera* mother's best friend – an ideal health food. *Food and Beverage World* 42(2), 33–34.

Ambrose, D.C.P., Naik, R. and Annamalai, S.J.K. (2013) A comparative study on the different drying methods of *Moringa oleifera* leaves for powder production. In: Sasikumar, B., Dinesh, R., Prasath, D., Biju, C.N. and Srinavasan, V. (eds) *National Symposium on Spices and Aromatic Crops, SYMSAC-VII, Postharvest Processing of Spices and Fruit Crops, 27–29 November, 2013, Madikeri, Karnataka. Souvenir and Abstracts.* Indian Society for Spices, Kozhikode, Kerala, India, p. 241. Available at: http://www.indianspicesociety.in/iss/pdf/SYMSAC%20VII%20-%20Souvenir.pdf (accessed 21 January 2016).

Ambrose, D.C.P., Naik, R. and Daunty, A.L. (2014) *Moringa oleifera* leaves – an excellent health food. In: Rajendran, A. (ed.) *The National Seminar on Emerging Trends in Biodiversity Conservation & Sustainable Utilisation (NSBCS-2014), held at Bharathiar University, Coimbatore, on 29–30 January 2014.* Bharathiar University, Coimbatore, India, pp. 96–97.

Ambrose, D.C.P., Naik, R. and Daunty, A.L. (2015) Investigation on the thin layer drying characteristics of *Moringa oleifera* leaves. In: *National Seminar on Emerging Solutions for Sustainable Agriculture and Food Processing, Punjab Agricultural University, Punjab, India, 23–25 February 2015.* Indian Society of Agricultural Engineers, New Delhi, p. 31.

Arun Kumar, P., Nirmala, R., Bhavya, E.P. (2013) Effect of different packaging and storage conditions on shelf-life of processed drumstick leaves. *International Journal of Agricultural Engineering* 6, 28–31.

Arabshahi, D.S., Devi, D.V. and Urooj, A. (2007) Evaluation of antioxidant activity of some plant extracts and their heat, pH and storage stability. *Food Chemistry* 100, 1100–1105.

Arun Prabhu, R., Rajan, A.P. and Santhalia, S. (2011) Comparative analysis of preservation techniques on *Moringa oleifera. International Journal of Agricultural and Food Science* 1, 12–22.

Bey, H.H. (2010) All Things Moringa. The Story of an Amazing Tree of Life. Published by www.allthingsmoringa.com. Available at: http://www.remediosnaturales.es/wp-content/uploads/2014/12/eBook-moringa-ingles.pdf (accessed 21 January 2016).

CIAE News (2014) Mechanical processing of *Moringa oleifera* leaves. *CIAE News* 23(2), 2.

Daunty, A.L. (2014) Investigation on mechanical processing of *Moringa oleifera* leaves in its fresh and dried form. Unpublished M. Tech. (Food Processing and Engineering) thesis, Karunya University, Coimbatore, India.

Fahey, J.W. (2005) *Moringa oleifera:* a review of the medical evidence for its nutritional, therapeutic, and prophylactic properties. Part 1. *Trees for Life Journal* 1:5. Available at: http://www.tfljournal.org/article.php/20051201124931586 (accessed 21 January 2016).

Foidl, N. and Paul, R. (2008) *Moringa oleifera.* In: Janick, J. and Paull, R.E. (eds) *The Encyclopedia of Fruit & Nuts.* CAB International, Wallingford, UK, pp. 509–512.

Foidl, N., Makkar, H.P.S and Becker, K. (2001) The potential of *Moringa oleifera* for agricultural and industrial uses. Available at: http://miracletrees.org/moringa-doc/the_potential_of_moringa_oleifera_for_agricultural_and_industrial_uses.pdf (accessed 21 January 2016).

Fuglie, L.J. (ed.) (1999) *The Miracle Tree:* Moringa oleifera*: Natural Nutrition for the Tropics.* Church World Service, Dakar. Data used in Table 14.1 available at: http://www.moringatree.co.za/analysis.html (accessed 21 January 2016).

Gyamfi, E.T., Kwarteng, I.K., Ansah, M.O., Anim, A.K., Ackah, M., Kpattah, L. and Bentil, N.O. (2011) Effects of processing on *Moringa oleifera. Proceedings of the International Academy of Ecology and Environmental Sciences* 1, 179–185.

Joshi, P. and Mehta, D. (2010) Effect of dehydration on the nutritive value of drumstick leaves. *Journal of Metabolomics and Systems Biology* 1(1), 5–9.

Karim, O.R., Kayode, R.M., Oyeyinka, S.A. and Oyeyinka, A.T. (2013) Proximate, mineral and sensory qualities of 'amala' prepared from yam flour fortified with moringa leaf powder. *Food Science and Quality Management* 12(2), 10–22.

Krishnamurthy, P.T., Vardarajalu, A., Wadhwani, A. and Patel, V. (2015) Identification and characterization of a potent anticancer fraction from the leaf extracts of *Moringa oleifera* L. *Indian Journal of Experimental Biology* 53, 98–103.

Mahmood, K.T., Mugal, T. and Ul Haq, I. (2010) *Moringa oleifera*: a natural gift – a review. *Journal of Pharmaceutical Science and Research* 2, 775–781.

Marcu, M.G. (2005) *Miracle Tree.* KOS Health Publications, La Canada, California.

Mishra, S.P., Singh, P. and Singh, S. (2012) Processing of *Moringa oleifera* for human consumption. *Bulletin of Environment, Pharmacology and Life Sciences* 2(1), 28–31.

Morton, J.F. (1991) The horseradish tree, *Moringa pterygosperma* (Moringaceae) – a boon to arid lands. *Economic Botany* 45, 318–333.

Premi, M., Sharma, H.K., Sarkar, B.C. and Singh, C. (2010) Kinetics of drumstick leaves (*Moringa oleifera*) during convective drying. *African Journal of Plant Science* 4, 391–400.

Price, M.L. (1985, rev. 2007) *The Moringa Tree.* ECHO Technical Note, ECHO Community, North Fort Myers, Florida. Revised version (2007) available at: http://miracletrees.org/moringa-doc/ebook_moringa.pdf (accessed 21 January 2016).

Richardson, A. (2009) Moringa oleifera – *Food, Medicine and Forage Crop.* Cited in Bey, H.H. (2010) *All Things on* Moringa. p. 23. Available at: www.allthingsmoringa.com (accessed 10 June 2016).

Salem, A.S., Salama, W.M., Hassanein, A.M. and El Ghandour, H.M.A. (2013) Enhancement of nutritional and biological values of Labneh by adding dry leaves of *Moringa oleifera* as innovative dairy products. *World Applied Sciences Journal* 22, 1594–1602.

Satwase, A.N., Pandhre, G.R., Sirsat, P.G. and Wade, Y.R. (2012) Studies on drying characteristic and nutritional composition of drumstick leaves by using sun, shadow, cabinet and oven drying methods. *Open Access Scientific Reports* 2, 584. Available at: http://www.omicsonline.org/scientific-reports/srep584.digital/srep584.html (accessed 21 January 2016).

Teye, G.A., Baffoe, F. and Teye, M. (2013) Effects of moringa (*Moringa oleifera*) leaf powder and dawadawa (*Parkia biglobosa*), on sensory characteristics and nutritional quality of frankfurter-type sausages – a preliminary study. *Global Advanced Research Journal of Agricultural Science* 2(1), 29–33.

Yameogo, C.W., Bengaly, M.D., Savadogo, A., Nikiema, P.A. and Traore, S.A. (2011) Determination of chemical composition and nutritional values *Moringa oleifera* leaves. *Pakistan Journal of Nutrition* 10, 264–268.

15 Oregano

K. Hüsnü Can Başer[1,2]* and Neşet Arslan[3]
[1]*Anadolu University, Eskişehir, Turkey;* [2]*Near East University, Nicosia, N. Cyprus;* [3]*Ankara University, Ankara, Turkey*

15.1 Botany

15.1.1 Introduction

The family Lamiaceae is composed of annual or perennial plants that are herbs or shrubs and are distributed mainly in the northern hemisphere and especially in the Mediterranean region. Their stems are generally square in cross section. The botanical description is as follows: Leaves simple or lobed, opposite, each pair at right angles to the previous one (decussate). Flowers bisexual and zygomorphic, emerging from the bottom of bracts, in dense clusters and verticillastrate. Bracts similar to leaves. Calyx 5 toothed, campanulate or tubular. Corolla tubular at base, bilabiate above. Upper lip 2, lower lip 3 toothed. Stamen 4, 2 fertile and 2 sterile; 2 with long and 2 with short filaments. Ovary superior, 2-celled and each cell 2-ovuled, style subterminal or ovary 4-parted and each lobe 1-ovuled and style gynobasic with 2-cleft apex. Fruits schizocarp usually with 4 dry nutlets (Ietswaart, 1982; Baytop, 1999; Zeybek and Zeybek, 2002; Fakılı, 2010).

Origanum is the most important genus in the Lamiaceae family. The genera *Thymus*, *Thymbra*, *Coridothymus* and *Satureja* are close to *Origanum*, and all five are commonly called 'kekik', and are used and traded similarly. In Turkey, 78 species, of which 39 are endemic, belong to these five genera (Table 15.1). They all smell similar owing to the presence of thymol and/or carvacrol in their essential oils.

The most important of the *Thymus* species is the cultivated *T. vulgaris*, which is not a native species in Turkey. However, *T. longicaulis*, *T. eigii*, *T. kotschyanus*, *T. pulvinatus* and *T. praecox* are native and are used locally or traded.

Of the *Satureja* species, *S. hortensis* (Bahçe sateri, Bakla kekiği) is cultivated and traded as a fresh herb for use in salads and soups. *S. cuneifolia*, *S. wiedemannia*, *S. thymbra*, *S. macrantha* and *S. spicigera* are collected from the wild and traded.

Of the *Coridothymus* species, *C. capitatus* (Spanish oregano, timari) is another traded oregano species. Among the *Thymbra* species, *T. spicata* (Karabaş kekiği, zahter) is wildcrafted and locally traded and in Turkey it ranks second to *Origanum* species (Başer *et al.*, 1993; Arslan, 1994; Ceylan, 1996; Arslan, 2002; Başer, 2002b; Sarı and Oğuz, 2002; Bayram, 2003).

*Corresponding author, e-mail: khcbaser@gmail.com

© CAB International 2016. *Leafy Medicinal Herbs: Botany, Chemistry, Postharvest Technology and Uses* (eds D.C.P. Ambrose *et al.*)

Table 15.1. Number of kekik[a] species in Turkey. From (Ietswaart, 1982; Başer *et al.*, 1993; Baytop, 1999; Arslan, 2002; Başer, 2002a,b; Zeybek and Zeybek, 2002; Sadıkoğlu, 2005; Fakılı, 2010.

Genera of kekik	Endemic species	Total species
Thymus	20	38
Origanum	15	23
Satureja	4	14
Thymbra	–	2
Coridothymus	–	1
Total species in all genera	39	78

[a]The name kekik includes all of the genera listed in the table.

Worldwide, *Thymus* (thyme), *Origanum* (oregano) and *Satureja* (savory) species are separately traded as commercial commodities; *C. capitatus* is known in world trade as Spanish oregano. *Thymbra* is not recognized in the world trade, but is used locally and traded under the name of wild thyme. This chapter places the greatest emphasis on *Origanum* species, which are referred to as either oregano or kekik.

Classification

The genus *Origanum* belongs to the family Lamiaceae. *Origanum* species are treated under ten sections: *Amaracus* (Gleditsch) Bentham, *Anatolicon* Bentham, *Brevifilamentum* Ietsw., *Longitubus* Ietsw., *Chilocalyx* (Briquet) Ietsw., *Majorana* (Miller) Bentham (*O. onites* L., *O. majorana* L.), *Campanulaticalyx* Ietsw., *Elongatispica* Ietsw., *Origanum* (*O. vulgare* L.) and *Prolaticorolla* Ietsw. Some 56 species have been characterized in these sections in the world. The most important species (including subspecies) are *O. vulgare*, *O. onites* and *O. majorana*. In Turkey, important species are *O. onites* (Izmir kekiği, Bilyalı kekik, Turkish oregano), *O. vulgare* L. subsp. *hirtum* (Link) Ietsw. (Istanbul kekiği, Kara kekik, Greek oregano) and *O. dubium* Boiss. (syn. *O. majorana* L. – Turkish type) (Alanya kekiği, beyaz kekik, yağ kekiği). *O. vulgare* is cultivated worldwide. It has six subspecies, namely, *O. vulgare* L. subsp. *glandulosum* (Desf.) Ietsw., *O. vulgare* subsp. *gracile* (K. Koch) Ietsw., *O. vulgare* subsp. *hirtum* (Link) Ietsw., *O. vulgare* L. subsp. *virens* (Hoffmanns. & Link) Ietsw., *O. vulgare* L. subsp. *viridulum* (Martrin-Donos) Nyman and *O. vulgare* L. subsp. *vulgare*. Four of these subspecies grow wild in Turkey and the most important of them commercially is *O. vulgare* subsp. *hirtum*.

Other commercially important *Origanum* species in Turkey are *O. minutiflorum* Schwartz et Davis (Tota kekiği, Toka kekiği, Yayla kekiği, Sütçüler Kekiği), an endemic species, and *O. syriacum* L. subsp. *bevanii* (Holmes) Greuter et Burdet (Suriye Kekiği, Tarsus kekiği, zahter). Some other *Origanum* species are also locally consumed as herbal tea or condiment. However, here we shall concentrate more on *O. onites*. It should also be noted here that *Lippia graveolens* Kunth. (Verbenaceae) is known in world trade as Mexican oregano (Ietswaart, 1982; Davis, 1988; Başer, 1993; Arslan, 2002; Başer, 2002b; Sadıkoğlu, 2005; Anonymous, 2011).

15.1.2 Origin

Most *Origanum* L. species are mainly distributed in the Mediterranean region, and being locally endemic naturally grow in countries neighbouring the Mediterranean sea, such as Turkey, Greece, Morocco, Algeria, Tunisia, Spain, Italy, Croatia, Albania, Cyprus, Syria, Lebanon, Israel, Egypt and Palestine. Some 75% of all *Origanum* species occur only in the eastern Mediterranean region (Fig. 15.1).

15.1.3 Location

The economically important *O. vulgare* is the most widely distributed *Origanum* species, growing naturally in Portugal, Spain, Albania, Greece, Croatia, Turkey, Iran, India, China, Afghanistan, Russia, Pakistan, Morocco, the Azores islands and the Canary islands. It has six subspecies and has been introduced to many countries of the world. It is the most widely distributed species in Turkey, and four subspecies occur in Turkey. These are: *O. vulgare* subsp. *gracile*, subsp. *hirtum* (Fig. 15.2), subsp. *viridulum* and subsp.

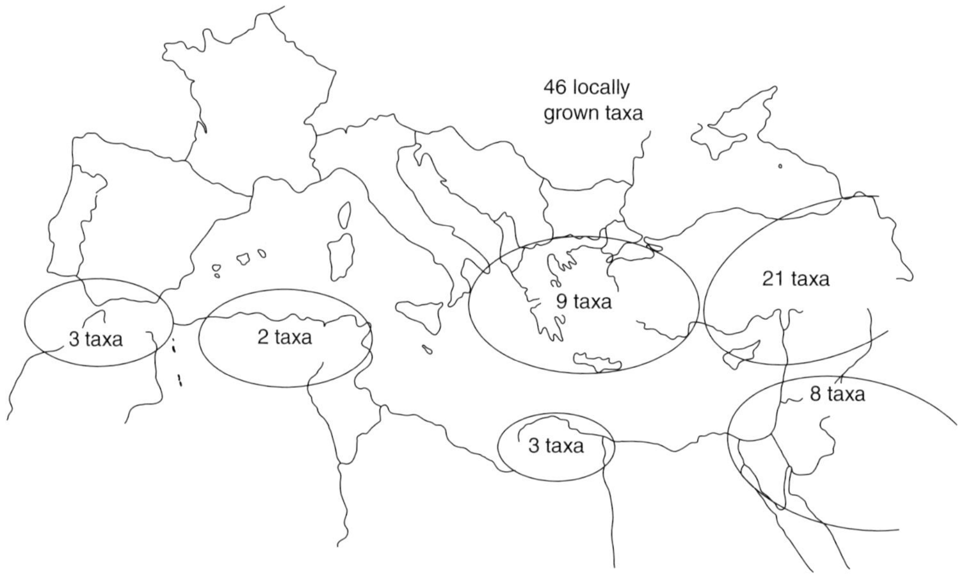

Fig. 15.1. Distribution of locally grown *Origanum* taxa in different Mediterranean countries/regions (Kokkini, 1996).

Fig. 15.2. Oregano – *Origanum vulgare* ssp. *hirtum*. Seedlings. Photo courtesy of Muzaffer Özdemir

vulgare. *O. onites*, in contrast, is distributed in Turkey, Greece, the Aegean islands and Sicily. Some *Origanum* species grow naturally outside the Mediterranean area (Ietswaart, 1982; Başer, 1993; Ceylan, 1996; Kokkini, 1996; Arslan, 2002).

15.1.4 Morphology

Origanum species are perennial and are usually 30–60 cm tall or, in some species, above 100 cm. The stems are four angled as in other Lamiaceae species. The plants are generally branched and the branches are somewhat hairy. The leaves are sessile in a couple of species and with stalks in others. The lamina is diverse in shape and size; while in some species its width and length are similar, in others its length is 2–2.5× greater than its width. The leaves have two types of glandular trichomes: sessile and stalked. These are also present on the stem, bract, calyx and corolla. The leaves are pale bluish green and leathery. The colour is due to the waxy cuticle covering the surface (Ietswaart, 1982; Sadıkoğlu, 2005; Fakılı, 2010). Generally, each stem and branch has a spike and these are of different shapes and sizes. The inflorescence is paniculate. The bracts are generally round or oval and of different sizes (Sadıkoğlu, 2005). The calyces are variable. The corolla is two toothed, the upper lip emarginate or with two short teeth, and the lower lip three toothed. There are four stamens. The style is quite variable in shape and size. The nutlet is small, ovoid and brown. The weight of 1000 seeds is 0.1–0.2 g (Ietswaart, 1982; Sadıkoğlu, 2005; Marquard and Malko, 2006; Fakılı, 2010).

15.2 Agronomy

15.2.1 Soil requirements

Origanum species grow in almost every type of soil, including dry, stony, lime and slightly alkaline soils, but they prefer loamy, alluvial, permeable, fast warming and humus-rich soils. Moist, cold, sandy, hard and lumpy soils are not suitable. However, some of the species do not follow these criteria, and can thrive in lime-rich and lime-poor soils. The soil pH should be neutral or slightly alkaline (Ceylan, 1996; Arslan, 2002; Sarı and Oğuz, 2002; Bayram, 2003).

Soils from natural growing areas of *O. onites* showed the following features: pH 6.03–7.99, sandy, limey, saltless or slightly salty, rich in organic matter, nitrogen and phosphorus and poor in potassium (Gönüz and Özgörücü, 1999).

15.2.2 Climatic requirements

As a Mediterranean plant, *Origanum* species grow in all areas under the Mediterranean influence or in neighbouring areas. They thrive in the temperate and subtropical belts of Europe and Asia, and can be found on sunny hills, open areas and roadsides in central Europe. Day length, light density, vegetation period, temperature and precipitation are important climatic factors for oregano, and it likes direct sunlight and is resistant to drought conditions.

O. vulgare is more resistant to cold than other species and *O. onites* is more sensitive to cold. Cold-resistant types are also available. *O. onites* shows seasonal dimorphism, and forms two basic leaf types that enable it to withstand extreme dry summer months. The plant develops small and short leaves in summertime, and bigger leaves in other months – an adaptation to changing environmental conditions (Ceylan, 1996; Arslan, 2002; Sarı and Oğuz, 2002; Bayram, 2003; Marquard and Malko, 2006).

15.2.3 Production

O. onites is the most widely cultivated oregano species in Turkey, although the cultivation of *O. vulgare* subsp. *hirtum* has just started. The domestication of some other *Origanum* species is ongoing. The cultivation of *O. onites* in the Izmir area started in 1990 as an alternative crop to the ailing tobacco farming industry and has become successful (Table 15.2). Cultivated areas reached 9428 ha in 2012, with a crop yield of 11,598 tons. Although oregano is cultivated in 19

Table 15.2. Annual production and export of oregano (kekik) in Turkey.[a] From TUIK, 2014.

				Export	
Year	Cultivation (ha)	Production (ton)	Yield (kg/da)	Amount (kg)	Revenue (US$)
1995	–	–	–	5,501,000	13,690,000
1999	–	–	–	7,640,000	16,556,000
2003	–	–	–	8,783,887	14,807,927
2004	5250.0	7,000	133	9,759,878	17,411,576
2005	4700.0	6,400	136	10,661,968	18,342,331
2006	5885.3	7,979	136	12,262,814	23,486,761
2007	6075.1	5,350	88	11,242,264	41,032,769
2008	8413.3	10,082	120	9,999,780	44,786,064
2009	8495.7	12,329	145	11,339,670	29,173,121
2010	8535.1	11,190	131	12,923,995	28,992,794
2011	7770.7	10953	141	13,261,353	30,903,286
2012	9428.3	11,598	123	14,003,696	40,582,079
2013	8913.7	13,658	153	14,813,000	56,324,000

[a]Production records have been kept only since 2004.

provinces, Denizli alone comprises 92% of the cultivation areas and is responsible for 86% of the production (TUIK, 2014).

O. vulgare and *O. majorana* (sweet marjoram, mercanköşk) are cultivated worldwide.

15.2.4 Cultivars

Domestication studies on *O. onites* started in 1970s at the Faculty of Agriculture of Ege University and two cultivars were eventually registered as 'Ceylan-2002' and 'Tayşi-2002'. They were developed by selective breeding and cloning techniques, but have not yet been used in cultivation. There are several cultivars of *O. vulgare* and *O. majorana* in Turkey and abroad. Two new cultivars of *O. vulgare* subsp. *hirtum*, called 'Tınmaz' and 'Başer', are in the process of registration at the Turkish Ministry of Food, Agriculture and Livestock.

15.2.5 Harvesting

No crop is expected in the first year of cultivation. However, if seedlings have been transplanted early in the field and show good development, one harvest may be possible. Harvest time is important for the quality of the crop. Harvesting is appropriate when 50% of the plants are flowering. Leaf fall is also a determinant of harvest time, and if the leaves start drying early, an early harvest is recommended. Depending on the climatic and maintenance conditions, two to six harvests a year may be possible. In non-irrigated fields, one harvest is made. For *O. onites* in irrigated fields, generally three harvests are possible, the first in mid-May, the second at the end of July and the third in mid-October. Depending on the extent of the field, harvesting is done manually, or by using a scythe, lawnmower or a special machine designed to harvest kekik. Harvesting height must be 10 cm above ground. Mid-September is suitable for seed harvesting (Ceylan, 1996; Bayram *et al.*, 1999; Arslan, 2002; Sarı and Oğuz, 2002; Marquard and Malko; 2006).

15.3 Postharvest Technology

15.3.1 Drying

Harvested kekik plants (kekiks) are dried. Stack height must not exceed 35–40 cm during drying, and stacks must be turned over frequently to enable good drying. Otherwise, darkening or rotting may occur. Artificial drying over a short period is ideal for quality, and is performed by blowing hot air not exceeding 35–40°C over the plants. In Turkey, both wildcrafted and cultivated kekiks are sun dried. The dried kekiks are beaten with wooden sticks or other such implements to separate the leaves from the stems. Sieving is then performed to complete the process. The fresh herbage yield is 1500–3000 kg/da (decare) from the second year onward. If the crop is well maintained, the dried herbage yield may go up to 800 kg/da; otherwise, it may go down to 150–200 kg/da; 3.5–4.5 kg fresh kekik yields 1 kg of dried kekik. The seed yield is 28–30 kg/da. Depending on the crop and the essential oil yield, 2–5 kg/da or more of essential oil can be obtained (Arslan, 1994, 2002; Ceylan, 1996; Sarı and Oğuz, 2002; Bayram, 2003; Marquard and Malko, 2006).

15.3.2 Value addition

Oregano oil, oregano water and oregano tea have been used for their regulatory properties on the gastrointestinal system, and for their antitussive and antimicrobial properties on the respiratory system. There is a monograph 'Origani herba' (Oregano herb), in the European Pharmacopoeia (EDQM, 2014) that requires the use of dried herbs of *O. onites* and *O. vulgare* subsp. *hirtum*, or a mixture of these) containing 2.5% essential oil with at least 60% carvacrol and thymol (which are found in the essential oil of oregano). As 80% of the immune system is regulated by the small intestines, oregano and carvacrol can be considered useful in immune system-related disorders (Furness *et al.*, 1999).

15.4 Uses

15.4.1 General uses

Kekik has been used as a spice or condiment in salads, soups, pizzas and other foods and food products. As noted above, it is a source of essential oil that contains carvacrol and thymol – two isomeric monoterpenoid phenols, as main constituents. These compounds have been shown to be antimicrobial and antioxidant. Many other activities of these constituents and of the essential oils that contain them have been reported (Başer, 2008).

15.4.2 Pharmacological uses

Oregano is mainly used for gastrointestinal system disorders in Turkish folk medicine. The following biological activities of carvacrol and carvacrol-bearing essential oils have been proven:

- activity against dental caries and periodontal disease (Botelho *et al.*, 2007, 2008);
- analgesic, antinociceptive, antiangiogenic and anti-inflammatory activities (Wagner *et al.*, 1986; Cingi *et al.*,1992; Aydin *et al.*, 1996; Demirci *et al.*, 2004; Xu *et al.*, 2006; Landa *et al.*, 2009; Ocana-Fuentes *et al.*, 2010; Yamada *et al.*, 2010; Doerner *et al.*, 2011; Joca *et al.*, 2012; Guimaraes *et al.*, 2012a,b; Liu *et al.*, 2012; Silva *et al.*, 2012; Spiering *et al.*, 2012; Boskabady and Jalali, 2013; Lima *et al.*, 2013; Pahlavan *et al.*, 2013);
- anticancer activity (Zeytinoglu *et al.*, 1998, 2003; Ipek *et al.*, 2003, 2005; Koparal and Zeytinoglu, 2003; Andersen, 2006; Karkabounas *et al.*, 2006; Mezzoug *et al.*, 2007; Ozkan and Erdogan, 2011; Aydin *et al.*, 2014; Subramaniyan *et al.*, 2014);
- anticonvulsant activity (Quintans-Junior *et al.*, 2010);
- antidepressant activity (Melo *et al.*, 2011);
- antidiabetic activity (Deng *et al.*,2013; Gul *et al.*, 2013; Ezhumalai *et al.*, 2014; Hajializadeh *et al.*, 2014; Hyun *et al.*, 2014);
- anti-elastase properties (Kacem and Meraihi, 2006);
- anti-gnawing properties against rodents (Ahn *et al.*, 1995);
- anti-genotoxic, antimutagenic and DNA-protective activities (Ipek *et al.*, 2005; Aydin *et al.*, 2005a,b; Horvathova *et al.*, 2007; Slamenova *et al.*, 2007; Carolina Vicuna *et al.*, 2010);
- antihistamine activity (Boskabady *et al.*, 2012);
- antihypercholesterolaemic activity (Case *et al.*, 1995; Ozdemir *et al.*, 2008; Akkol *et al.*, 2009);
- anti-lipoxygenase activity (Albano *et al.*, 2012);
- antimicrobial activity (Rideal *et al.*, 1930; Martin-Smith and Khatoon, 1963; Sticher, 1977; Knobloch *et al.*, 1986; Helander *et al.*, 1998; Ultee *et al.*, 1999, 2002; Lambert *et al.*, 2001; Pauli, 2001; Baricevic *et al.*, 2002; Burt *et al.*, 2005, 2007; Kristinsson *et al.*, 2005; Bayramoglu *et al.*, 2006; Ben Arfa *et al.*, 2006; Demirci *et al.*, 2006; Di Pasqua *et al.*, 2006; Friedman *et al.*, 2006; Gill and Holley, 2006a,b; Neri *et al.*, 2006; Nostro *et al.*, 2006; Soylu *et al.*, 2006; Veldhuizen *et al.*, 2006; Botelho *et al.*, 2007; Guilen *et al.*, 2007; Lopez *et al.*, 2007; Martinez-Romero *et al.*, 2007; Bakkali *et al.*, 2008; Nowotarska *et al.*, 2014; Zabka *et al.*, 2014; Zanini *et al.*, 2014);
- antioxidant and free-radical scavenging activity (Prieto *et al.*, 2007; Kosar *et al.*, 2008; Ozkan *et al.*, 2010; Yin *et al.*, 2012; Alinkina *et al.*, 2013; Inanc and Maskan, 2013; Slamenova *et al.*, 2013; Carbone-Howell *et al.*, 2014; Yousefzadi *et al.*, 2014);
- anti-parasitic activity against, e.g. protozoa, *Schistosoma*, *Leishmania*, *Trypanosoma*, *Plasmodium* (Force *et al.*, 2000; Tampieri *et al.*, 2003; Morillas Marquez *et al.*, 2004; Tasdemir *et al.*, 2006; Hussain *et al.*, 2011; Farias-Junior *et al.*, 2012);
- anti-platelet activity (Son *et al.*, 2005);
- antispasmodic activity (Van Den Boucke and Limli, 1980, 1982; Goze *et al.*, 2010);
- antitussive and bronchodilatory activities (Van Den Broucke and Lemli,

1981; Boskabady and Jandaghi, 2003; Boskabady *et al.*, 2003, 2005; Hudaib and Aburjai, 2007);

- antiviral activity (Dunkic *et al.*, 2010; Pilau *et al.*, 2011; Santoyo *et al.*, 2014);
- cardioprotective activity (Magyar *et al.*, 2004; Earley *et al.*, 2010; Santos *et al.*, 2011; Yu *et al.*, 2013);
- cognitive-enhancing activity (Azizi *et al.*, 2012);
- deodorant activity (Varel and Miller, 2001a,b; Varel, 2002; Varel *et al.*, 2004, 2006);
- effectiveness in honey bee diseases (Ozkirim, 2006; Ozkirim *et al.*, 2012);
- effects on female hormones (Trabace *et al.*, 2011);
- food preservative and feed *additive* activity (Bassett, 2000; Losa, 2001; Krimpen and Binnendijk, 2001; Moller, 2001; Onibala *et al.*, 2001; Botsoglou *et al.*, 2002a,b; Bozkurt and Baser, 2002a,b; Alcicek *et al.*, 2003, 2004; Busquet *et al.*, 2006; Cabuk *et al.*, 2006a,b; Horosova *et al.*, 2006; Oviedo-Rondon *et al.*, 2006; Si *et al.*, 2006; Garcia *et al.*, 2007; Jang *et al.*, 2007; Janz *et al.*, 2007; Doyle *et al.*, 2008; Ahmadifar *et al.*, 2011; Tunc and Duman, 2011; Giannenas *et al.*, 2012; Zodrow *et al.*, 2012; Belda-Galbis *et al.*, 2013, 2014; Castillo *et al.*, 2014; Gormez and Diler, 2014; Jouki *et al.*, 2014; Kamimura *et al.*, 2014; Otoni *et al.*, 2014; Pirgozliev *et al.*, 2014; Quintero *et al.*, 2014; Ramos *et al.*, 2014; Upadhyay *et al.*, 2014);
- inhibitor of gastrointestinal contraction, choleretic and antidiarrhoeal activity (Cingi *et al.*, 1992; Tabata *et al.*, 1993; Aydin *et al.*, 1997d; Skocibusic and Bezic, 2003; Oliveira *et al.*, 2012);
- hepatoprotective/anti-hepatotoxic activity (Canbek *et al.*, 2008; Slamenova *et al.*, 2008, 2011, 2013; Uyanoglu *et al.*, 2008, 2011; Aristatile *et al.*, 2009, 2011, 2013, 2014);
- hypotensive (essential oil) and hypertensive activity (aromatic water) (Aydin *et al.*, 1997a,b,c,d, 2007);
- immunomodulatory activity (Chan *et al.*, 2005);
- insecticidal and acaricidal activity (Karpouhtsis *et al.*, 1998; Aslan *et al.*, 2005; Erler, 2005; Park *et al.*, 2005; Sampson *et al.*, 2005; Dietrich *et al.*, 2006; Cetin *et al.*, 2007, 2009, 2010; Choi *et al.*, 2007; Coşkun *et al.*, 2008; Vucinic *et al.*, 2011; Koc *et al.*, 2013; Tabanca *et al.*, 2013; Tong *et al.*, 2013; Yilmaz and Tunaz, 2013; Ma *et al.*, 2014);
- nematicidal activity (Oka *et al.*, 2000; Tsao and Yu, 2000; Lei *et al.*, 2010; Ntalli *et al.*, 2010, 2011; Faria *et al.*, 2013);
- neuroprotective activity (Peters *et al.*, 2012; Celik *et al.*, 2013; Zhong *et al.*, 2013);
- active in prevention of obesity (Cho *et al.*, 2012);
- phytotoxic, herbicidal and allelopathic activity (Dudai *et al.*, 1999; Taban *et al.*, 2013; Vasilakoglou *et al.*, 2013); and
- wound healing activity (Altiok *et al.*, 2010; Suntar *et al.*, 2011).

15.5 Summary

Oregano has been found to have many beneficial properties as a medicinal herb. In recent years, after the ban of the use of antibiotics in animal feed, essential oils, and especially oregano oil rich in carvacrol, have been studied as alternatives to antibiotics. Their use in animal feed has been shown to improve health, yield and quality of poultry and pigs, and many such products are on the market. In some of these, essential oil is micro-encapsulated for easy availability to the animals (Franz *et al.*, 2010; Wallace *et al.*, 2010).

References

Ahmadifar, E., Falahatkar, B. and Akrami, R. (2011) Effects of dietary thymol–carvacrol on growth performance, hematological parameters and tissue composition of juvenile rainbow trout, *Oncorhynchus mykiss*. *Journal of Applied Ichthyology* 27, 1057–1060.

Ahn, Y.J., Lee, S.B., Okubo, T. and Kim, M. (1995) Antignawing factor of crude-oil derived from *Thujopsis dolabrata* S-et-Z var. *hondai* sawdust against mice. *Journal of Chemical Ecology* 21, 263–271.
Akkol, E.K., Avci, G., Kucukkurt, I., Keles, H., Tamer, U., Ince, S. and Yesilada, E. (2009) Cholesterol-reducer, antioxidant and liver protective effects of *Thymbra spicata* L. var. *spicata*. *Journal of Ethnopharmacology* 126, 314–319.
Albano, S.M., Sofia Lima, A., Graca Miguel, M., Pedro, L.G., Barroso, J.G. and Cristina Figueiredo, A. (2012) Antioxidant, anti-5-lipoxygenase and antiacetylcholinesterase activities of essential oils and decoction waters of some aromatic plants. *Records of Natural Products* 6(1), 35–48.
Alcicek, A., Bozkurt, M. and Cabuk, M. (2003) The effect of an essential oil combination derived from selected herbs growing wild in Turkey on broiler performance. *South African Journal of Animal Science* 33, 89–94.
Alcicek, A., Bozkurt, M. and Cabuk, M. (2004) The effect of a mixture of herbal essential oils, an organic acid or a probiotic on broiler performance. *South African Journal of Animal Science* 34, 217–222.
Alinkina, E.S., Misharina, T.A. and Fatkullina, L.D. (2013) Antiradical properties of oregano, thyme, and savory essential oils. *Applied Biochemistry and Microbiology* 49, 73–78.
Altiok, D., Altiok, E. and Tihminlioglu, F. (2010) Physical, antibacterial and antioxidant properties of chitosan films incorporated with thyme oil for potential wound healing applications. *Journal of Materials Science – Materials in Medicine* 21, 2227–2236.
Andersen, A. (2006) Final report on the safety assessment of sodium *p*-chloro-*m*-cresol, *p*-chloro-*m*-cresol, chlorothymol, mixed cresols, *m*-cresol, *o*-cresol, *p*-cresol, isopropyl cresols, thymol, *o*-cymen-5-ol, and carvacrol. *International Journal of Toxicology* 25(Suppl 1), 29–127.
Anonymous (2011) Türkiye Bitkileri Veri Servisi (TÜBİVES) (Turkish Plant Data Service). Available at: http://web.archive.org/web/20071028083719/http://turkherb.ibu.edu.tr/ (accessed 25 January 2016).
Aristatile, B., Al-Numair, K.S., Veeramani, C. and Pugalendi, K.V. (2009) Antihyperlipidemic effect of carvacrol on D-galactosamine-induced hepatotoxic rats. *Journal of Basic and Clinical Physiology and Pharmacology* 20, 15–27.
Aristatile, B., Al-Numair, K.S., Al-Assaf, A.H. and Pugalendi, K.V. (2011) Pharmacological effect of carvacrol on D-galactosamine-induced mitochondrial enzymes and DNA damage by single-cell gel electrophoresis. *Journal of Natural Medicines* 65, 568–577.
Aristatile, B., Al-Assaf, A.H. and Pugalendi, K.V. (2013) Carvacrol suppresses the expression of inflammatory marker genes in D-galactosamine-hepatotoxic rats. *Asian Pacific Journal of Tropical Medicine* 6, 205–211.
Aristatile, B., Al-Assaf, A.H. and Pugalendi, K.V. (2014) Carvacrol ameliorates the Ppar-A and cytochrome P450 expression on D-galactosamine induced hepatotoxicity rats. *African Journal of Traditional Complementary and Alternative Medicines* 11, 118–123.
Arslan, N. (1994) *Tütün, ilaç ve baharat bitkileri ders notları.* (Unpublished Lecture Notes), Ankara University, Ankara.
Arslan, N. (2002) *Kokulu bitkiler yetiştirme ve ıslahı ders notları.* (Unpublished Lecture Notes), Ankara University, Ankara.
Aslan, I., Calmasur, O., Sahin, F. and Caglar, O. (2005) Insecticidal effects of essential plant oils against *Ephestia kuehniella, Lasioderma serricorne* and *Sitophilus granarius. Zeitschrift für Pflanzenkrankheiten und Pflanzenschutz* 112, 257–267.
Aydin, E., Turkez, H. and Keles, M.S. (2014) The effect of carvacrol on healthy neurons and N2a cancer cells: some biochemical, anticancerogenicity and genotoxicity studies. *Cytotechnology* 66, 149–157.
Aydin, S., Ozturk, Y., Beis, R. and Baser, K.H.C. (1996) Investigation of *Origanum onites, Sideritis congesta* and *Satureja cuneifolia* oils for analgesic activity. *Phytotherapy Research* 10, 342–344.
Aydın, S., Ozturk, Y. and Baser, K.H.C. (1997a) Cardiovascular actions of kekik (*Origanum onites* L.) essential oil. In: Coşkun, M.(ed.) *Proceedings of the 11th Symposium on Plant Originated Crude Drugs (11th BIHAT), Ankara, 22–24 May 1996.* Üniversitesi Eczacılık Fakültesi Yayınları No. 75, Ankara University, Ankara, pp. 332–338.
Aydin, S., Ozturk, Y. and Baser, K.H.C. (1997a) Cardiovascular actions of kekik (*Origanum onites* L.) aqueous distillate which accumulates under the essential oil. In: Coşkun, M. (ed.) *Proceedings of the 11th Symposium on Plant Originated Crude Drugs (11th BIHAT), Ankara, 22–24 May 1996.* Ankara Üniversitesi Eczacılık Fakültesi Yayınları No. 75, Ankara University, Ankara, pp. 339–344.
Aydin, S., Ozturk, Y. and Baser, K.H.C. (1997b) Choleretic actions of kekik (*Origanum onites* L.) aqueous distillate which accumulates under the essential oil. In: Coşkun, M. (ed.) *Proceedings of the*

11th Symposium on Plant Originated Crude Drugs (11th BIHAT), Ankara, 22–24 May 1996. Ankara Üniversitesi Eczacılık Fakültesi Yayınları No. 75, Ankara University, Ankara, pp. 345–351.

Aydin, S., Baser, K.H.C. and Ozturk, Y. (1997d) The chemistry and pharmacology of *Origanum* (kekik) water. In: Franz, C., Máthé, A. and Buchbauer, G. (eds) *Proceedings of the 27th International Symposium on Essential Oils, September 8–11, 1996. Essential Oils: Basic and Applied Research*. Allured Publishing, Vienna, pp. 52–60.

Aydin, S., Basaran, A.A. and Basaran, N. (2005a) The effects of thyme volatiles on the induction of DNA damage by the heterocyclic amine IQ and mitomycin C. *Mutation Research* 581, 43–53.

Aydin, S., Basaran, A.A. and Basaran, N. (2005b) Modulating effects of thyme and its major ingredients on oxidative DNA damage in human lymphocytes. *Journal of Agricultural and Food Chemistry* 53, 1299–1305.

Aydin, Y., Kutlay, O., Ari, S., Duman, S., Uzuner, K. and Aydin, S. (2007) Hypotensive effects of carvacrol on the blood pressure of normotensive rats. *Planta Medica* 73, 1365–1371.

Azizi, Z., Ebrahimi, S., Saadatfar, E., Kamalinejad, M. and Majlessi, N. (2012) Cognitive-enhancing activity of thymol and carvacrol in two rat models of dementia. *Behavioural Pharmacology* 23, 241–249.

Bakkali, F., Averbeck, S., Averbeck, D. and Idaomar, M. (2008) Biological effects of essential oils –a review. *Food and Chemical Toxicology* 46, 446–475.

Baricevic, D. and Bartol, T. (2002) The biological/pharmacological activity of the *Origanum* genus. In: Kintzios, S.E. (ed.) *Oregano, the Genera Origanum and Lippia*. Taylor and Francis, London and New York, pp. 177–213.

Başer, K.H.C. (1993) Essential oils of Anatolian Labiatae: a profile. *Acta Horticulture* 333, 217–238.

Başer, K.H.C. (2002a) The Turkish *Origanum* species. In: Kintzios, S.E. (ed.) *Oregano, the Genera* Origanum *and* Lippia. Taylor and Francis, London and New York, pp. 109–126.

Başer, K.H.C. (2002b) Aromatic biodiversity among the flowering plant taxa of Turkey. *Pure and Applied Chemistry* 74, 527–545.

Başer, K.H.C. (2008) Biological and pharmacological activities of carvacrol and carvacrol bearing essential oils. *Current Pharmaceutical Design* 14, 3106–3120.

Başer, K.H.C., Özek, T., Tümen, G. and Sezik, E. (1993) Composition of the essential oils of Turkish *Origanum* species with commercial importance. *Journal of Essential Oil Research* 5, 619–623.

Bassett, R. (2000) Oregano's positive impact on poultry production. *World Poultry* 16, 31–34.

Bayram, E. (2003) *Kekik Yetiştiriciliği*. Ege Üniversitesi Tarımsal Araştırma ve Uygulama Merkezi Bülteni, Yayım Bülteni No. 42, Izmir, Turkey.

Bayram, E., Geren, H., Ceylan, A. and Özay, N. (1999) İzmir kekiği (*Origanum onites* L.) 'nde farkli biçim ve yüksekliğin verim ve kaliteye etkisi. In: Türkiye III: Tarla Bitkileri Kongresi, 15–18 Kasım 1999, Adana, Turkey, pp. 222–226.

Bayramoglu, E.E., Gulumser, G. and Karaboz, I. (2006) Ecological and innovative fungicide for leather industry: essential oil of *Origanum minutiflorum*. *JALCA* 101, 96–104.

Baytop, T. (1999) *Türkiye'de Bitkiler ile Tedavi*. Nobel Tıp Yayınevleri, İstanbul, Turkey.

Belda-Galbis, C.M., Pina-Perez, M.C., Leufven, A., Martinez, A. and Rodrigo, D. (2013) Impact assessment of carvacrol and citral effect on *Escherichia coli* K12 and *Listeria innocua* growth. *Food Control* 33, 536–544.

Belda-Galbis, C.M., Leufven, A., Martinez, A. and Rodrigo, D. (2014) Predictive microbiology quantification of the antimicrobial effect of carvacrol. *Journal of Food Engineering* 141, 37–43.

Ben Arfa, A., Combes, S., Preziosi-Bellov, L., Gontard, N. and Chalier, P. (2006) Antimicrobial activity of carvacrol related to its chemical structure. *Letters in Applied Microbiology* 43, 149–154.

Boskabady, M.H. and Jalali, S. (2013) Effect of carvacrol on tracheal responsiveness, inflammatory mediators, total and differential WBC count in blood of sensitized guinea pigs. *Experimental Biology and Medicine* 238, 200–208.

Boskabady, M.H. and Jandaghi, P. (2003) Relaxant effects of carvacrol on guinea pig tracheal chains and its possible mechanisms. *Pharmazie* 58, 661–663.

Boskabady, M.H., Ramazani, M. and Tabei, T. (2003) Relaxant effects of different fractions of essential oil from *Carum copticum* on guinea pig tracheal chains. *Phytotherapy Research* 17, 1145–1149.

Boskabady, M.H., Jandaghi, R., Kiani, S. and Hasanzadeh, L. (2005) Antitussive effect of *Carum copticum* in guinea pigs. *Journal of Ethnopharmacology* 97, 79–82.

Boskabady, M.H., Tabanfar, H., Gholamnezhad, Z. and Sadeghnia, H.R. (2012) Inhibitory effect of *Zataria multiflora* Boiss and carvacrol on histamine (H1) receptors of guinea-pig tracheal chains. *Fundamental and Clinical Pharmacology* 26, 609–620.

Botelho, M.A., Nogueira, N.A.P., Bastos, G.M., Forseca, S.G.C., Lemos, T.L.G. and Matos, F.J.A. (2007) Antimicrobial activity of the essential oil from *Lippia sidoides*, carvacrol and thymol against oral pathogens. *Brazilian Journal of Medical and Biological Research* 40, 349–356.

Botelho, M.A., Rao, V.S., Montenegro, D., Menezes Bandeira, M.A., Cruz Fonseca, S.G., Pinto Nogueira, N.A., Ribeiro, R.A. and Castro Brito, G.A. (2008) Effects of a herbal gel containing carvacrol and chalcones on alveolar bone resorption in rats on experimental periodontitis. *Phytotherapy Research* 22, 442–449.

Botsoglou, N.A., Floron-Paneri, P., Christaki, E., Fletouris, D.J. and Spais, A.B. (2002a) Effect of dietary oregano essential oil on performance of chickens and on iron-induced lipid peroxidation of breast, thigh and abdominal fat tissues. *British Poultry Science* 43, 223–230.

Botsoglou, N.A., Christaki, E., Fletouris, D.J., Florou-Paneri, P. and Spais, A.B. (2002b) The effect of dietary oregano essential oil on lipid oxidation in raw and cooked chicken during refrigerated storage. *Meat Science* 62, 259–265.

Bozkurt, M. and Başer, K.H.C. (2002a) The effect of commercial organic acid, probiotic and essential oil mixture at two levels on the performance of broilers. In: *First European Symposium on Bioactive Secondary Plant Products in Veterinary Medicine, 4–5 October 2002, Vienna, Austria.*

Bozkurt, M. and Başer, K.H.C. (2002b) The effect of antibiotic, mannan oligosaccharide and essential oil mixture on the laying hen performance. In: *First European Symposium on Bioactive Secondary Plant Products in Veterinary Medicine, 4–5 October 2002, Vienna, Austria.*

Busquet, M., Calsamiglia, S., Ferret, A. and Kamel, C. (2006) Plant extracts affect *in vitro* rumen microbial fermentation. *Journal of Dairy Science* 89, 761–771.

Burt, S.A., Vlielander, R., Haagsman, H.P. and Veldhuizen, E.J.A. (2005) Increase in activity of essential oil components carvacrol and thymol against *Escherichia coli* O157:H7 by addition of food stabilizers. *Journal of Food Protection* 68, 919–926.

Burt, S.A., van der Zee, R., Koets, A.P., de Graaff, A.M., van Knapen, F. and Gaastra, W. (2007) Carvacrol induces heat shock protein 60 and inhibits synthesis of flagellin in *Escherichia coli* O157:H7. *Applied and Environmental Microbiology* 73, 4484–4490.

Cabuk, M., Bozkurt, M., Alcicek, A., Akbas, Y. and Kucukyilmaz, K. (2006a) Effect of a herbal essential oil mixture on growth and internal organ weight of broilers from young and old breeder flocks. *South African Journal of Animal Science* 36, 135–141.

Cabuk, M., Bozkurt, M., Alcicek, A., Catli, A.U. and Başer, K.H.C. (2006b) Effect of a dietary essential oil mixture on performance of laying hens in the summer season. *South African Journal of Animal Science* 36, 215–221.

Canbek, M., Uyanoglu, M., Bayramoglu, G., Senturk, H., Erkasap, N. and Koken, T. (2008) Effects of carvacrol on defects of ischemia reperfusion in the liver. *Phytomedicine* 15, 447–452.

Carbone-Howell, A.L., Stebbins, N.D. and Uhrich, K.E. (2014) Poly(anhydride-esters) comprised exclusively of naturally occurring antimicrobials and EDTA: antioxidant and antibacterial activities. *Biomacromolecules* 15, 1889–1895.

Carolina Vicuna, G., Stashenko, E.E. and Luis Fuentes, J. (2010) Chemical composition of the *Lippia origanoides* essential oils and their antigenotoxicity against bleomycin-induced DNA damage. *Fitoterapia* 81, 343–349.

Case, G.I., He, L., Mo, H.B. and Elson, C.E. (1995) Induction of geranyl pyrophosphate pyrophosphatase activity by cholesterol-suppressive isoprenoids. *Lipids* 30, 357–359.

Castillo, S., Perez-Alfonso, C.O., Martinez-Romero, D., Guillen, F., Serrano, M. and Valero, D. (2014) The essential oils thymol and carvacrol applied in the packing lines avoid lemon spoilage and maintain quality during storage. *Food Control* 35, 132–136.

Celik, F., Gocmez, C., Bozkurt, M., Kaplan, I., Kamasak, K., Akil, E., Dogan, E., Guzel, A. and Uzar, E. (2013) Neuroprotective effects of carvacrol and pomegranate against methotrexate-induced toxicity in rats. *European Review for Medical and Pharmacological Sciences* 17, 2988–2993.

Cetin, H., Erler, F. and Yanikoglu, A. (2007) A comparative evaluation of *Origanum onites* essential oil and its four major components as larvicides against the pine processionary moth. *Thaumetopoea wilkinsoni* Tams. *Pest Management Science* 63, 830–833.

Cetin, H., Cilek, J.E., Aydin, L. and Yanikoglu, A. (2009) Acaricidal effects of the essential oil of *Origanum minutiflorum* (Lamiaceae) against *Rhipicephalus turanicus* (Acari: Ixodidae). *Veterinary Parasitology* 160, 359–361.

Cetin, H., Cilek, J.E., Oz, E., Aydin, L., Deveci, O. and Yanikoglu, A. (2010) Acaricidal activity of *Satureja thymbra* L. essential oil and its major components, carvacrol and gamma-terpinene against adult *Hyalomma marginatum* (Acari: Ixodidae). *Veterinary Parasitology* 170, 287–290.

Ceylan, A. (1996) *Tıbbi Bitkiler-II (Uçucu yağ bitkileri).* Ege Üniversitesi Ziraat Fakültesi Yayınları No. 481, Izmir Turkey.
Chan, A.S.L., Pang, H.H., Yip, E.C.H., Tam, Y.K. and Wong, Y.H. (2005) Carvacrol and eugenol differentially stimulate intracellular Ca^{2+} mobilization and mitogen-activated protein kinases in Jurkat T-cells and monocytic THP-1 cells. *Planta Medica* 71, 634–639.
Cho, S., Choi, Y., Park, S. and Park, T. (2012) Carvacrol prevents diet-induced obesity by modulating gene expressions involved in adipogenesis and inflammation in mice fed with high-fat diet. *Journal of Nutritional Biochemistry* 23, 192–201.
Choi, I.-H., Kim, J., Shin, S.-C. and Park, I.-K. (2007) Nematicidal activity of monoterpenoids against the pine wood nematode (*Bursaphelenchus xylophilus*). *Russian Journal of Nematology* 15, 35–40.
Cingi, M.I., Kirimer, N., Sarikardasoglu, I., Cingi, C. and Başer, K.H.C. (1992) Pharmacological activities of the essential oils of *Origanum onites* and *Origanum minutiflorum.* In: Başer, K.H.C. (ed.) *Proceedings of the 9th Symposium on Plant Drugs.* Anadolu Universitesi Yayınları No. 641, Anadolu University Press, Eskişehir, Turkey, pp. 10–15.
Coşkun, S., Giriskin, O., Kurkcuoglu, M., Giriskin, A.O., Kirimer, N. and Başer, K.H.C. (2008) Acaricidal efficacy of *Origanum onites* L. essential oil against *Rhipicephalus turanicus* (Ixodidae). *Parasitology Research* 103, 259–261.
Davis, P.H. (1982) *Flora of Turkey and Aegean Islands, Vol. 7.* Edinburgh University Press, Edinburgh, UK.
Davis, P.H. (1988) *Flora of Turkey and Aegean Islands, Vol. 10.* Edinburgh University Press, Edinburgh, UK.
Demirci, F., Paper, D.H., Franz, G. and Başer, K.H.C. (2004) Investigation of the *Origanum onites* L. essential oil using the chorioallantoic membrane (CAM) assay. *Journal of Agricultural and Food Chemistry* 52, 251–254.
Demirci, B., Baser, K.H.C., Tabanca, N. and Wedge, D.E. (2006) Characterization of volatile constituents of *Haplopappus greenei* and studies on the antifungal activity against phytopathogens. *Journal of Agricultural and Food Chemistry* 54, 3146–3150.
Deng, W., Lu, H. and Teng, J. (2013) Carvacrol attenuates diabetes-associated cognitive deficits in rats. *Journal of Molecular Neuroscience* 51, 813–819.
Dietrich, G., Dolan, M.C., Peralta-Cruz, J., Schmidt, J., Piesman, J. and Eisen, R.J. (2006) Repellent activity of fractioned compounds from *Chamaecyparis nootkatensis* essential oil against nymphal *Ixodes scapularis* (Acari: Ixodidae). *Journal of Medical Entomology* 43, 957–961.
Di Pasqua, R., Hoskins, N., Betts, G. and Mauriello, G. (2006) Changes in membrane fatty acids composition of microbial cells induced by addition of thymol, carvacrol, limonene, cinnamaldehyde and eugenol in the growing media. *Journal of Agricultural Food Chemistry* 54, 2745–2749.
Doerner, J.F., Hatt, H. and Ramsey, I.S. (2011) Voltage- and temperature-dependent activation of TRPV3 channels is potentiated by receptor-mediated PI(4,5)P-2 hydrolysis. *Journal of General Physiology* 137, 271–288.
Doyle, P.S., Sajid, M., O'Brien, T., Dubois, K., Engel, J.C. and Mackey, Z.B. (2008) Drugs targeting parasite lysosomes. *Current Pharmaceutical Design* 14, 889–900.
Dudai, N., Poljakoff-Mayber, A., Mayer, A.M., Putievsky, E. and Lerner, H.R. (1999) Essential oils as allelochemicals and their potential use as bioherbicides. *Journal of Chemical Ecology* 25, 1079–1089.
Dunkic, V., Bezic, N., Vuko, E. and Cukrov, D. (2010) Antiphytoviral activity of *Satureja montana* L. ssp *variegata* (Host) P.W. Ball essential oil and phenol compounds on CMV and TMV. *Molecules* 15, 6713–6721.
Earley, S., Gonzales, A.L. and Garcia, Z.I. (2010) A dietary agonist of transient receptor potential cation channel V3 elicits endothelium-dependent vasodilation. *Molecular Pharmacology* 77(4), 612–620.
Erler, F. (2005) Fumigant activity of six monoterpenoids from aromatic plants in Turkey against the two stored-product pests confused flour beetle, *Tribolium confusum* and Mediterranean flour moth, *Ephestia kuehniella. Zeitschrift für Pflanzenkrankheiten und Pflanzenschutz* 112, 602–611.
EDQM (2014) *European Pharmacopoeia*, 8th edn. European Directorate for the Quality of Medicines – Council of Europe, Strasbourg, France.
Ezhumalai, M., Radhiga, T. and Pugalendi, K.V. (2014) Antihyperglycemic effect of carvacrol in combination with rosiglitazone in high-fat diet-induced type 2 diabetic C57BL/6J mice. *Molecular and Cellular Biochemistry* 385, 23–31.
Fakılı, O. (2010) *Türkiye'de Kekik Adı Ile Anılan Bitkiler Konusunda Yapılan Çalışmaların Envanteri.* Ç.Ü. Fen Bilimleri Enstitüsü Yüksek Lisans Tezi, Adana, Turkey.

Faria, J.M.S., Barbosa, P., Bennett, R.N., Mota, M. and Cristina Figueiredo, A. (2013) Bioactivity against *Bursaphelenchus xylophilus*: nematotoxics from essential oils, essential oils fractions and decoction waters. *Phytochemistry* 94, 220–228.

Farias-Junior, P.A., Rios, M.C., Moura, T.A., Almeida, R.P., Alves, P.B., Blank, A.F., Fernandes, R.P. and Scher, R. (2012) Leishmanicidal activity of carvacrol-rich essential oil from *Lippia sidoides* Cham. *Biological Research* 45, 399–402.

Force, M., Sparks, W.S. and Ronzio, R.A. (2000) Inhibition of enteric parasites by emulsified oil of oregano *in vivo*. *Phytotherapy Research* 14, 213–214.

Franz, C., Başer, K.H.C. and Windisch, W. (2010) Aromatic plants, essential oils and volatiles in animal feeding – a European perspective. *Flavour and Fragrance Journal* 25, 327–340.

Friedman, M., Henika, P.R., Levin, C.E. and Mandrell, R.E. (2006) Antimicrobial wine formulations active against the foodborne pathogens *Escherichia coli* O157:H7 and *Salmonella enterica*. *Journal of Food Science* 71, M245–M251.

Furness, J.B., Kunze, W.A. and Clerc, N. (1999) Nutrient tasting and signaling mechanisms in the gut. II. The intestine as a sensory organ: neural, endocrine, and immune responses. *American Journal of Physiology* 277, G922–G928.

Garcia, V., Catala-Gregori, P., Madrid, J., Hernandez, F., Megias, M.D. and Andrade-Montemayor, H.M. (2007) Potential of carvacrol to modify *in vitro* rumen fermentation as compared with monensin. *Animal* 1, 675–680.

Giannenas, I., Triantafillou, E., Stavrakakis, S., Margaroni, M., Mavridis, S., Steiner, T. and Karagouni, E. (2012) Assessment of dietary supplementation with carvacrol or thymol containing feed additives on performance, intestinal microbiota and antioxidant status of rainbow trout (*Oncorhynchus mykiss*). *Aquaculture* 350, 26–32.

Gill, A.O. and Holley, R.A. (2006a) Disruption of *Escherichia coli*, *Listeria monocytogenes* and *Lactobacillus sakei* cellular membranes by plant essential oils aromatics. *International Journal of Food Microbiology* 108, 1–9.

Gill, A.O. and Holley, R.A. (2006b) Inhibition of membrane bound ATPases of *Escherichia coli* and *Listeria monocytogenes* by plant oil aromatics. *International Journal of Food Microbiology* 111, 170–174.

Gönüz, A. and Özgörücü, B. (1999)An investigation on the morphology, anatomy and ecology of *Origanum onites* L. *Turkish Journal of Botany* 23, 19–32.

Gormez, O. and Diler, O. (2014). *In vitro* antifungal activity of essential oils from *Thymbra*, *Origanum*, *Satureja* species and some pure compounds on the fish pathogenic fungus, *Saprolegnia parasitica*. *Aquaculture Research* 45, 1196–1201.

Goze, I., Alim, A., Cetinus, S.A., Cetin, A., Durmus, N., Atas, A.T. and Vural, N. (2010). *In vitro* antimicrobial, antioxidant, and antispasmodic activities and the composition of the essential oil of *Origanum acutidens* (Hand.-Mazz.) Ietswaart. *Journal of Medicinal Food* 13, 705–709.

Guilen, F., Zapata, P.J., Martinez-Romero, D., Castillo, S., Serrano, M. and Valero, D. (2007) Improvement of the overall quality of table grapes stored under modified atmosphere packaging in combination with natural antimicrobial compounds. *Journal of Food Science* 72, S185–S90.

Guimaraes, A.G., Silva, F.V., Xavier, M.A., Santos, M.R.V., Oliveira, R.C.M., Oliveira, M.G.B., De Souza, C.C. and Quintans-Junior, L.J. (2012a) Orofacial analgesic-like activity of carvacrol in rodents. *Zeitschrift fur Naturforschung C* 67, 481–485.

Guimaraes, A.G., Xavier, M.A., de Santana, M.T., Camargo, E.A., Santos, C.A., Brito, F.A., Barreto, E.O., Cavalcanti, S.C., Antoniolli, A.R., Oliveira, R.C. and Quintans-Junior, L.J. (2012b) Carvacrol attenuates mechanical hypernociception and inflammatory response. *Naunyn Schmiedeberg's Archives of Pharmacology* 385, 253–263.

Gul, A.S.D., Fadillioglu, E., Karabulut, I., Yesilyurt, A. and Delibasi, T. (2013) The effects of oral carvacrol treatment against H_2O_2 induced injury on isolated pancreas islet cells of rats. *Islets* 5, 149–155.

Hajializadeh, Z., Nasri, S., Kaeidi, A., Sheibani, V., Rasoulian, B. and Esmaeili-Mahani, S. (2014) Inhibitory effect of *Thymus caramanicus* Jalas on hyperglycemia-induced apoptosis in *in vitro* and *in vivo* models of diabetic neuropathic pain. *Journal of Ethnopharmacology* 153, 596–603.

Helander, I.M., Alakomi, H.-L., Latva-Kala, K., Mattila-Sandholm, T., Pol, I. and Smid, E.J. (1998) Characterization of the action of selected essential oil components on Gram-negative bacteria. *Journal of Agricultural and Food Chemistry* 46, 3590–3595.

Horosova, K., Bujnakova, D. and Kmet, V. (2006) Effect of oregano essential oil on chicken lactobacilli and *E. coli*. *Folia Microbiology* 5, 278–280.

Horvathova, E., Turcaniova, V. and Slamenova, D. (2007) Comparative study of DNA-damaging and DNA-protective effects of selected components of essential plant oils in human leukemic cells K562. *Neoplasma* 54, 478–483.
Hudaib, M. and Aburjai, T. (2007) Volatile components of *Thymus vulgaris* L. from wild-growing and cultivated plants in Jordan. *Flavour and Fragrance Journal* 22, 322–327.
Hussain, A.I., Anwar, F., Rasheed, S., Nigam, P.S., Janneh, O. and Sarker, S.D. (2011) Composition, antioxidant and chemotherapeutic properties of the essential oils from two *Origanum* species growing in Pakistan. *Revista Brasileira de Farmacognosia* 21, 943–952.
Hyun, T.K., Kim, H.-C. and Kim, J.-S. (2014) Antioxidant and antidiabetic activity of *Thymus quinquecostatus* Celak. *Industrial Crops and Products* 52, 611–616.
Ietswaart, J.H. (1982) *Origanum*. In: Davis, P.H. (ed.) *Flora of Turkey and the East Aegean Islands, Vol. 7*. Edinburgh University Press, Edinburgh, UK, pp. 297–313.
Inanc, T. and Maskan, M. (2013) Testing the antioxidant effect of essential oils and BHT on corn oil at frying temperatures: a response surface methodology. *Journal of the American Oil Chemists Society* 90, 1845–1850.
Ipek, E., Ayaz Tuylu, B. and Zeytinoglu, H. (2003) Effects of carvacrol on sister chromatid exchanges in human lymphocyte cultures. *Cytotechnology* 43, 145–148.
Ipek, E., Zeytinoglu, H., Okay, S., Tuylu, B.A., Kurkcuoglu, M. and Başer, K.H.C. (2005) Genotoxicity and antigenotoxicity of *Origanum* oil and carvacrol evaluated by Ames *Salmonella*/microsomal test. *Food Chemistry* 93, 551–556.
Jang, I.S., Ko, Y.H., Kang, S.Y. and Lee, C.Y. (2007) Effect of commercial essential oil on growth performance digestive enzyme activity and intestinal microflora population in broiler chickens. *Animal Feed Science and Technology Journal* 134, 304–315.
Janz, J.A.M., Morel, P.C.H., Wilkinson, B.H.P. and Purchas, R.W. (2007) Preliminary investigation of the effects of low-level dietary inclusion of fragrant essential oils and oleoresins on pig performance and pork quality. *Meat Science* 75, 350–355.
Joca, H.C., Cruz-Mendes, Y., Oliveira-Abreu, K., Moreno Maia-Joca, R.P., Barbosa, R., Lemos, T.L., Lacerda Beirao, P.S. and Leal-Cardoso, J.H. (2012) Carvacrol decreases neuronal excitability by inhibition of voltage-gated sodium channels. *Journal of Natural Products* 75, 1511–1517.
Jouki, M., Yazdi, F.T., Mortazavi, S.A., Koocheki, A. and Khazaei, N. (2014) Effect of quince seed mucilage edible films incorporated with oregano or thyme essential oil on shelf life extension of refrigerated rainbow trout fillets. *International Journal of Food Microbiology* 174, 88–97.
Kacem R. and Meraihi, Z. (2006) Effects of essential oil extracted from *Nigella sativa* L. seeds and its main components on human neutrophil elastase activity. *Yakugaku Zasshi* 126, 301–305.
Kamimura, J.A., Santos, E.H., Hill, L.E. and Gomes, C.L. (2014) Antimicrobial and antioxidant activities of carvacrol microencapsulated in hydroxypropyl-beta-cyclodextrin. *LWT – Food Science and Technology* 57, 701–709.
Karkabounas, S., Kostoula, O.K., Daskalou, T., Veltsistas, P., Karamousis, M. and Zelovitis, I. (2006) Anticarcinogenic and antiplatelet effects of carvacrol. *Experimental Oncology* 28, 121–125.
Karpouhtsis, I., Pardali, E., Feggou, E., Kokkini, S., Scouras, Z.G. and Mavragani-Tsipidou, P. (1998) Insecticidal and genotoxic activities of oregano essential oils. *Journal of Agricultural and Food Chemistry* 46, 1111–1115.
Knobloch, K., Weigand, H., Weis, N., Schwarm, H.M. and Vigenschow, H. (1986) Action of terpenoids on energy metabolism. In: Brunke, E.J. (ed.) *Progress in Essential Oil Research*. Walter de Gruyter, Berlin, pp. 429–445.
Koc, S., Oz, E., Cinbilgel, I., Aydin, L. and Cetin, H. (2013) Acaricidal activity of *Origanum bilgeri* PH Davis (Lamiaceae) essential oil and its major component, carvacrol against adults *Rhipicephalus turanicus* (Acari: Ixodidae). *Veterinary Parasitology* 193, 316–319.
Kokkini, S. (1996) Taxonomy, diversity and distribution of *Origanum* species. In: Padulosi, S. (ed.) *Proceedings of the IPGRI International Workshop on Oregano, 8–12 May 1996, CIHEAM, Valenzano, Bari, Italy*. International Plant Genetic Resources Institute, Rome, pp. 2–12.
Koparal, A.T. and Zeytinoglu, M. (2003) Effects of carvacrol on a human non-small cell lung cancer (NSCLC) cell line, A549. *Cytotechnology* 43, 149–154.
Kosar, M., Demirci, B., Demirci, F. and Başer, K.H.C. (2008) Effect of maturation on the composition and biological activity of the essential oil of a commercially important *Satureja* species from Turkey: *Satureja cuneifolia* Ten. (Lamiaceae). *Journal of Agricultural and Food Chemistry* 56, 2260–2265.

Krimpen, M.V. and Binnendijk, G.P. (2001) *Ropadiar® as Alternative for Antimicrobial Growth Promoter in Diets of Weanling Pigs.* Rapport Praktijkonderzoek Veehouderij, European Food Safety Authority, Brussels.

Kristinsson, K.G., Magnusdottir, A.B., Petersen, H. and Hermansson, A. (2005) Effective treatment of experimental acute otitis media by application of volatile fluids into the ear canal. *The Journal of Infectious Diseases* 191, 1876–1880.

Lambert, R.J.W., Skandamis, P.N., Coote, P.J. and Nychas, G.-J.E. (2001) A study of the minimum inhibitory concentration and mode of action of oregano essential oil, thymol and carvacrol. *Journal of Applied Microbiology* 91, 453–462.

Landa, P., Kokoska, L., Pribylova, M., Vanek, T. and Marsik, P. (2009) *In vitro* anti-inflammatory activity of carvacrol: inhibitory effect on COX-2 catalyzed prostaglandin E-2 biosynthesis. *Archives of Pharmacal Research* 32, 75–78.

Lei, J., Leser, M. and Enan, E. (2010) Nematicidal activity of two monoterpenoids and SER-2 tyramine receptor of *Caenorhabditis elegans. Biochemical Pharmacology* 79, 1062–1071.

Lima, M.d.S., Quintans-Junior, L.J., de Santana, W.A., Kaneto, C.M., Pereira Soares, M.B. and Villarreal, C.F. (2013) Anti-inflammatory effects of carvacrol: evidence for a key role of interleukin-10. *European Journal of Pharmacology* 699, 112–117.

Liu, Y., Song, M., Che, T.M., Bravo, D. and Pettigrew, J.E. (2012) Anti-inflammatory effects of several plant extracts on porcine alveolar macrophages *in vitro. Journal of Animal Science* 90, 2774–2783.

Lopez, P., Sanchez, C., Battle, R. and Nerin, C. (2007) Vapor-phase activities of cinnamon, thyme, and oregano essential oils and key constituents against foodborne microorganisms. *Journal of Agricultural and Food Chemistry* 55, 4348–4356.

Losa, R. (2001) The use of essential oils in animal nutrition. In: Brufau, J. (ed.) *Feed Manufacturing in the Mediterranean Region. Improving Safety: From Feed to Food.* CIHEAM-IAMZ, Zaragota, pp. 39–44 (Cahiers Options Méditerranéennes Volume 54), 3rd Conference of Feed Manufacturers of the Mediterranean, 2000/03/2224, Reus, Spain. Centre International de Hautes Etudes Agronomiques Méditerranéennes, Paris.

Ma, W.-B., Feng, J.-T., Jiang, Z.-L., Wu, H., Ma, Z.-Q. and Zhang, X. (2014) Fumigant activity of eleven essential oil compounds and their selected binary mixtures against *Culex pipiens pallens* (Diptera: Culicidae). *Parasitology Research* 113, 3631–3637.

Magyar, J., Szentandrassy, N., Banyasz, T., Fulop, L., Varro, A. and Nanasi, P.P. (2004) Effects of terpenoid phenol derivatives on calcium current in canine and human ventricular cardiomyocytes. *European Journal of Pharmacology* 487, 29–36.

Marquard, R.A and Malko, A. (2006) Dost. In: Heyland, K.-U., Hanus, H. and Keller, E.R. (eds) *Handbuch des Pflanzenbaues 4.* Ulmer, Stuttgart, pp. 465–468.

Martinez-Romero, D., Guillen, F., Valverde, J.M., Bailen, G., Zapata, P. and Serrano, M. (2007) Influence of carvacrol on survival of *Botrytis cinerea* inoculated in table grapes. *International Journal of Food Microbiology* 115, 144–148.

Martin-Smith, M. and Khatoon, T. (1963) Biological activity of the terpenoids and their derivatives. In: Jucker, E. (ed.) *Progress in Drug Research, Vol. IV.* Birkhauser, Basel/Stuttgart, Germany, pp. 279–346.

Melo, F.H., Moura, B.A., de Sousa, D.P., de Vasconcelos, S.M., Macedo, D.S., Fonteles, M.M., Viana, G.S. and de Sousa, F.C. (2011) Antidepressant-like effect of carvacrol (5-isopropyl-2-methylphenol) in mice: involvement of dopaminergic system. *Fundamental and Clinical Pharmacology* 25, 362–367.

Mezzoug, N., Elhadri, A., Dallouh, A., Amkiss, S., Skali, N.S. and Abrini, J. (2007) Investigation of the mutagenic and antimutagenic effects of *Origanum compactum* essential oil and some of its constituents. *Mutation Research* 629, 100–110.

Moller, T. (2001) *Studies on the Effect of an Oregano-oil-addition to Feed Towards Nutrient Digestibilities, N-balance as Well as Towards the Parameters of Microbial Activity in the Alimentary Tract of Weaned Piglets.* European Food Safety Authority, Brussels.

Morillas Marquez, F., Navarro Moll, C., Montilla Herrera, P., Perez Galindo, P., Morales Yuste, M. and Martin Sanchez, J. (2004) Activity of the monoterpene derivatives carvacrol, linalool and alpha-terpineol, obtained from aromatic plants, on *Leishmania infantum.* In: Santaiago, M.-C. (ed.) *Multidisciplinarity for Parasites, Vectors and Parasitic Diseases. Proceedings of the IX European Multicolloquium of Parasitology. Volume 2. Articles of Free Oral Papers and Posters. Valencia, Spain, July 18–23, 2004.* Medimond, Pianoro (Bologna), Italy, pp. 93–96.

Neri, F., Mari, M. and Brigati, S. (2006) Control of *Penicillium expansum* by plant volatile compounds. *Plant Pathology* 55, 100–105.

Nostro, A., Blanco, A.R., Cannatelli, M.A., Enea, V., Flamini, G. and Morelli, I. (2006) Susceptibility of methicillin-resistant staphylococci to oregano essential oil, carvacrol and thymol. *FEMS Microbiological Letters* 230, 191–195.

Nowotarska, S.W., Nowotarski, K.J., Friedman, M. and Situ, C. (2014) Effect of structure on the interactions between five natural antimicrobial compounds and phospholipids of bacterial cell membrane on model monolayers. *Molecules* 19, 7497–7515.

Ntalli, N.G., Ferrari, F., Giannakou, I. and Menkissoglu-Spiroudi, U. (2010) Phytochemistry and nematicidal activity of the essential oils from 8 Greek Lamiaceae aromatic plants and 13 terpene components. *Journal of Agricultural and Food Chemistry* 58, 7856–7863.

Ntalli, N.G., Ferrari, F., Giannakou, I. and Menkissoglu-Spiroudi, U. (2011) Synergistic and antagonistic interactions of terpenes against *Meloidogyne incognita* and the nematicidal activity of essential oils from seven plants indigenous to Greece. *Pest Management Science* 67, 341–351.

Ocana-Fuentes, A., Arranz-Gutierrez, E., Senorans, F.J. and Reglero, G. (2010) Supercritical fluid extraction of oregano (*Origanum vulgare*) essentials oils: anti-inflammatory properties based on cytokine response on THP-1 macrophages. *Food and Chemical Toxicology* 48, 1568–1575.

Oka, Y., Nacar, S., Putievsky, E., Ravid, U., Yaniv, Z. and Spiegel, Y. (2000) Nematicidal activity of essential oils and their components against the root-knot nematode. *Phytopathology* 90, 710–715.

Oliveira, I.S., da Silva, F.V., Viana, A.F.S.C., dos Santos, M.R.V., Quintans-Junior, L.J., Martins, M.d.C.C., Moreira, P.H. and Oliveira, R.d.C.M. (2012) Gastroprotective activity of carvacrol on experimentally induced gastric lesions in rodents. *Naunyn-Schmiedebergs Archives of Pharmacology* 385, 899–908.

Onibala, J.S.I.T., Gunther, K.D. and Meulen, U.T. (2001) *Effects of Essential Oil of Spices as Feed Additives on the Growth and Carcass Characteristics of Growing-Finishing Pigs. Sustainable Development in the Context of Globalization and Locality: Challenges and Options for Networking in Southeast Asia.* European Food Safety Authority, Brussels.

Otoni, C.G., Pontes, S.F.O., Medeiros, E.A.A. and Soares, N.d.F.F. (2014) Edible films from methylcellulose and nanoemulsions of clove bud (Syzygium *aromaticum*) and oregano (*Origanum vulgare*) essential oils as shelf life extenders for sliced bread. *Journal of Agricultural and Food Chemistry* 62, 5214–5219.

Oviedo-Rondon, E.O., Clemente-Hernandez, S., Salvador, F., Williams, R. and Losa, R. (2006) Essential oils on mixed coccidia vaccination and infection in broilers. *International Journal of Poultry Science* 5, 723–730.

Ozdemir, B., Ekbul, A., Topal, N.B., Sarandol, E., Sag, A., Başer, K.H.C., Cordan, J., Gullulu, S., Tuncel, E., Baran, I. and Aydinlar, A. (2008) Effects of *Origanum onites* on endothelial function and serum biochemical markers in hyperlipidaemic patients. *Journal of International Medical Research* 36, 1326–1334.

Ozkan, A. and Erdogan, A. (2011) A comparative evaluation of antioxidant and anticancer activity of essential oil from *Origanum onites* (Lamiaceae) and its two major phenolic components. *Turkish Journal of Biology* 35, 735–742.

Ozkan, G., Baydar, H. and Erbas, S. (2010) The influence of harvest time on essential oil composition, phenolic constituents and antioxidant properties of Turkish oregano (*Origanum onites* L.). *Journal of the Science of Food and Agriculture* 90, 205–209.

Ozkirim A. (2006) The detection of antibiotic resistance in the American and European foulbrood diseases of honey bees (*Apis mellifera L.*). PhD thesis, Hacettepe University, Ankara.

Ozkirim, A., Keskin, N., Kurkcuoglu, M. and Başer, K.H.C. (2012) Evaluation of some essential oils as alternative antibiotics against American foulbrood agent *Paenibacillus larvae* on honey bees *Apis mellifera* L. *Journal of Essential Oil Research* 24, 465–470.

Pahlavan, Y., Sepehri, G., Sheibani, V., Khaki, M.A., Gojazadeh, M., Pahlavan, B. and Pahlavan, F. (2013) Study the antinociceptive effect of intracerebroventricular injection of aqueous extract of *Origanum vulgare* leaves in rat: possible involvement of opioid system. *Iranian Journal of Basic Medical Sciences* 16, 1109–1113.

Park, B.-S., Choi, W.-S., Kim, J.-H., Kim, K.-H. and Lee, S.-E. (2005) Monoterpenes from thyme (*Thymus vulgaris*) as potential mosquito repellents. *Journal of American Mosquito Control Association* 21, 80–83.

Pauli, A. (2001) Antimicrobial properties of essential oil constituents. *International Journal of Aromatherapy* 11, 126–133.

Peters, M., Trembovler, V., Alexandrovich, A., Parnas, M., Birnbaumer, L., Minke, B. and Shohami, E. (2012) Carvacrol together with TRPC1 elimination improve functional recovery after traumatic brain injury in mice. *Journal of Neurotrauma* 29, 2831–2834.

Pilau, M.R., Alves, S.H., Weiblen, R., Arenhart, S., Cueto, A.P. and Lovato, L.T. (2011) Antiviral activity of the *Lippia graveolens* (Mexican oregano) essential oil and its main compound carvacrol against human and animal viruses. *Brazilian Journal of Microbiology* 42, 1616–1624.

Pirgozliev, V., Bravo, D. and Rose, S.P. (2014) Rearing conditions influence nutrient availability of plant extracts supplemented diets when fed to broiler chickens. *Journal of Animal Physiology and Animal Nutrition* 98, 667–671.

Prieto, J.M., Jacopini, P., Cioni, P. and Chericoni, S. (2007) *In vitro* activity of the essential oils of *Origanum vulgare, Satureja montana* and their main constituents in peroxynitrite-induced oxidative processes. *Food Chemistry* 104, 889–895.

Quintans-Junior, L.J., Guimaraes, A.G., Araujo, B.E.S., Oliveira, G.F., Santana, M.T., Moreira, F.V., Santos, M.R.V., Cavalcanti, S.C.H., Lucca Junior, W.D., Bothelho, M.A. *et al.* (2010) Carvacrol, (–)-borneol and citral reduce convulsant activity in rodents. *African Journal of Biotechnology* 9, 6566–6572.

Quintero, R.I., Galotto, M.J., Rodriguez, F. and Guarda, A. (2014). Preparation and characterization of cellulose acetate butyrate/organoclay nanocomposites produced by extrusion. *Packaging Technology and Science* 27, 495–507.

Ramos, M., Beltran, A., Peltzer, M., Valente, A.J.M. and Garrigos, M.d.C. (2014) Release and antioxidant activity of carvacrol and thymol from polypropylene active packaging films. *LWT – Food Science and Technology* 58, 470–477.

Rideal, E.K., Sciver, A. and Richardson, N.E.G. (1930) An investigation into the germicidal powers and capillary activities of certain pure constituents of essential oils. *Perfume and Essential Oil Record* 21, 341–344.

Sadıkoğlu, N. (2005) Türkiye'de ihracatı yapılan kekik türleri üzerinde farmasötik botanik araştırmalar. Ph.D. thesis, İ.Ü.Sağlık Bilimleri Enstitüsü, Istanbul, Turkey.

Sampson, B.J., Tabanca, N., Kirimer, N., Demirci, B., Başer, K.H.C. and Khan, I.A. (2005) Insecticidal activity of 23 essential oils and their major compounds against adult *Lipaphis pseudobrassicae* (Davis) (Aphididae: Homoptera). *Pest Management Science* 61, 1122–1128.

Santos, M.R.V., Moreira, F.V., Fraga, B.P., de Sousa, D.P., Bonjardim, L.R. and Quintans-Junior, L.J. (2011) Cardiovascular effects of monoterpenes: a review. *Revista Brasileira de Farmacognosia* 21, 764–771.

Santoyo, S., Jaime, L., Garcia-Risco, M.R., Ruiz-Rodriguez, A. and Reglero, G. (2014) Antiviral properties of supercritical CO_2 extracts from oregano and sage. *International Journal of Food Properties* 17, 1150–1161.

Sarı, A.O. and Oğuz, B. (2002) *Kekik*. Tarım ve Köyişleri Bakanlığı Ege Tarımsal Araştırma, Enstitüsü Yayın No. 82, Izmir, Turkey.

Si, W., Gong, J., Chanas, C., Cui, S., Yu, H., Caballero, C. and Friendship, R.M. (2006) *In vitro* assessment of antimicrobial activity of carvacrol, thymol and cinnamaldehyde towards *Salmonella* serotype Typhimurium DT104: effects of pig diets and emulsification in hydrocolloids. *Journal of Applied Microbiology* 101, 1282–1291.

Silva, F.V., Guimaraes, A.G., Silva, E.R.S., Sousa-Neto, B.P., Machado, F.D.F., Quintans-Junior, L.J., Arcanjo, D.D.R. and Oliveira, R.C.M. (2012) Anti-inflammatory and anti-ulcer activities of carvacrol, a monoterpene present in the essential oil of oregano. *Journal of Medicinal Food* 15, 984–991.

Skocibusic, M. and Bezic, N. (2003) Chemical composition and antidiarrhoeal activities of winter savory (*Satureja montana* L.) essential oil. *Pharmaceutical Biology* 41, 622–626.

Slamenova, D., Horvathova, E., Sramkova, M. and Marsalkova, L. (2007) DNA protective effects of two components of essential plant oils carvacrol and thymol on mammalian cells cultured *in vitro*. *Neoplasma* 54, 108–112.

Slamenova, D., Horvathova, E., Marsalkova, L. and Wsolova, L. (2008) Carvacrol given to rats in drinking water reduces the level of DNA lesions induced in freshly isolated hepatocytes and testicular cells by H_2O_2. *Neoplasma* 55, 394–399.

Slamenova, D., Horvathova, E., Chalupa, I., Wsolova, L. and Navarova, J. (2011) *Ex vivo* assessment of protective effects of carvacrol against DNA lesions induced in primary rat cells by visible light excited methylene blue (VL plus MB). *Neoplasma* 58, 14–19.

Slamenova, D., Kozics, K., Hunakova, L., Melusova, M., Navarova, J. and Horvathova, E. (2013) Comparison of biological processes induced in HepG2 cells by *tert*-butyl hydroperoxide (*t*-BHP) and hydroperoxide (H_2O_2): the influence of carvacrol. *Mutation Research* 757, 15–22.

Son, D.-J., Park, Y.-H., Kim, Y.-M., Chung, N.-H. and Lee, H.-S. (2005) Antiplatelet activity of *Thujopsis dolabrata* var. *hondai*-derived component against platelet aggregation. *Journal of Microbiology and Biotechnology* 15, 425–427.
Soylu, E.M., Soylu, S. and Kurt, S. (2006) Antimicrobial activities of the essential oils of various plants against tomato late blight disease agent *Phytophthora infestans*. *Mycopathologia* 161, 119–128.
Spiering, R., van der Zee, R., Wagenaar, J., Kapetis, D., Zolezzi, F., van Eden, W. and Broere, F. (2012) Tolerogenic dendritic cells that inhibit autoimmune arthritis can be induced by a combination of carvacrol and thermal stress. *PLoS ONE* 7(9), e46336.
Sticher, O. (1977) Plant mono-, di- and sesquiterpenoids with pharmacological or therapeutical activity. In: Wagner, H. and Wollf, P. (eds) *New Natural Products and Plant Drugs with Pharmacological, Biological or Therapeutical Activity*. Springer, Berlin, pp. 137–176.
Subramaniyan, J., Krishnan, G., Balan, R., Divya, M.G.J., Ramasamy, E., Ramalingam, S., Veerabathiran, R., Thandavamoorthy, P., Mani, G.K. and Thiruvengadam, D. (2014) Carvacrol modulates instability of xenobiotic metabolizing enzymes and downregulates the expressions of PCNA, MMP-2, and MMP-9 during diethylnitrosamine-induced hepatocarcinogenesis in rats. *Molecular and Cellular Biochemistry* 395, 65–76.
Suntar, I., Kupeli Akyol, E., Keles, H., Oktem, A., Başer, K.H.C. and Yesilada, E. (2011) A novel wound healing ointment: a formulation of *Hypericum perforatum* oil and sage and oregano essential oils based on traditional Turkish knowledge. *Journal of Ethnopharmacology* 134, 89–96.
Taban, A., Saharkhiz, M.J. and Hadian, J. (2013) Allelopathic potential of essential oils from four *Satureja* spp. *Biological Agriculture and Horticulture* 29, 244–257.
Tabanca, N., Bernier, U.R., Ali, A., Wang, M., Demirci, B., Bythe, E.K., Khan, S.I., Başer, K.H.C. and Khan, I.A. (2013) Bioassay-guided investigation of two monarda essential oils as repellents of yellow fever mosquito *Aedes aegypti*. *Journal of Agricultural of Food Chemistry* 61, 8573–8580.
Tabata, M., Honda, G., Sezik, E. and Yesilada, E. (1993) *A Report on Traditional Medicine and Medicinal Plants in Turkey (1990, 1991)*. Faculty of Pharmaceutical Sciences, Kyoto University, Kyoto, Japan, pp. 11–21.
Tampieri, M.P., Galuppi, R., Carelle, M.S., Macchioni, F., Cioni, P.L. and Morelli, I. (2003) Effect of selected essential oils and pure compounds on *Saprolegnia parasitica*. *Pharmaceutical Biology* 41, 584–591.
Tasdemir, D., Kaiser, M., Demirci, F. and Başer, K.H.C. (2006) Essential oil of Turkish *Origanum onites* L. and its main components, carvacrol and thymol show potent antiprotozoal activity without cytotoxicity. *Planta Medica* 72, 1006.
Tong, F., Gross, A.D., Dolan, M.C. and Coats, J.R. (2013) The phenolic monoterpenoid carvacrol inhibits the binding of nicotine to the housefly nicotinic acetylcholine receptor. *Pest Management Science* 69(7), 775–780.
Trabace, L., Zotti, M., Morgese, M.G., Tucci, P., Colaianna, M., Schiavone, S., Avato, P. and Cuomo, V. (2011) Estrous cycle affects the neurochemical and neurobehavioral profile of carvacrol-treated female rats. *Toxicology and Applied Pharmacology* 255, 169–175.
Tsao, R. and Yu, Q. (2000). Nematicidal activity of monoterpenoid compounds against economically important nematodes in agriculture. *Journal of Essential Oil Research* 12, 350–354.
TUIK (2014) Bitkisel Üretim İstatistikleri, Türkiye İstatistik Kurumu [Crop Production Statistics, Turkish Statistical Institute]. Available at: https://biruni.tuik.gov.tr/bitkiselapp/bitkisel.zul (accessed 25 January 2016).
Tunc, S. and Duman, O. (2011) Preparation of active antimicrobial methyl cellulose/carvacrol/montmorillonite nanocomposite films and investigation of carvacrol release. *LWT – Food Science and Technology* 44, 465–472.
Ultee, A., Kets, E.P.W. and Smid, E.J. (1999) Mechanisms of action of carvacrol in the food-borne pathogen *Bacillus cereus*. *Applied and Environmental Microbiology* 65, 4606–4610.
Ultee, A., Bennik, M.H.J. and Moezelaar, R. (2002) The phenolic hydroxyl group of carvacrol is essential for action against the food-borne pathogen *Bacillus cereus*. *Applied and Environmental Microbiology* 68, 1561–1568.
Upadhyay, A., Upadhyaya, I., Mooyottu, S., Kollanoor-Johny, A. and Venkitanarayanan, K. (2014) Efficacy of plant-derived compounds combined with hydrogen peroxide as antimicrobial wash and coating treatment for reducing *Listeria monocytogenes* on cantaloupes. *Food Microbiology* 44, 47–53.

Uyanoglu, M., Canbek, M., Aral, E. and Başer, K.H.C. (2008) Effects of carvacrol upon the liver of rats undergoing partial hepatectomy. *Phytomedicine* 15, 226–229.

Uyanoglu, M., Canbek, M., Ceyhan, E., Senturk, H., Bayramoglu, G., Gunduz, O., Ozen, A. and Turgak, O. (2011) Preventing organ injury with carvacrol after renal ischemia/reperfusion. *Journal of Medicinal Plants Research* 5, 72–80.

Van Den Broucke, C.O. and Lemli, J.A. (1980) Antispasmodic activity of *Origanum compactum*. *Planta Medica* 38, 317–331.

Van Den Broucke, C.O. and Lemli, J.A. (1981) Pharmacological and chemical investigation of thyme liquid extracts. *Planta Medica* 41, 129–135.

Van Den Broucke, C.O. and Lemli, J.A. (1982) Antispasmodic activity of *Origanum compactum* 2. Antagonistic effects of thymol and carvacrol. *Planta Medica* 45, 188–190.

Varel, V.H. (2002) Carvacrol and thymol reduce swine waste odor and pathogens: stability of oils. *Current Microbiology* 44, 38–43.

Varel, V.H. and Miller, D.N. (2001a) Effect of carvacrol and thymol on odor emissions from livestock wastes. *Water Science and Technology* 44, 143–148.

Varel, V.H. and Miller, D.N. (2001b) Plant-derived oils reduce pathogens and gaseous emissions from stored cattle waste. *Applied and Environmental Microbiology* 67, 1366–1370.

Varel, V.H., Miller, D.N. and Lindsay, A.D. (2004). Plant oils thymol and eugenol affect cattle and swine waster emissions differently. *Water Science and Technology* 50, 207–213.

Varel, V.H., Miller, D.N. and Berry, E.D. (2006) Incorporation of thymol into corncob granules for reduction of odor and pathogens in feedlot cattle waste. *Journal of Animal Science* 84, 481–487.

Vasilakoglou, I., Dhima, K., Paschalidis, K. and Ritzoulis, C. (2013) Herbicidal potential on *Lolium rigidum* of nineteen major essential oil components and their synergy. *Journal of Essential Oil Research* 25, 1–10.

Veldhuizen, E.J.A., Tjeerdsma-Van Bokhoven, J.L.M., Zweijtzer, C., Burt, S.A. and Haagsman, H.P. (2006) Structural requirements for the antimicrobial activity of carvacrol. *Journal of Agricultural and Food Chemistry* 54, 1874–1879.

Vucinic, M., Nedeljkovic-Trailovic, J., Trailovic, S., Ivanovic, S., Milovanovic, M. and Krnjaic, D. (2011) Carvacrol importance in veterinary and human medicine as ecologic insecticide and acaricide [Karvakrol kao ekoloski insekticid i akaricid od znacaja za humanu i veterinarsku medicinu]. *Veterinarski Glasnik* 65, 433–441.

Wagner, H., Wierer, M. and Bauer, R. (1986) *In vitro* inhibition of prostaglandin biosynthesis by essential oils and phenolic compounds. *Planta Medica* 52, 184–187.

Wallace, R.J., Oleszek, W., Franz, C., Hahn, I., Başer, K.H.C., Mathe, A. and Teichmann, K. (2010) Dietary plant bioactives for poultry health and productivity. *British Poultry Science* 51, 461–487.

Xu, H., Delling, M., Jun, J.C. and Clapham, D.E. (2006) Oregano, thyme and clove derived flavors and skin sensitizers activate specific TRP channels. *Nature Neuroscience* 9, 628–635.

Yamada, T., Ueda, T., Ugawa, S., Ishida, Y., Imayasu, M., Koyama, S. and Shimada, S. (2010) Functional expression of transient receptor potential vanilloid 3 (TRPV3) in corneal epithelial cells: involvement in thermosensation and wound healing. *Experimental Eye Research* 90, 121–129.

Yilmaz, Y.B. and Tunaz, H. (2013) Fumigant toxicity of some plant essential oils and their selected monoterpenoid components against adult American cockroach, *Periplaneta americana* (Dictyoptera: Blattidae). *Turkiye Entomoloji Dergisi* 37, 319–328.

Yin, Q.-H., Yan, F.-X., Zu, X.-Y., Wu, Y.-H., Wu, X.-P., Liao, M.-C., Deng, S.W., Yin, L.I. and Zhuang, Y.-Z. (2012) Anti-proliferative and pro-apoptotic effect of carvacrol on human hepatocellular carcinoma cell line HepG-2. *Cytotechnology* 64, 43–51.

Yousefzadi, M., Riahi-Madvar, A., Hadian, J., Rezaee, F., Rafiee, R. and Biniaz, M. (2014) Toxicity of essential oil of *Satureja khuzistanica*: *in vitro* cytotoxicity and anti-microbial activity. *Journal of Immunotoxicology* 11, 50–55.

Yu, W., Liu, Q. and Zhu, S. (2013) Carvacrol protects against acute myocardial infarction of rats via anti-oxidative and anti-apoptotic pathways. *Biological and Pharmaceutical Bulletin* 36, 579–584.

Zabka, M., Pavela, R. and Prokinova, E. (2014) Antifungal activity and chemical composition of twenty essential oils against significant indoor and outdoor toxigenic and aeroallergenic fungi. *Chemosphere* 112, 443–448.

Zanini, S.F., Silva-Angulo, A.B., Rosenthal, A., Rodrigo, D. and Martinez, A. (2014) Effect of citral and carvacrol on the susceptibility of *Listeria monocytogenes* and *Listeria innocua* to antibiotics. *Letters in Applied Microbiology* 58, 486–492.

Zeybek, N. and Zeybek, U. (2002) *Farmasötik Botanik, Değiştirilmiş 3.* Baskı, Ege Üniversitesi Eczacılık Fakültesi Yayınları, Bornova, İzmir, Turkey.
Zeytinoglu, M., Aydin, S., Ozturk, Y. and Başer, K.H.C. (1998) Inhibitory effects of carvacrol on DMBA induced pulmonary tumorigenesis in rats. *Acta Pharmaceutica Turcica* 40, 93–98.
Zeytinoglu, H., Incesu, Z. and Başer, K.H.C. (2003) Inhibition of DNA synthesis by carvacrol in mouse myoblast cells bearing a human *N-Ras* oncogene. *Phytomedicine* 10, 292–299.
Zhong, Z., Wang, B., Dai, M., Sun, Y., Sun, Q., Yang, G. and Bian, L. (2013) Carvacrol alleviates cerebral edema by modulating AQP4 expression after intracerebral hemorrhage in mice. *Neuroscience Letters* 555, 24–29.
Zodrow, K.R., Schiffman, J.D. and Elimelech, M. (2012) Biodegradable polymer (PLGA) coatings featuring cinnamaldehyde and carvacrol mitigate biofilm formation. *Langmuir* 28, 13993–13999.

16 Parsley

Ghazi Daradkeh[1,2] and Musthafa Mohamed Essa[1]*
[1]*Sultan Qaboos University, Muscat, Oman;* [2]*Hamad Medical Corporation, Doha, Qatar*

16.1 Botany

16.1.1 Introduction

Parsley (*Petroselinum crispum*) is a herb belonging to the Apiaceae (formerly Umbelliferae) family (Fig. 16.1). It is native to the Mediterranean region where it is found in the wild form. It is mostly grown outdoors and is seasonally harvested (Navazio, 2012). Parsley is a leafy vegetable, rich in many biologically active compounds, and its name (*Petroselinum*) is derived from the Greek for 'rock celery'; it can be distinguished from other leafy green herbs by its unique aroma. In sunny areas with suitable environmental conditions – in a humid soil with a pH of 5.3–7.3 – parsley may grow up to 60–120 cm tall (Navazio, 2012).

Parsley is sensitive to water stress, especially if it is planted in the summer and at the end of spring, and to increase production and improve quality, a permanent source of water should be provided. Both growth stage and parsley type determine the susceptibility of the plants to water stress (Petropoulos *et al.*, 2006; Najla *et al.*, 2012).

Fresh parsley has been reported to have a storage life of 1–2 months at 0°C and 95–100% RH (Cantwell, 2001) and of over 12 days in a cold store at 0–2°C and RH 95–97% (Lisiewska *et al.*, 1997). However, at 18–20°C and 85–90% RH, it can only be stored for 3 days (Lisiewska *et al.*, 1997). As already noted, parsley can be cold stored, but it is sensitive to chilling injury.

16.1.2 History

The botanical name *Petroselinum* is derived from the Greek words 'petros', meaning stone (it grows on rocky hillsides) and 'selinon' (parsley or celery). Parsley is mentioned in Greek historical records as being used for cheering sportsmen by wearing crowns made of parsley; wreaths made from parsley were also used to adorn graves. Parsley was also used in Roman rituals. There are reports of it being sprinkled over dead bodies to remove the smell too. Parsley is used in the Jewish celebration of the Passover as well. It is mentioned as one of the plants in the gardens of Charlemagne and Catherine de Medici, and there is a rumour that parsley was popularized in France by Catherine de Medici.

*Corresponding author, e-mail: drmdessa@squ.edu.om

© CAB International 2016. *Leafy Medicinal Herbs: Botany, Chemistry, Postharvest Technology and Uses* (eds D.C.P. Ambrose *et al.*)

Fig. 16.1. A parsley plant and leaves.

16.1.3 Location

Parsley is believed to be native to southern Europe, but it is now found throughout the world. It has been grown in Britain since at least the 16th century. Parsley has now become naturalized throughout Europe, North America, the West Indies, Algeria and Lebanon.

16.1.4 Morphology

There are three varieties of parsley: curly-leaved or common parsley, Italian or flat-leaved parsley and root (Hamburg) parsley, which is grown for its edible root. The leaves are compound, alternately arranged, and are divided into two to three leaflets and the plant can grow to over 1 m tall, and as an annual (in tropical regions) or a biennial crop (in temperate areas). The typical flowering period is in the warmer months and the ideal temperature for pollination and seed production is 29–30°C (Teuscher, 2005).

The roots are a faint yellow colour and carrot shaped. They can grow up to 20 cm in length and 5 cm in width. Hamburg root parsley has larger roots and is commonly used in European cuisine (Teuscher, 2005).

16.2 Chemistry

16.2.1 Chemical (nutritional) composition

Parsley is a 'powerhouse' of nutrition, and is rich in B vitamins, vitamin C, β-carotene and zinc; it is an important dietary component for strengthening bone due to its high content of boron and fluorine, and also contains iron and calcium in an absorbable form (Table 16.1).

16.2.2 Phytochemistry

Parsley has anti-inflammatory, antimicrobial, diuretic and hypoglycaemic properties due to its content of essential oil and phenolic compounds (Taiz and Zeiger, 1998). Yoshikawa *et al.* (2000) reported several flavone glycosides with oestrogenic activity from the aerial parts of parsley, along with a new monoterpene glucoside, petroside.

The leaves contain 0.04–0.4% of volatile oil, and this includes as major constituents α-pinene, β-pinene, myrcene, β-phellandrene, 1,3,8-*p*-menthatriene and myristicin (Charles, 2004). The aroma of parsley is due to the presence of terpenes, which are toxic to many insects. Myristicin (see Fig. 16.2) is a toxic phenylpropene/allylbenzene compound (also known as 5-methoxysafrole) and has hallucinogenic properties, acting as a psychoactive at high intake levels (Hallström and Thuvander, 1997).

The seeds contain 2–8% of volatile oil and 13–22% of fixed oil, and the major compounds found in the volatile oil are α-pinene, β-pinene, myristicin, elemicin, 2,3,4,5-tetramethoxy-allylbenzene and apiol (Charles, 2004). Apiol, a phenylpropene (also known as apiole and as dimethoxysafrole; see Fig. 16.3), is responsible for the

Table 16.1. Nutritional value of parsley for one serving (60 g). Adapted from SELFNutritionData (2016); original source USDA National Nutrient Database for Standard Reference, Release 21 (USDA ARS, 2008).

Nutrient	Amount		
Calories	21.6	**Vitamins**	
From carbohydrate	13.3	Vitamin A	5055 IU
From fat	4.0	Vitamin B	
From protein	4.3	Thiamine (B_1)	0.1 mg
From alcohol	0.0	Riboflavin (B_2)	0.1 mg
Carbohydrates		Niacin (B_3)	0.8 mg
Total carbohydrate	3.8 g	Vitamin B_6	0.1 mg
Dietary fibre	2.0 g	Pantothenic acid (B_5)	0.2 mg
Starch	~[a]	Folate (B_9)	91.2 mcg
Sugar	0.5 g	Vitamin B_{12}	0.0 mcg
Fats and fatty acids		Vitamin C	79.8 mg
Total fat	0.5 g	Vitamin D	~
Saturated fat	0.1 g	Vitamin E (α-tocopherol)	0.4 mg
Monounsaturated fat	0.2 g	Vitamin K	984 mcg
Polyunsaturated fat	0.1 g	Choline	7.7 mg
Omega-3 fatty acids	4.8 mg	Betaine	~
Omega-6 fatty acids	69.0 mg	**Minerals**	
Protein and amino acids		Calcium	82.8 mcg
Protein	1.8 g	Iron	3.7 mg
Sterols		Magnesium	30.0 mg
Cholesterol	0.0 mg	Phosphorus	34.8 mg
Phytosterols	3.0 mg	Potassium	332 mg
Other constituents		Sodium	33.6 mg
Alcohol	0.0 mg	Zinc	0.1 mg
Ash	1.3 g	Copper	0.1 mg
Caffeine	0.0 mg	Manganese	0.1 mg
Theobromine	0.0 mg	Selenium	0.1 mcg
Water	52.6 g	Fluoride	~

[a]Signifies missing or incomplete data.

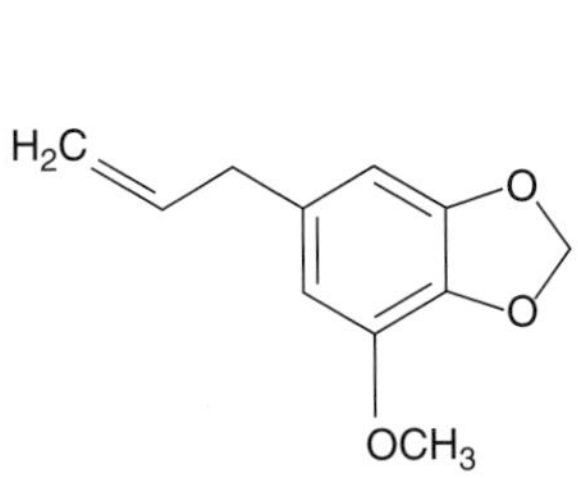

Fig. 16.2. The chemical structure of myristicin. Adapted from Zhang *et al.*, 2006.

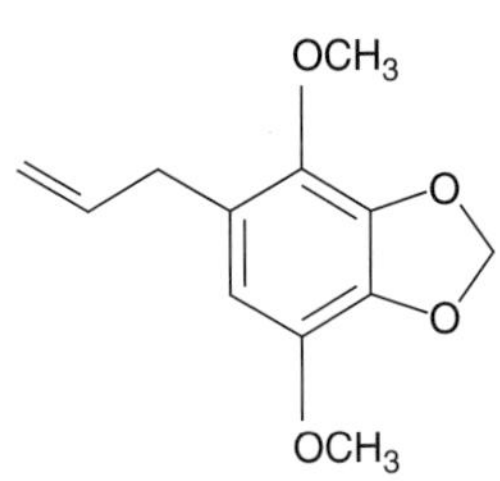

Fig. 16.3. The chemical structure of apiol. Adapted from Zhang *et al.*, 2006.

abortifacient properties of parsley, and the herb may be used to treat menstrual disorders (Castleman, 2009).

Parsley roots contain 0.2–0.75% of essential oil, which has as its main components terpinolene, apiol and myristicin (Orav *et al.*, 2003), while apiin (apigenin-7-apioglucoside; see Fig. 16.4) makes up to 0.2–1.6% of the roots (Taiz and Zeiger, 1998).

Table 16.2 compares the constituents of commercial samples of parsley leaf oil and seed oil.

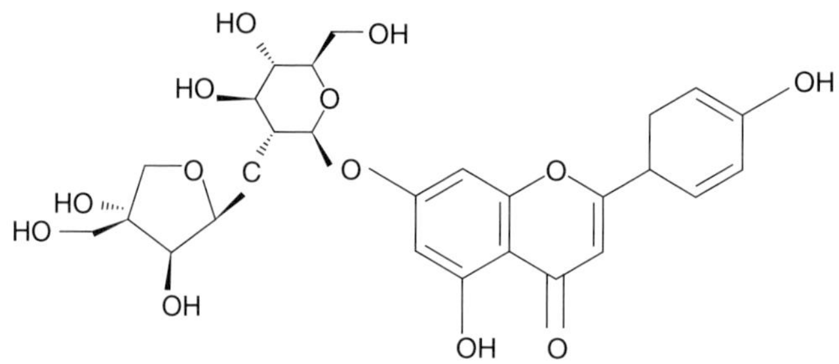

Fig. 16.4. The chemical structure of apiin. From Hostetler *et al.*, 2013.

Table 16.2. Composition of commercial samples of parsley essential oil. Adapted from Charles, 2004.

Compounds	Leaf oil (%)	Seed oil (%)
Apiol	0.27	18.32
Elemicin	2.71	4.84
1,3,8-*p*-Mentha-triene	16.41	0.12
Myrcene	4.24	0.22
Myristicin	11.92	39.65
α-Phellandrene	0.51	0.12
β-Phellandrene	6.48	2.14
α-Pinene	26.42	15.73
β-Pinene	18.04	10.01
Sabinene	1.1	0.64
Terpinolene	2.52	0.01
2,3,4,5-Tetramethoxy-allylbenzene	0.72	7.82

16.3 Postharvest Technology

16.3.1 Extraction of phenolic compounds

Plant phenolics have several health benefits, including: antioxidant, anti-inflammatory, antimicrobial, antitumour and hepatoprotective activities (Rice-Evans *et al.*, 1996; Middleton *et al.*, 2000; Hinneburg *et al.*, 2006). They vary in structure from monomers to complex polymeric tannins, and their isolation from plants involves various steps (including sample grinding, extraction, pre-concentration, hydrolysis and derivatization) that it is important to get right. Luthria and co-workers have investigated the process in dried parsley flakes. First (Luthria *et al.*, 2006), they evaluated the influence of extraction solvents and techniques, and the number of extraction cycles on the quantity of phenolic compounds obtained, and determined that pressurized liquid extraction (PLE), used with four extractions and 50:50 ethanol:water, was the best method. Second, Luthria (2008) examined the influence of six additional PLE parameters (particle size, extraction temperature and pressure, flush volume, static time, and solid:solvent ratio), and showed that the phenolic compounds obtained were influenced by temperature, particle size and solid:solvent ratio, with temperature having the major effect on the phenolic profile (Luthria, 2008). When the extraction temperature was higher, malonyl apiin was partially degraded to acetylapiin and apiin, while flush volume showed marginal influence on the extracted yield (Luthria, 2008).

16.3.2 Minimal processing

Fresh plant food can either be minimally processed or not further processed at all prior to consumption. In a test of the storage of fresh parsley, leaves were sealed in polyethylene bags and stored at 4°C for 12 days. Quality characteristics colour, appearance (succulence and firmness) and aroma (odour, taste) were evaluated on a scale of 1 to 5 on the 1st, 5th, 8th and 12th days. The scores were used to produce life curves, from which it was estimated that minimally processed parsley could stay fresh for a period of 23 days (Cătunescu *et al.*, 2012).

16.4 Uses

16.4.1 Traditional/medicinal benefits

The medicinal benefits of parsley have been long known, and Hippocrates classified parsley as a diuretic. Wine boiled with parsley was recommended in medieval times as a treatment for arthritis and chest pain. Parsley was recommended to alleviate menstrual symptoms in the 17th century (Lis-Balchin, 2006), and was documented as a laxative, diuretic and quinine substitute by the US Pharmacopeia in 1850 (Castleman, 2009). Parsley infusions were used to treat and regulate menstrual pain by Colombian immigrants in London (Ceuterick *et al.*, 2008). Parsley has also been reported as beneficial for postmenopausal women.

The use of both parsley seed powder and parsley juice have been reported for the stimulation of hair growth when used to massage the scalp, and also for the treatment and prevention of insect bites (Charles, 2004).

16.4.2 Pharmacological uses

Antioxidant properties

Parsley was reported to protect cells by decreasing the ageing process due to its antioxidant properties, with the major contributors to its antioxidant activity being myristicin and apiol (Chevallier, 1996; Taiz and Zeiger 1998). Parsley essential oil was found to have β-carotene bleaching and free radical scavenging activities. Apiol contributes more free radical scavenging activity than myristicin, even though it is present at a lower concentration (contrast Fig. 16.3, which shows the methoxy electron donor groups of apiol, with the structure of myristicin shown in Fig. 16.2) (Zhang *et al.*, 2006).

The high content of flavonoids such as apiin, other apigenin glycosides and luteolin, tocopherol, ascorbic acid and essential oils in parsley could encourage antioxidant activity and might decrease any harm caused by oxidation (Nielsen *et al.*, 1999). Myristicin, which is found in parsley oil, activates glutathione-*S*-transferase, which catalyses the action of glutathione in fighting against oxidized molecules (Ozsoy-Sacan *et al.*, 2006; Kolarovic *et al.*, 2010).

Wong and Kitts (2006) reported on the antioxidant (and antibacterial) activities of methanol and water extracts of freeze-dried and irradiated parsley leaves and stems.

Anti-inflammatory properties

Parsley has been traditionally used for the treatment of allergies and autoimmune and chronic inflammatory disorders (Yousofi *et al.*, 2012). Tissue damage, neuropathological diseases and autoimmune disorders may result from chronic activation of the immune system, and myristicin oil from parsley can reduce the immune inflammation by inhibiting nitric oxide, cytokine production and the release of inflammatory proteins (Lee and Park, 2011). Yousofi *et al.* (2012) investigated the suppressive effects of parsley essential oil on mouse splenocytes and macrophages, and found that it could suppress nitric oxide production and the immune functions of macrophages.

Antimicrobial effects

Parsley plays an important role in the defence mechanisms against microbes such as bacteria and fungi. For example, Manderfeld *et al.* (1997) reported that photoactive coumarins from fresh and freeze-dried parsley leaves protected against various human pathogens and food spoilage organisms in a photobiological assay.

The addition of fresh parsley leaves to 'Kareish' cheese reduced the amount of yeast present within 2 h. Additionally parsley extracts showed a significant inhibitory activity against *Staphylococcus aureus* and antibacterial activity against other microbial flora in the cheese (Wahba *et al.*, 2010).

Wong and Kitts (2006) reported on the antibacterial activities of methanol and water extracts of freeze-dried and irradiated parsley leaves and stems against *Bacillus subtilis* and *Escherichia coli*.

Diuretic effect

The diuretic effect of parsley has long been noticed in traditional medicine. More recently, Marczal *et al.* (1997) studied the phenol ether components of the diuretic effect of parsley. Also, in rat *in vivo* and *in vitro* experiments, Kreydiyyeh and Usta (2002) found that parsley extracts increased the urine output/day by inhibiting Na^+-K^+-ATPase, thus leading to an increased K^+ concentration in the kidney lumen that leads to an osmotic water flow into the lumen and diuresis.

The German Commission E has accepted the use of parsley for the treatment of kidney stones (Charles, 2004), most likely because the consumption of parsley tea can increase urine output, although it is recommended not to exceed three cups of parsley seed tea a day (Kreydiyyeh and Usta, 2002).

Hypoglycaemic effect

The hypoglycaemic effect of parsley has been reported by various researchers. Studies in other plants suggested that terpenoids have the ability to enhance insulin for the stimulation of glucose disposal and exert their antidiabetic actions via α-glucosidase modulation, a typical extra-pancreatic mechanism (Luo *et al.*, 1999; Kumar *et al.*, 2011).

It has also been reported that coumarins and flavonoid glucosides (as found in parsley) act in the scavenging or quenching of free radicals (Anand *et al.*, 1981), and that parsley, also being a good source of vitamin C, would also be effective in preventing the non-enzymatic glycosylation of proteins (Afkhami-Ardekani *et al.*, 2003).

Hepatoprotective effect

Ozsoy-Sacan *et al.* (2006) showed that parsley has a hepatoprotective effect on the liver tissue of streptozotocin (STZ)-induced diabetic rats by decreasing blood glucose and lipid peroxidation, along with elevating the level of liver glutathione.

Anti-platelet aggregating effect

Gadi *et al.* (2009) reported that parsley extract inhibited *in vitro* and *ex vivo* platelet aggregation and prolonged bleeding time in rats. Later, Chaves *et al.* (2011) have isolated and identified the flavonoids apigenin, apigenin-7-*O*-glucoside (cosmosiin) and apiin, and the coumarin 2″,3″-dihydroxyfuranocoumarin (oxypeucedanin hydrate) from aqueous extracts of the leaves of flat-leaved parsley. The extract, and apigenin and cosmosiin, all interfered with haemostasis-inhibiting platelet aggregation in human platelets.

Contraindications

Parsley is best avoided by pregnant women because its myristicin and apiol content may stimulate the uterus. The availability of these components in the leaves, stalks and roots is lower than in the seed oil, so these are safe to consume. Those with allergies to plants in the Apiaceae family should avoid all parsley components and constituents (Castleman, 2009; Fig. 16.5).

Drug interactions

Parsley is rich in vitamin K, which interferes with warfarin, so those who are taking the blood thinner warfarin should monitor parsley intake closely (Heck *et al.*, 2000).

Parsley should also be avoided by those on diuretic drugs, because it will exacerbate the diuretic action of the drugs, thus increasing urine output and possibly causing too much fluid loss, leading to dehydration, dizziness and hypotension.

In studies on mice, Jakovljvic *et al.* (2002) have reported on the effect of parsley juice on pharmacodynamic activity of drugs involving cytochrome P450 in their metabolism (the hypnotic pentobarbital and the analgesics paracetamol and aminopyrine).

16.5 Summary

Parsley is used as a table garnish worldwide, but its heath-promoting uses are often ignored, like its valuable medicinal effects. It has great demand in the food and cosmetic industries, and the expansion of its cultivation is of prime importance.

Acknowledgement

The support provided by Internal grant from CAMS, Sultan Qaboos University (IG/AGR/FOOD/14/01) is highly acknowledged. The authors state that there is no conflict of interest in this article.

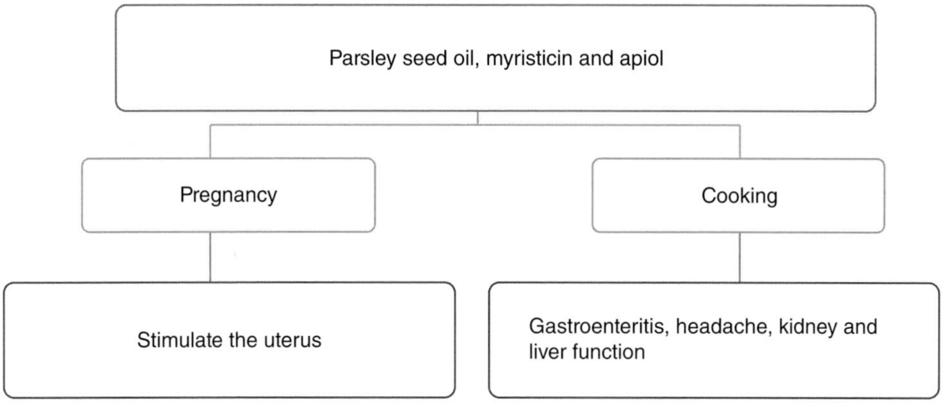

Fig. 16.5. Contraindications for parsley seed oil and seed oil components. From/after Castleman, 2009.

References

Afkhami-Ardekani, M., Vahidi, A.R., Borjian, L. and Borjian, L. (2003) Effect of vitamin C supplement on glycosylated hemoglobin in patients with type 2 diabetes. *Journal of Shah Sad University* 10, 15–18.

Anand, N.K., Sharma, N.D. and Gupta, S.R. (1981) Coumarins from *Apium petroselinum* seeds. *National Academy Science Letters (India)* 4, 249–251.

Cantwell, M. (2001) *Properties and Recommended Conditions for Storage of Fresh Fruit and Vegetables.* Available at: http://postharvest.ucdavis.edu/files/109107.pdf (accessed 26 January 2016).

Castleman, M. (2009) Chapter 5: The healing herbs – Parsley. In: Castleman, M. *The New Healing Herbs: The Essential Guide to More Than 125 of Nature's Most Potent Herbal Remedies.* Rodale, Emmaus, Pennsylvania, pp. 354–357.

Cătunescu, G.M., Tofană, M., Mureşan, C., David, A. and Stănilă, S. (2012) Sensory evaluation of minimally processed parsley (*Petroselinum crispum*), dill (*Anethum graveolens*) and lovage (*Levisticum officinale*) stored at refrigeration temperatures. *Bulletin UASVM Agriculture* 69, 205–212.

Ceuterick, M., Vanderbroek, I., Tony, B. and Pieroni, A. (2008) Cross-cultural adaptation in urban ethnobotany: the Columbian folk pharmacopeia in London. *Journal of Ethnopharmacology* 120, 342–359.

Charles, D.J. (2004) Parsley. In: Peter, K.V. (ed.) Handbook of Herbs and Spices, Volume 2. Woodhead Publishing, Abington, Cambridge, UK and CRC Press, Boca Raton Florida, pp. 231–234.

Chaves, D.S., Frattani, F.S., Assafim, M., Almeida, A.P. de, Zingali, R.B. and Costa, S.S. (2011) Phenolic chemical composition of *Petroselinum crispum* extract and its effect on haemostasis. *Natural Product Communications* 6, 961–964.

Chevallier, A. (1996) *The Encyclopedia of Medicinal Plants.* Dorling Kindersley, London.

Gadi, D., Bnouham, M., Aziz, M., Ziyyat, A., Legssyer, A., Legrand, C., Lafeve, F.F. and Mekhfi, H. (2009) Parsley extract inhibits *in vitro* and *ex vivo* platelet aggregation and prolongs bleeding time in rats. *Journal of Ethnopharmacology* 125, 170–174.

Hallström, H. and Thuvander, A. (1997) Toxicological evaluation of myristicin. *Natural Toxins* 5, 186–192.

Heck, A., DeWitt, B. and Lukes, A. (2000) Potential interactions between alternative therapies and warfarin. *American Journal of Health-System Pharmacy* 57, 1221–1227.

Hinneburg, I., Dorman, H.J.D. and Hiltunen, R. (2006) Antioxidant activities of extracts from selected culinary herbs and spices. *Food Chemistry* 97, 122–129.

Hostetler, G.L., Riedl, K.M. and Schwartz, S.J. (2013) Effects of food formulation and thermal processing on flavones in celery and chamomile. *Food Chemistry* 141, 1406–1411.

Jakovljvic, V., Raskovic, A., Popovic, M. and Sabo, J. (2002) The effect of celery and parsley juice on pharmacodynamic activity of drugs involving cytochrome P450 in their metabolism. *European Journal of Drug Metabolism and Pharmacokinetics* 27, 153–156.

Kolarovic, J., Popovic, M., Zlinská, J., Trivic, S. and Vojnovic, M. (2010) Antioxidant activities of celery and parsley juices in rats treated with doxorubicin. *Molecules* 15, 6193–6204.

Kreydiyyeh, S.I. and Usta, J. (2002) Diuretic effect and mechanism of action of parsley. *Journal of Ethnopharmacology* 79, 353–357.

Kumar, S., Narwal, S., Kumar, V. and Prakash, O. (2011) α-Glucosidase inhibitors from plants: a natural approach to treat diabetes. *Pharmacological Reviews* 5(9), 19–29.

Lee, J. and Park, W. (2011) Anti-inflammatory effect of myristicin on RAW 264.7 macrophages stimulated with polyinosinic-polycytidylic acid. *Molecules* 16, 7132–7142.

Lis-Balchin, M. (2006) Parsley oil. In: Lis-Balchin, M. *Aromatherapy Science: A Guide for Healthcare Professionals*, 1st edn. Pharmaceutical Press, London, pp. 267–269.

Lisiewska, Z., Kmieccik, W. and Budnik, A. (1997) Effect of condition and time of storage on technological quality changes of parsley leaves. *Folia Horticulturae* 9, 21–29.

Luo, J., Cheung, J., Yevich, E.M., Clark, J.P., Tsai, J., Lapresca, P., Ubillas, R.P., Fort, D.M., Carlson, T.J., Hector, R.F. *et al.* (1999) Novel terpenoid-type quinones isolated from *Pycnanthus angolensis* of potential utility in the treatment of type 2 diabetes. *The Journal of Pharmacology and Experimental Therapeutics* 288, 529–534.

Luthria, D.L. (2008) Influence of experimental conditions on the extraction of phenolic compounds from parsley (*Petroselinum crispum*) flakes using a pressurized liquid extractor. *Food Chemistry* 107, 745–752.

Luthria, D.L., Mukhopadhyay, S. and Kwansa, A.L. (2006) A systematic approach for extraction of phenolic compounds using parsley (*Petroselinum crispum*) flakes as model substrate. *Journal of Science of Food and Agriculture* 86, 1350–1358.

Manderfeld, M.M., Schafer, H.W., Davidson, P.M. and Zottola, E.A. (1997) Isolation and identification of antimicrobial furocoumarins from parsley. *Journal of Food Protection* 60, 72–77.

Marczal, G., Balogh, M. and Verzar-Petri, G. (1997) Phenolether components of diuretic effect in parsley I. *Acta Agronomica Academiae Scientiarum Hungaricae* 26, 7–13.

Middleton, E., Kandaswami, C. and Theoharides, T.C. (2000) The effects of plant flavonoids on mammalian cells: implications for inflammations, heart disease, and cancer. *Pharmacology Reviews* 52, 673–751.

Najla, S., Sanoubar, R. and Murshed, R. (2012) Morphological and biochemical changes in two parsley varieties upon water stress. *Physiology and Molecular Biology of Plants* 18, 133–139.

Navazio, J. (2012) Apiaceae. In: Navazio, J. *The Organic Seed Grower: A Farmer's Guide to Vegetable Seed Production*. Chelsea Green Publishing, White River Junction, Vermont, pp. 108–111.

Nielsen, S.E., Young, J.F., Daneshvar, B., Lauridsen, S.T., Knuthsen, P., Sandstrom, B. and Dragsted, L.O. (1999) Effect of parsley (*Petroselinum crispum*) intake on urinary apigenin excretion, blood antioxidant enzymes and biomarkers for oxidative stress in human subjects. *The British Journal of Nutrition* 81, 447–455.

Orav, A., Kailas, T. and Jegorova, A. (2003) Composition of the essential oil of dill, celery, and parsley from Estonia. *Proceedings of the Estonian Academy of Sciences: Chemistry* 52, 147–154.

Ozsoy-Sacan, O., Yanardag, R., Orak, H., Ozgey, Y., Yarat, A. and Tunali, T. (2006) Effects of parsley (*Petroselinum crispum*) extract versus glibornuride on the liver of streptozotocin-induced diabetic rats. *Journal of Ethnopharmacology* 104, 175–181.

Petropoulos, S.A., Akoumianakis, C.A. and Passam, H.C. (2006) Evaluation of turnip-rooted parsley (*Petroselinum crispum* ssp. *tuberosum*) for root and foliage production under a warm Mediterranean climate. *Scientia Horticulturae* 109, 282–287.

Rice-Evans, C.A., Miller, N.J. and Paganga, G. (1996) Structure–antioxidant activity relationships of flavonoids and phenolic acids. *Free Radical Biology and Medicine* 20, 933–956.

SELFNutritionData (2016) Nutrition Facts and Analysis for Parsley, raw. Available at: http://nutritiondata.self.com/facts/vegetables-and-vegetable-products/2513/2 (accessed 26 January 2016).

Taiz, L. and Zeiger, E. (1998) Chapter 13. Secondary metabolites and plant defense. In: Taiz, L. and Zeiger, E. *Plant Physiology*, 2nd edn. Sinauer Associates, Sunderland, Massachusetts, pp. 320–344.

Teuscher, E. (2005) *Medicinal Spices: A Handbook of Culinary Herbs, Spices, Spice Mixtures and their Essential Oils*. (1st edn). Medpharm Scientific, Stuttgart, Germany and CRC Press, Boca Raton, Florida.

USDA ARS (2008) *USDA National Nutrient Database for Standard Reference*, Release 21. Nutrient Data Laboratory, Agricultural Research Service, US Department of Agriculture, Beltsville, Maryland. Available via: http://www.ars.usda.gov/News/docs.htm?docid=18880 (accessed 26 January

2016). *Nutrient Data Laboratory home page* available at: http://www.ars.usda.gov/ba/bhnrc/ndl (accessed 26 January 2016).
Wahba, N.M., Ahmed, A.S. and Ebraheim, Z.Z. (2010) Antimicrobial effects of pepper, parsley, and dill and their roles in the microbiological quality enhancement of traditional Egyptian Kareish cheese. *Foodborne Pathogens and Disease* 7, 411–418.
Wong, P. and Kitts, D. (2006) Studies on the dual antioxidant and antibacterial properties of parsley (*Petroselinum crispum*) and cilantro (*Coriandrum sativum*) extracts. *Food Chemistry* 97, 505–515.
Yoshikawa, M., Uemura, T., Shimoda, H., Kishi, A., Kawahara, Y. and Matsuda, H. (2000) Medicinal foodstuffs. XVIII. Phytoestrogens from the aerial part of *Petroselinum crispum* Mill (parsley) and structures of 6-acetylapiin and a new monoterpene glycoside petroside. *Chemical and Pharmaceutical Bulletin (Tokyo)* 48, 1039–1044.
Yousofi, A., Daneshmandi, S., Soleimani, N., Bagheri, K. and Karimi, M.H. (2012) Immunomodulatory effect of parsley (*Petroselinum crispum*) essential oil on immune cells: mitogen-activated splenocytes and peritoneal macrophages. *Immunopharmacology and Immunotoxicology* 34, 303–308.
Zhang, H., Chen, F., Xi, W. and Yao, H. (2006). Evaluation of antioxidant capacity of parsley (*Petroselinum crispum*) essential oil and identification of its antioxidant constitutents. *Food Research International* 39, 833–839.

17 Patchouli

H.G. Ramya*
Punjab Agricultural University, Ludhiana, India

17.1 Botany

17.1.1 Introduction

Patchouli (*Pogostemon cablin*) is a small bushy perennial herb (see Fig. 17.1), which has odorous leaves and has been acclaimed as a valuable aromatic plant. The oil obtained from patchouli is a chief constituent in several perfumes as it imparts a rich scent. Raw patchouli oil itself can be used as an alternative for exotic perfumes. The oil also has good properties for use in perfuming soaps (Vijaykumar, 2004).

The herb is mainly grown for its essential oil, which can be obtained from the leaves, and also, in very small quantities, from the tender part of its stem. The oil is extracted from dried leaves of patchouli by the steam distillation technique. About 2.5–3.5% of high-quality oil with significant economic value can be obtained from shade dried patchouli leaves.

Popular cultivated patchouli varieties include cv 'Java' and 'Singapore' (named after their country of origin), from which distinctive quality oil of specified chemical composition and odorific value can be obtained. In contrast, the oils from cv 'Johore' and 'Malaysia' have a harsher odour. The varieties 'Johore', 'Singapore' and 'Indonesia' are those that are commonly cultivated; the best quality oil can be obtained from 'Johore' while a high yield of oil can be obtained from the other two varieties (Sarwar *et al.*, 1983; TNAU, 2016).

Taxonomically, the plant is classified as follows:

Kingdom: Plantae
Order: Lamiales
Family: Lamiaceae (Mint family)
Class: Magnoliopsida
Subclass: Asteridae
Genus: *Pogostemon*
Species: *Pogostemon cablin* (Blanco) Benth.

The vernacular names of the plant in various countries are given in Table 17.1.

In recent times, growing interest in patchouli fragrance has led to its widespread cultivation throughout tropical Asia. The plant was introduced to India during 1941 in Madhya Pradesh, Tamil Nadu, Kerala and Karnataka. Field experiments have shown patchouli oil of superior quality can be produced from patchouli herbage (leaves and tender twigs) grown under the weather conditions

*Corresponding author, e-mail: ramyarinda@gmail.com

© CAB International 2016. *Leafy Medicinal Herbs: Botany, Chemistry, Postharvest Technology and Uses* (eds D.C.P. Ambrose *et al.*)

Fig. 17.1. A patchouli plant.

Table 17.1. Vernacular names of the patchouli in various countries.

Country	Vernacular name(s)
China	Guang huo xiang
France	Patchouli
India	Pacha, sugandhi pandi (Gujarati); pachapat, patchouli (Bengali); pachauli (Hindi); pachetene (Kannada); pachi (Sanskrit); pachila, kattam (Malayalam); panch (Marathi)
Indonesia	Nilam (Aceh); Nilam Wangi (general); Singalon (Batak)
Korea	Hyangdulkkaephul
Malaysia	Dhalum wangi, nilam, tilam wangi
Philippines	Kabling (Tagalog); kadlum (Bikol, Bisaya, Sulu); katluen (Bisaya)
Spanish	Cablan, pachuli
Thailand	Phimsen (Bangkok)
UK	Patchouli

of Bangalore (ACHS, 2012). India could play an important role in the production of this economically valuable aromatic oil, if proper care is taken in the cultivation of patchouli, and would then have considerable scope for entering the world market (NABARD, 2007).

17.1.2 History/origin

Botanically, patchouli is identified as *P. cablin* (Blanco) Benth., and it belongs to the family Lamiaceae. It is indigenous and native to the Philippines. The plant was first described from the Philippines in 1837 as *Mentha cablin* Blanco by Francisco Manuel Blanco, a Spanish priest, in *Flora de Filipinas* (Flora of the Philippines, published between 1873 and 1877). The word *cablin* is similar to the word *cablan*, the vernacular name of the species in Spanish, and indeed also to *kabling*, the name in the Philippines. Soon after, in 1848, it was renamed *P. cablin* by Bentham. In 1845, the botanist Pelletier-Sautelet described this species grown in a hothouse in France as *P. patchouli*. J.D. Hooker designated the Indian patchouli plant, which is indigenous to the Western Ghats, as *P. heyneanus* Benth., and that in Sri Lanka as *P. patchouli*.

17.1.3 Location

Patchouli grows wild in the Philippines, Malaysia, Indonesia and Singapore. In India, it is cultivated in the coastal regions of Karnataka, Tamil Nadu, Maharashtra, Madhya Pradesh, Gujarat, West Bengal and Assam.

17.1.4 Morphology

Patchouli is a perennial bushy herb with smooth, opposite, ovate to oblong, leathery leaves with dentate margins, which are pale to purplish green in colour. It has a square stem that grows up to 1 m tall. It has a peculiar and distinctive aroma when it is rubbed. It has whitish flowers tinged with purple that grow in axillary as well as terminal spikes (Grieve, 2012).

17.2 Chemistry

17.2.1 Chemical composition

The various attributes in patchouli essential oil make it unique and able to be advantageously used for a range of purposes from medicinal to cosmetic. The main traits of patchouli oil are that it is slightly sticky, thick, viscous and very slow to volatilize. High quality oil has an intangible, wine-like, floral and sweet 'top note'. This top note becomes more obvious as the oil ages (about a year on from distillation). The 'body note' is rich, intensely sweet, woody, balsamic and earthy (Shankaranarayan, 2002). The chemical constituents of the oil that are the principal odour-intensive components are patchouli alcohol and norpatchoulenol.

Other components of patchouli essential oil include α-bulnesene, α-, β- and δ-patchoulene, α- and δ-guaiene, β-elemene and seychellene (Tierra, 1939). About 92% of patchouli oil is composed of compounds that have only a meagre influence on its aroma. Sesquiterpenes constitute 40–45% of the oil, of which patchouli camphor or patchouli alcohol or patchoulol represent 35–40%. Patchouli alcohol is odourless, although one or more compounds of patchoulol are possibly responsible for the unique odour of the oil. Also, a crystalline portion of norsesquiterpenic alcohol popularly known as patchoulenol has been separated that has been found to be the actual odour or aroma carrier of patchouli oil. The oil obtained from patchouli leaves is difficult to synthesize or to replace by an alternative, so the natural oil from patchouli remains the only source (NBH, undated).

17.2.2 Phytochemistry

A survey of the literature has ascertained that patchouli oil contains over 70 chemical compounds. The prime components are patchouli alcohol, pogostol, seychellene, nor-patchoulinol, patchoulipyridine, ethylchavicol, limonene, pinene and *p*-methoxycinnamaldehyde (Guan *et al.*, 1994; Daniel, 2006; Skaria *et al.*, 2007), and the remaining constituents are phytochemicals extracted from patchouli during the oil distillation process (Park *et al.*, 1998; Luo *et al.*, 2002; Kiuchi *et al.*, 2004; Sundaresan *et al.*, 2009). Patchouli essential oil goes well with other oils, for example, vetiver, sandalwood, geranium, lavender and cedarwood derivatives, and clove oil, thereby enabling the manufacturing of strong perfumes of enduring odour. Patchouli oil is extensively used in the perfumery industry, and is exploited in the production of scents, cosmetics, body lotions, shaving lotions, toiletry products, tobacco products, detergents and incense sticks.

The essential oil constituents of *P. cablin* (patchouli) and *P. travancoricus* Bedd. var. *travancoricus* were examined using gas chromatography (GC) and GC/mass spectrometry (MS). Eleven compounds were recognized from *P. cablin* oil and 13 from *P. travancoricus* var. *travancoricus* oil. Both species shared compounds such as α- and β-patchoulene, patchouli alcohol (patchoulol), β-caryophyllene, α-guaiene, seychellene and selinene, though they were quantitatively less in *P. travancoricus* var. *travancoricus*. The essential oil of *P. cablin* cultivated in Vietnam was investigated by capillary GC/MS, and about 32–38% of the patchouli oil was accounted for patchouli alcohol. Ten additional compounds were also identified that included α-bulnesene and α-guaiene as the main components (Dũng *et al.*, 1989).

17.3 Postharvest Technology

17.3.1 Processing

Drying of patchouli herbage

Drying is possibly the oldest, most widespread and diverse of chemical engineering unit operations. Over 400 types of dryers have been reported in the literature while over 100 distinct types are commonly available. Drying competes with distillation as the most energy-intensive unit operation due to the high latent heat of vaporization and the inherent inefficiency of using hot air as the most common drying medium. It is a complex operation involving transient transfer of heat and mass, along with several other processes, such as physical or chemical transformations, which may cause changes in product quality as well as in the mechanisms of heat and mass transfer (Mujumdar and Devahastin, 2000).

Dried herbs have great importance not only for culinary purposes, but also for medicinal uses. Harvesting and drying of herbs are not complicated. The major issue that must be kept in mind during the gathering and drying of herbs is the presence of volatile oil, and the most significant part of the patchouli plant is stored in the leafy material, although the tender twigs also give the plant its exceptional aroma and fragrance. It is the essential oil that is the chief element that must be conserved during the drying process (Joykumar, 1997; Parziale, 1998).

Shade drying of patchouli herbage is carried out by spreading the fresh herbage on stainless steel trays and drying in a well-ventilated room. The preliminary drying bed thickness is kept at 100 mm and the leaves should be turned regularly. The weight loss of samples should be recorded every hour using a sensitive electronic balance and the moisture content at any given time estimated accurately by the toluene distillation method (Farooqui *et al.*, 2001). Shade drying is continued until there is no more weight loss in samples as indicated by constant consecutive weight readings. At this point, the herbage can be assumed to be dried to its stable equilibrium moisture content of around 11–12% (wet basis, w.b.) from 80% (w.b.) at the beginning of drying. The process is dependent upon the incidence of sunshine and atmospheric humidity. Sufficient drying is of great importance for acquiring the highest yield and superior quality of oil. Appropriately dried leaves develop a typical patchouli odour that is less obvious in fresh herbage.

At several localities, fermentation is a common procedure, but it generates an uninvited consequence in that the oil develops a mouldy odour. Hence, it is advisable to prevent any such fermentation (Anitha, 2008).

Pallavi *et al.* (2006) investigated the drying behaviour of patchouli herbage under several methods: in the shade, in a tray dryer and in a trouble-free batch-type tray dryer commonly known as the ASTRA model developed by the Centre For Sustainable Technologies of the Indian Institute of Sciences at Bangalore, India. This is an agricultural waste-based crop dryer based on the principle of a fuel-efficient wood stove. Drying bed thickness was maintained consistently in all drying methods and the herbage was dried from 80% (w.b.) initial to 11–12% (w.b.) final moisture content. Under climatic conditions of Bangalore (21.0–24.4°C; 40–81% R.H.), patchouli required 54 h drying time in the shade, while in the ASTRA dryer, it required just 14 h. In a convectional (electrical) tray dryer, the drying time at 30, 40, 50, 60 and 70°C was 13, 12, 11, 7 and 6 h, respectively. Freshly harvested patchouli herbage was dried at a lower temperature by a forced flow dryer under shade-drying conditions in well-ventilated room (mean ambient air temperature 33°C and 70% RH). Lower temperatures of 40 and 45°C were selected after trials that aimed to attain the required moisture content at the lowest possible temperature and in the shortest drying time. The airflow maintained in the dryer was 37.93 m^3/min. The initial moisture content of the sample was noted by the toluene distillation method. The type of drying and the temperature of the drying air influence the quantity and quality of the active elements present in aromatic and medicinal plants, and in order to attain a better yield of patchouli oil in a shorter drying time, mechanical drying of the herbage is an excellent option. Accordingly, it was found that drying at 45°C temperature in a forced flow dryer would be a good method for drying patchouli as it is rapid and gives oil recovery of 2.60%, which is better than that obtained by shade drying (Ambrose *et al.*, 2013).

Processing of patchouli oil by steam distillation

Patchouli cultivation is considered to be farmer friendly as the plant is easy to handle compared with other aromatic plants. Additionally, once it is dried and properly preserved, the herbage can be used for oil extraction at leisure. The essential oil is found in all parts of the herbage including the root, but the top leaves and tender twigs have the maximum quality oil (Vijayakumar, 2004). Generally, dry leaves stored for 4–6 months produce more oil with a superior aroma.

Steam distillation is the most usual method for the extraction of essential oils from aromatic plants. Even though there are various extraction techniques (hydrodistillation, microwave distillation, supercritical fluid extraction, ultrasound extraction), from the consumer standpoint, steam distillation remains the ideal process for the extraction of essential oils. The ease of the technique presents a guarantee of purity as the process uses only water (Tannous *et al.*, 2012). At present, the extraction efficiency of patchouli oil is very low, about 2.5–3.5%, probably due to improper postharvesting practices at various stages, such as herbage drying and the distillation method used, so it is hoped that the use of steam distillation will improve this situation.

The apparatus for steam distillation comprises a boiler, distillation entity, condenser and receiver. The distillation entity is usually made up of mild steel with a perforated bottom on to which the herbage is loaded for distillation. The herbage needs to be evenly/tightly packed so that steam channels do not form during distillation and result in a low essential oil yield. The level of water in the boiler is inspected regularly so that it is kept at the appropriate level. The maintenance of pressures approximating to 1.4–3.5 kg/m^2 produce superior quality oil as more cell walls rupture at this pressure. The distillation interval varies from 6 to 8 h. The condenser of the apparatus cools the vapours received from the distillation entity. This is made of several tubes made up of stainless steel or copper and mounted inside a jacket, and it has an inlet and outlet for the circulation of cooling water. The hot vapours of steam and essential oil are cooled in the condensing tubes and the condensate moves out into the receiver. The essential oil vapour and the spent steam that come out of the distillation entity will be condensed back to the liquid phase in the water-cooled condenser and the condensate will be received into the receiver tank. The steam distillation process results in separate products: the liquid distillate that contains the volatile, water-soluble parts of the herbage known as the 'hydrosol' and the volatile water-insoluble material of the herbage which is the essential oil (Fig. 17.2) (Ramya, 2010; Ramya *et al.*, 2013).

The condensate in the receiver tank should be allowed to stand for a sufficient time (overnight) so that the patchouli oil separates out as far as possible from the water layer. After appropriate separation in the receiver, the essential oil must be further separated from the water phase using a separating funnel. As it is lighter than water and insoluble, the oil floats on the apex of the receiver and only water is drained off. The oil will be still turbid, and all traces of moisture present should be removed by adding anhydrous sodium sulfate at 20–30 g/l and maintaining the distillate blend for 4–5 h. Following this, the oil should be filtered through a Whatman filter paper to obtain a clear essential oil. Any moisture present in the essential oil can encourage polymerization and lead to loss in quality. The clear essential oil that is obtained should be stored in a cool dry, dark place in air-tight aluminium containers or coloured glass bottles filled up to the rim (depending on the quantity of oil to be stored). Normally 60 kg of oil/ha is obtained in a year (Jain, 1978; Anonymous, 2011; Horticulture Portal, 2013).

Fig. 17.2. Patchouli essential oil.

Oil recovery

When patchouli herbage is harvested, shade dried and steam distilled, an average essential oil yield of 2.5% may be considered satisfactory in commercial distillation (Leung and Foster, 1996). Many studies have been conducted to establish the best method for essential oil recovery. Factors that can affect the quality of essential oil are extraction time, pressure, temperature and sample mass or bed thickness. The quality of the oil produced depends on the time of extraction – the longer the extraction time, the higher the content of patchouli alcohol (Yahya and Yunus, 2013).

Demand for and economics of patchouli oil

The majority of the patchouli oil in the world is manufactured in the Philippines, Indonesia, Malaysia and Singapore, with Indonesia distributing most of the world's requirement, which is around 1500 t a year. India's production is about 1 t, while its requirement borders around 220 t annually (Sarwar *et al.*, 1983). The world's production of patchouli oil is around 800 t/annum. Java produces about two thirds of this, followed by China and Malaysia. Patchouli farming in India is scanty, but it has expanded in the last 5 years to cover around 600 ha, accounting for the production of 20 t of oil per annum (Kumar *et al.*, 1986). Industrial cultivation of the crop in India was first attempted by Tata Oil Mills in 1942, and the requirement for patchouli oil is expanding more rapidly than that for most of the other essential oils.

17.3.2 Value addition

Perfumery

The perfume or fragrance industry is one of the ancient industries of the world. In the early days, the perfumer used only natural essential oils for perfume production, which was viewed as an art. With time, due to the scarcity of natural essential oils, the price of oils started rising. With the advent of organic chemistry, synthetic perfumery materials became affordable. Meanwhile, the use of perfumes in household products became more widespread and the requirement for industrial perfumes increased significantly. This benefited perfumers, as they now had more alternatives for ingredients at an economic cost, and enforced progress on improving the chemistry of perfumery ingredients, including their quality and profiles (Ranade and Paranape, 2000; Singh *et al.*, 2002).

Patchouli oil has an outstanding base note in the perfumery and fragrance industry, and is highly valued for its exceptional fragrance, and the perfume industry has exploited patchouli oil in a number of the world's luxury perfumes for its soulful, sensual, woody and voluptuous notes. There is no alternative synthetic chemical to substitute for the oil of patchouli, which further boosts its value and its eminence in the perfumery market. There is a great demand for it in soaps, scents, body lotions, detergents, tobacco and incense manufacturing. Patchouli oil is also used extensively in perfumed and scented industrial commodities such as paper towels, laundry detergents and air fresheners (Tac Ethno, 2012).

Incense

Incense sticks are generally prepared from aromatic plants and essential oils extracted from plants or animal sources. When they are burned, they emit a very captivating and appealing fragrance (Borah *et al.*, 2003). In prehistoric times, only naturally fragrant resins or woods such as sandalwood, and patchouli, were used for incense. Modern fragrance production allows practically any fragrance to be duplicated, but patchouli is still an important component in East Asian incense. The incense is familiar for its sturdy fragrance, and incense sticks made using patchouli essential oils are known to have antifungal, anti-inflammatory, antiseptic and antidepressant properties. Patchouli incense sticks have also been recognized to generate aphrodisiac effects when burned. There is a scope for utilizing the patchouli herbage left after steam distillation (the spent charge) for incense stick production after drying and macerating it into appropriate particle size powder (Ramya, 2010). At present, the spent charge is not used except for the production of biogas by anaerobic digestion and for manure (Anonis, 2007).

Cosmetics

Patchouli essential oil has been exploited in several natural cosmetics products such as aromatic bath gel, aromatic soaps, body lotions, cosmetics fragrances and so forth (Grieve, 2012).

Deodorant

The strong sweet, spicy and musky aroma of patchouli essential oil masks body odour, but it should be used in a diluted form as the aroma can be very strong to some people's olfactory senses. Some people find the aroma astonishing, while others are rather irritated by it.

Household uses

The dried leaves and oil of patchouli are used for pot-pourri and to scent fabric products. In Asia and South America, patchouli oil is frequently mixed with anise and clove as a breath sweetener (Wikipedia, 2009). Besides its medicinal and perfumery uses, it can act as an insect repellent (Yoshihiro *et al.*, 1992; Organic Facts, 2014). Notably, the patchouli herbage is claimed to be an effective repellent against the Formosan subterranean termite (Zhu *et al.*, 2003).

Flavourings

Patchouli oil is widely used as a flavour ingredient in major food products and in alcoholic and non-alcoholic beverages. A low concentration (2 mg/kg) of oil is used for flavouring beverages, frozen dry desserts, candy, baked items, gelatin, meat and meat products.

17.4 Uses

17.4.1 General medicinal uses

Patchouli oil has been identified as helping in the treatment of eczema, dermatitis, psoriasis and sores. It also relieves constipation and can be used as a fleeting antidote or soothing agent against snake and insect bites.

17.4.2 Pharmacological uses

There are numerous benefits of patchouli oil ranging from skincare, medicinal uses to aromatherapy. It performs different pharmacological activities including use as an antidepressant, anti-inflammatory, antifungal, antiseptic, astringent, diuretic and sedative. The insecticidal and insect repellent properties of the oil have been well known for many years, and it has been used to protect clothes and fabrics from insects. It is now viewed as one of the most adaptable and versatile essential oils in the global market (Ichikawa *et al.* 1989; NABARD, 2007; Karimi 2014; Organic Facts, 2014). The uses of the oil have been summarized and discussed on the Organic Facts website, and much of the information below (and some of that given above) is based on that information (see Organic Facts, 2014).

Antimicrobial activity

Patchouli essential oil is effective against Gram-positive organisms (acting as an antibacterial) and quite a few dermatophytes (skin, hair and nail fungi) collected from wound and skin infections (acting as an antifungal. It can also be used as a substitute medication for skin infections that are caused by strains of bacteria that have developed resistance to antibiotics (Osawa *et al.*, 1990). Patchouli oil has also shown antibacterial activity against periodontopathic bacteria, including *Actinobacillus*, *Capnocytophaga*, *Fusobacterium* and *Eikenella* (Kongkathip *et al.*, 2009). The oil has been found to be quite helpful against fungal development and infectivity in infections such as athlete's foot (Organic Facts, 2014).

Aromatherapy

Patchouli oil is used for massages and aromatherapy. Massaging with the oil can boost the mood, control the emotions and calm the nerves, thus alleviating stress. In circumstances of dermatitis and depression, it is the best universal remedy (Chakrapani, 2013).

Skin problems, healing activity and use as an antiseptic

The oil is being used for healing a variety of skin problems such as psoriasis, dermatitis and eczema. The oil has antiseptic attributes that heals the wounds comparatively fast, and prevents the spread of infection. It has anti-ageing properties as well, and so it makes a good skincare product as it keeps the skin young. It is also good for acne, scars, burns, cuts, stretch marks and other skin-related problems. Frequent use of patchouli oil can hide scars and lasting marks from measles and pox, keep a check on dry skin and wrinkles, and promote a soft facial appearance (Ichikawa *et al.*, 1989; Karimi, 2014).

Relief of depression

People experiencing depression can benefit from the use of patchouli oil as it is regarded as a mood lifter. It relieves the feelings of sadness, so is commonly used in aromatherapy (see above). It boosts the mood, alleviates disappointment and calms down anxiety and anger, maybe by having an impact on hormones and chemical reactions in the body.

Anti-inflammatory and anti-irritant activity

Patchouli oil soothes inflammation and irritation due to fever and relieves the fever itself. It is useful in treating a wide range of problems, including internal inflammation such as caused by arthritis and gout (Ichikawa *et al.*, 1989). The essential oil has long been used in traditional Chinese medicine to treat inflammatory illness.

Antiviral activity

Wu *et al.* (2011) found that patchouli alcohol had an inhibitory action against influenza A (H2N2) in *in vitro*, *in vivo* and *in silico* studies. It was also found to have a direct effect against influenza virus A/PR/8/34 (H1N1) in *in vitro* tests (Kiyohara *et al.*, 2012).

Use as an aphrodisiac

Patchouli oil is also able to cure various sexual problems, such as lack of libido, erectile malfunctions and frigidity, and has been used as an aphrodisiac for centuries (Ichikawa *et al.*, 1989).

Constriction activity

In crude drug screening tests, patchouli oil was active in promoting K+ contracture of guinea pig taenia coli (smooth mucle) *in vitro*, and might therefore be useful in averting sagging skin and slackening of muscle tissue in old age, and in helping to stop haemorrhaging (Ichikawa *et al.*, 1989).

Diuretic activity

Patchouli oil regulates the frequency of urination and the quantity of urine, and in doing so alleviates weight loss, reduces blood pressure, improves the appetite, lowers cholesterol, eliminates toxins from the body and decreases the possibilities of development of gall stones and kidney stones, and problems such as gout (Ichikawa *et al.*, 1989).

Insecticidal activity

The insecticidal attributes of patchouli oil have been documented since antiquity. In India, the oil has long been used to keep moths and other insects away from linens, woollen shawls and rugs (Chakrapani, 2013). It is used in sprays, body lotions, fumigants, vaporizers and incense sticks or mixed with water to rinse clothes and bed linen to chase off mosquitoes, ants, bedbugs, lice, fleas, flies and moths (only a few drops are necessary). Gokulakrishnan *et al.* (2013) reported the pupicidal and repellent activities of chemical compounds from patchouli essential oil against medically important human vector mosquitoes.

17.5 Summary

Patchouli oil is familiar for its persistent spicy aroma, and is extensively used in oriental perfumes, incense and joss sticks, commonly mixed with jasmine and sandalwood. Post-harvest management of patchouli herbage contributes significantly to obtaining a good yield and quality of essential oil. Although there are numerous techniques for extraction of the oil from patchouli herbage, steam distillation is the ideal process from the consumer standpoint. Patchouli can be constructively exploited in aromatherapy, perfumery, cosmetics, incense stick production and the food flavouring industries.

References

ACHS (2012) Essential oil of patchouli: *Pogostemon cablin* (syn. *Pogostemon patchouli*). American College of Healthcare Sciences, Portland, Oregon. Available at: http://www.achs.edu/mediabank/files/achs_patchouli_monograph.pdf (accessed 17 May 2012).

Ambrose, D.C.P., Annamalai, S.J.K. and Naik, R. (2013) Effect of drying on the volatile oil yield of patchouli. *Indian Journal of Science and Technology* 6, 5559–5562.

Anitha, M. (2008) Agro-processing of patchouli for efficient essential oil extraction. M.Tech thesis, University of Agricultural Sciences, Bangalore, India.

Anonis, D.P. (2007) Woody notes in perfumery: patchouli in fragrances, Part II. *Perfumer and Flavorist* 31(11), 36.

Anonymous (2011) Patchouli – Booklet No. 494. Medicinal Plants: MPS-21. Formerly available at: www.inseda.org/Additional%20material/CD%20/Patchouli-494.doc (accessed 2 May 2011).

Borah, R.C., Talukdar, A., Kataky, J.C.S, Unni, B.G., Modi, M.K. and Deka, P.C. (2003) Bio-prospecting of commercially important plants. Paper presented at: *National Symposium on 'Biochemical Approaches for Utilization and Exploitation of Commercially Important Plants' Organized at Jorhat, Assam, India, 12–14 November 2003*, pp. 290–293.

Chakrapani, P., Venkatesh, K., Chandra Sekhar Singh, B., Arun Jyothi, B., Prem Kumar, Amareshwari, P. and Roja Rani, A. (2013) Phytochemical, pharmacological importance of patchouli (*Pogostemon cablin* Benth.) an aromatic medicinal plant. *International Journal of Pharmaceutical Sciences Review and Research* 2, 7–15.

Daniel, M. (2006) *Medicinal Plants: Chemistry and Properties*. Science Publishers, Enfield, New Hampshire.

Dũng, N.X., Leclercq, P.A., Thai, T.H. and Moi, L.D. (1989) Chemical composition of patchouli oil from Vietnam. *Journal of Essential Oil Research* 1, 99–100.

Farooqui, A., Vasundhara, A.M. and Srinivasappa, K.N. (2001) *A Guide to the Cultivation of Commercially Important Aromatic Crops*. Division of Horticulture, University of Agricultural Sciences and Association for Promotion of Medicinal and Aromatic crops, Bangalore, India.

Gokulakrishnan, J., Kuppuswamy, E., Shanmugam, D., Appavu, A. and Kaliyamurthy, K. (2013) Pupicidal and repellent activities of *Pogostemon cablin* essential oil chemical compounds against medically important human vector mosquitos. *Asian Pacific Journal of Tropical Disease* 3, 26–31.

Grieve, M. (2012) Patchouli. Botanical.com: A Modern Herbal. Available at: http://www.botanical.com/botanical/mgmh/p/patcho15.html (accessed 15 June 2012).

Guan, L., Quan, L.H., Xu, L.Z. and Cong, P.Z. (1994) Chemical constituents of *Pogostemon cablin* (Blanco) Benth. *Zhongguo Zhong Yao Za Zhi* 19, 355–356, 383.

Horticulture Portal (2013) Crop encyclopedia: Patchouli *(Pogostemon patchouli)*. Society for Advancement of Horticulture, Indiranagar, Lucknow, India. Available at: http://hortportal.org/crop.php?alpha=P (accessed 9 January 2013).

Ichikawa, K., Kinoshita, T. and Sankawa, U. (1989) The screening of Chinese crude drugs for calcium antagonist activity: identification of active principles from the aerial part of *Pogostemon cablin* and the fruits of *Prunus mume*. *Chemical and Pharmaceutical Bulletin (Tokyo)* 37, 345–348.

Jain, P.C. (1978) Potentialities of growing patchouli *(Pogostemon patchouli)*.*Indian Perfumer* 22(1), 47–55.

Joykumar, N. (1997) Studies on the drying characteristics of some important flowers. M.Sc. thesis, University of Agricultural Sciences, Bangalore, India.

Karimi, A. (2014) Characterization and antimicrobial activity of patchouli essential oil extracted from *Pogostemon cablin* (Blanco) Benth.(Lamiaceae). *Advances in Environmental Biology* 8, 2301–2309.

Kiuchi, F., Matsuo, K., Ito, M., Qui, T.K. and Honda, G. (2004) New sesquiterpene hydroperoxides with trypanocidal activity from *Pogostemon cablin*. *Chemical and Pharmaceutical Bulletin (Tokyo)* 52, 1495–1496.

Kiyohara, H., Ichino, C., Kawamura, Y., Nagai, T., Sato, N. and Yamada, H. (2012) Patchouli alcohol: *in vitro* direct anti-influenza virus sesquiterpene in *Pogostemon cablin* Benth. *Journal of Natural Medicines* 66, 55–61.

Kongkathip, N., Sam-ang, P., Kongkathip, B., Pankaew, Y., Tanasombat, M. and Udomkusonsri, P. (2009) Development of patchouli extraction with quality control and isolation of active compounds with antibacterial activity. *Kasetsart Journal (Natural Science)* 43, 519–525.

Kumar, A., Gauniyal, A.K. and Virmani, O.P. (1986) Cultivation of *Pogostemon patchouli* for its oil. *CROMAP (Current Research on Medicinal and Aromatic Plants)* 8(2), 79–86.

Leung, A.Y. and Foster, S. (1996) *Encyclopedia of Common Natural Ingredients Used in Food, Drugs and Cosmetics.*, 2nd edn. Wiley, Hoboken, New Jersey.

Luo, J., Guo, X. and Feng, Y. (2002) Constituents analysis on volatile oil of *Pogostemon cablin* from different collection time cultivated in Hainan. *Zhong Yao Cai* 25(1), 21–23.

Mujumdar, A.S. and Devahastin, S. (2000) Fundamental principles of drying. In: Devahastin, S. (ed.) *Mujumdar's Practical Guide to Industrial Drying*. Exergex Corporation, Montreal, Canada, pp. 1–22.

NABARD (2007) Model Bankable Projects: Patchouli. National Bank for Agriculture and Rural Development, Mumbai, India. Available at: http://farmextensionmanager.com/English/Agribusiness%20opportunities/Medicinal%20plant%20sector/patchuoli.htm (accessed 27 January 2016).

NBH (undated) Patchouli. National Horticulture Board, Gurgaon, Haryana, India. Available at: http://nhb.gov.in/report_files/patchouli/PATCHOULI.htm (accessed 27 January 2016).

Organic Facts (2014) Health Benefits of Patchouli Essential Oil. Organic Information Services, Andheri East, Mumbai, India. Available at: http://www.organicfacts.net/health-benefits/essential-oils/health-benefits-of-patchouli-essential-oil.html (accessed 5 May 2014).

Osawa, K., Matsumoto, T., Maruyama, T., Takiguchi, T., Okuda, K. and Takazoe, I. (1990) Studies of the antibacterial activity of plant extracts and their constituents against periodontopathic bacteria. *The Bulletin of Tokyo Dental College* 31(1), 17–21.

Pallavi, G.S., Palanimuthu, V., Chandru, R. and Ranganna, B. (2006) Study of tray drying behaviour of aromatic patchouli herbage and its effect on essential oil recovery. Paper presented at: *National Seminar on '40th Annual Convention & Symposium of ISAE' Organized by TNAU, Coimbatore, India, January 19–21.*

Park, E.J., Park, H.R., Lee, J.S., Kim, J. and Licochalcone, A (1998) An inducer of cell differentiation and cytotoxic agent from *Pogostemon cablin*. *Planta Medica* 64, 46.

Parziale, E. (1998) How to Harvest, Dry and Store Herbs. Available at: http://earthnotes.tripod.com/harvest.htm (accessed 12 January 2011).

Ramya, H.G. (2010) Study on steam distillation of patchouli (*Pogostemon cablin* Benth.) and utilization of the by-product spent leaves. M.Tech. thesis, University of Agricultural Sciences, Bangalore, India.

Ramya, H.G., Palanimuthu, V. and Singla Rachna (2013) An introduction to patchouli (*Pogostemon cablin* Benth.) – a medicinal and aromatic plant: its importance to mankind. *CIGR Journal* 15, 243–250.

Ranade, G.S. and Paranape, S. (2000) Odour quality and profiles of perfumery ingredients. *Indian Perfumer* 44, 183–189.

Sarwar, M., Narayana, M.R. and Virmani, O.P. (1983) *Patchouli and Its Cultivation in India*. Farm Bulletin No. 17, Central Institute of Medicinal and Aromatic Plants, Lucknow, India.

Shankaranarayan, V. (2002) Patchouli constituents and its usage in perfumery. *Indian Perfumer* 46, 313–314.

Singh, M., Sharma, S. and Ramesh, S. (2002) Herbage, oil yield and oil quality of patchouli (*Pogostemon cablin* (Blanco) Benth.) influenced by irrigation, organic mulch and nitrogen application in semi-arid tropical climate. *Industrial Crops and Products* 16, 101–107.

Skaria, B.P., Joy, P.P., Mathew, S., Mathew, G., Joseph, A. and Joseph, R. (2007) *Aromatic Plants: Horticulture Science Series – 1.* New India Publishing Agency, New Delhi.

Sundaresan, V., Singh, S.P. and Mishra A.N. (2009) Composition and comparison of essential oils of *Pogostemon cablin* (Blanco) Benth. (patchouli) and *Pogostemon travancoricus* Bedd. var. *travancoricus*. *Journal of Essential Oil Research* 21, 220–222.

Tac Ethno (2012) *Pogostemon patchouli*, Patchouli. TAC Ethnobotanical Database. Available at: http://www.tacethno.com/pogostemon-patchouli-patchouli.html (accessed 7 September 2012).

Tannous, P., Juliani, R., Wang, M. and Simon, J. (2004) *Water Balance in Hydrosol Production via Steam Distillation: Case Study using Lavandin (Lavandula × intermedia).* New Use Agriculture and Natural Plant Products and ASNAPP Program, Department of Plant Biology and Plant Pathology, Rutgers, The State University of New Jersey, New Brunswick, New Jersey.

Tierra, M. (1939) *Planetary Herbology: An Integration of Western Herbs into the Traditional Chinese and Ayurvedic Systems.* Edited and supplemented by Frawley, D. (1988). Published in 1992 by Lotus Press, Twin Lakes, Wisconsin.

TNAU (2016) Horticulture: Aromatic Crops: Patchouli. TNAU Agrictech Portal, Tamil Nadu Agricultural University, Coimbatore, India. Available at: http://agritech.tnau.ac.in/horticulture/horti_aromatic%20crops_patchouli.html (accessed 27 January 2016).

Vijayakumar, K. (2004) Patchouli and India –a great leap forward. In: *Proceedings of National Seminar of Prospectus and Potentials of Medicinal and Aromatic Crops, Held at Bangalore, 18–19 June 2004*, pp. 106–107.

Wikipedia (2009) Patchouli. Available at: http://en.wikipedia.org/wiki/patchouli (accessed 25 June 2009).

Wu, H., Li, B., Wang, X., Jin, M. and Wang, G. (2011) Inhibitory effect and possible mechanism of action of patchouli alcohol against influenza A (H2N2) virus. *Molecules* 16, 489–501.

Yahya, A.O. and Yunus, R.M. (2013) Influence of sample preparation and extraction time on chemical composition of steam distillation derived patchouli oil. *Procedia Engineering* 53, 1–6.

Yoshihiro, H., Katsuhiko, T. and Toi, N. (1992) An additional constituent occurring in the oil from a patchouli cultivar. *Flavour and Fragrance Journal* 7, 333–335.

Zhu, B.C.-R., Henderson, G., Yu, Y. and Laine, R.A. (2003) Toxicity and repellency of patchouli oil and patchouli alcohol against Formosan subterranean termites *Coptotermes formosanus* Shiraki (Isoptera: Rhinotermitidae). *Journal of Agricultural and Food Chemistry* 51, 4585–4588.

18 Rosemary

Milda E. Embuscado*
McCormick & Company, Hunt Valley, Maryland, USA

18.1 Botany

18.1.1 Introduction

Rosemary, scientific name *Rosmarinus officinalis* L., belongs to the mint family (Lamiaceae) together with oregano, thyme, basil and lavender. It is one of the most important herbs used for culinary purposes, as well as being used in alternative and herbal medicine. Rosemary is a fragrant woody plant that is native to the Mediterranean region and is an essential element of the Mediterranean diet, along with many other spices and herbs. Rosemary leaves are used as a culinary condiment that provides excellent flavour for meats such as chicken, beef, lamb and fish. The leaves are also used as teas and extracts are used in beverages or as flavourings.

The name rosemary was derived from the Latin words *ros* meaning dew and *marinus* meaning sea, thus 'dew of the sea'. The plant is a perennial evergreen shrub with minty needle-like leaves that measure 0.8–1.6 in (2–3 cm) and it grows well in sandy soil or on dry and rocky slopes near the sea in full sun – which is probably the reason why it is called 'dew of the sea'. Rosemary has pretty pink, white, blue or purple flowers and the leaves are green on top and whitish underneath with short, dense, woolly hairs (Fig. 18.1). The shrub can reach 5 ft (1.5 m) in height when it grows upright, or it can grow sprawling on rocky slopes. The common or vernacular names of rosemary are: the compass plant, polar plant, compass weed, Rosmarinus coronarium, incensier, sea dew, ros maris, rosmarine, rosemarie, guardrobe, kelil, romanyi, rosmarin, rozemarjin, harilikrosmarin, rohtorosmarini, rozmarin, ramerino, mannenrû, rozmaryn, alecrim, romero, dumero and biberiye.

18.1.2 History/origin

Rosemary has been in use for thousands of years, starting from ancient times when the Greeks and Romans used it for its mystical and healing powers. The gardens of the ancient Greeks and Romans contained rosemary plants for protection from evil spirits, and it is said that rosemary grows only in the gardens of the righteous. The Roman army brought rosemary with them on their travels, and so its influence spread to Britain, then to the rest of Europe and, ultimately, reached the New World. The various uses of

*Corresponding author, e-mail: milda_embuscado@mccormick.com

© CAB International 2016. *Leafy Medicinal Herbs: Botany, Chemistry, Postharvest Technology and Uses* (eds D.C.P. Ambrose *et al.*)

Fig. 18.1. A rosemary shoot and flowers (Wikimedia Commons, 2014a).

rosemary for its healing powers and its ability to increase blood circulation and strengthen blood vessels were mentioned in ancient European history. Rosemary has also been associated with improving memory and the heart. Here is a brief summary of its history:

1. At around 5000 BC, references to rosemary were found in Cuneiform writing on stone tablets. This form of writing is one of the earliest known systems, and employed reed or grass as the writing instrument and clay or stone tablets as the writing medium.
2. In *c.*40–90, Pedanius Dioscorides, a Greek physician, phamacologist and botanist, who practised in Rome during the time of Nero, wrote *De Materia Medica*, one of the most influential herbal books. This book, which was written in five volumes, also contained remedies provided by other herbs that were used by the ancient Greeks and Romans.
3. In *c.*1525, Rycharde Banckes printed the book *Banckes' Herbal in England*, in which healing and other powers of rosemary were discussed. The title page contained the following passage 'Here begynneth a newe mater, the whichesheweth and treateth of ye vertues and proprytes of herbes, the whiche is called anHerball'.
4. In 1652 and 1653, Nicholas Culpepper, an English botanist, herbalist and physician, wrote *The English Physician* and the *Complete Herbal*, which contributed to the vast knowledge of the pharmacological properties of herbs. Dr Culpepper spent most of his life cataloguing hundreds of medicinal herbs and devoted himself to using herbs to treat his patients.

There are numerous inclusions or citations of rosemary in historical events, in poems and writings, and in practices and traditions through legends and myths. Here are a few of these:

1. Rosemary is known as the 'herb of remembrance' and was used by early Europeans in the graves of their loved ones so that they would not be forgotten. This was also a practice of the ancient Egyptians, who placed sprigs of rosemary in the tomb to remember the dead and used it in bouquets of funeral flowers.
2. The ancient Greek scholars wore the rosemary sprigs in their hair while they were studying because it was believed to enhance concentration. Rosemary sprigs were also placed under a student's pillow the night before exams because it was believed to improve memory during sleep.
3. Queen Isabella of Hungary used rosemary extracted by alcohol to treat gout. In Banckes' book, rosemary leaves that had been extracted using boiled water were applied as a poultice on the leg that 'be blown with gout'. Rosemary tea was also said to be 'for much worth against allevils of the body' according to Banckes. Dr Culpepper stated that 'the (rosemary) water is an admirable cure-all remedy of all kinds of cold, loss of memory, headache, coma'.
4. Rosemary was also used in cosmetics as recommended by Banckes in the following excerpt from his book: 'boyle the leves in whytewyne and wasshe thy face therwith ... thou shalt have a fayre face; make thee a box of the wood (rosemary) and smell to it and it shall preserve thy youthe'. Gervase Markham (1568–1637) regarded the many excellent uses of rosemary water for improving the

complexion and the heart, the brain and the whole body, 'cleansing away the spots of the face' making 'a man look young'.

From well-known literary writings, rosemary was part of these stories, as in the Sleeping Beauty, who was said to have been awoken from her sleep by Prince Charming when a rosemary sprig brushed over her cheek. In Shakespeare's *Hamlet*, Ophelia says 'There's rosemary, that's for remembrance'; and in *Romeo and Juliet*, it was quoted that Juliet was honoured at her burial with rosemary for remembrance. Rosemary has been a symbol of loyalty, fidelity, love and remembrance, and it is still used in Europe today when brides wear a rosemary sprig in their hair during their wedding ceremonies.

18.1.3 Location

Rosemary is grown commercially in France, Italy, Spain, Morocco and Tunisia. The major producers are in France, Spain and Morocco. There are several countries in Europe that cultivate rosemary on organic farms. It is also grown in the USA, Mexico, in almost all regions of South Africa and in some regions in Asia. In addition, the herb can also be found under cultivation in countries such as Austria, Belgium, Bulgaria, Denmark, Finland, Germany, Greece, Hungary, Iceland, the Netherlands, Portugal, Romania, Sweden and the UK.

Rosemary is drought tolerant, easy to grow and relatively pest resistant. It grows well in sandy to clay loam soil with good drainage and in full sun, and best in neutral to alkaline conditions. There are a number of cultivars available but for commercial purposes the following are the most important: 'Camphor-borneol' grown in Spain, '1,8 Cineole' grown in Tunisia and 'Verbenone' grown in France. Rosemary is a hardy, temperate plant that can tolerate frost and grows well at temperatures of 20–25° C. It is very adaptable.

Reliable production and consumption figures are not available because, in records, rosemary is lumped together with other aromatic herbs (see Fig. 18.2) such as basil, capers, horseradish, hyssop, juniper berry, marjoram, mint, oregano, parsley, savory, tarragon, tejpath (*Cinnamomum tamala*), thyme, sage and others. The global export of aromatic herbs has been steady and slightly increasing since 2003 compared with spices, whose export increased dramatically over the same time period. The spices include major ones such as allspice, annatto seed, aniseed, asafoetida, cambodge (Malabar tamarind), caraway, cardamom (large and small), cassia, cinnamon, cloves, coriander, cumin, dill, dried chillies, dried ginger, fennel, fenugreek, greater galanaga (or galangal, *Alpinia galangal*), kokam (*Garcinia indica*), lovage, mace, mustard, nutmeg, paprika, pepper, pepper long, pomegranate seeds, poppy, saffron, star anise, sweet flag (*Acorus calamus*), tamarind, turmeric, vanilla, etc. As precise data is not available, it was recommended by the Codex Alimentarius Commission (2013) that there is a need to collect this type of data because rosemary provides functionality in addition to providing aroma and taste in foods, and it also has a big impact on the identity of any food. Hence, impediments to the availability of vital statistics on the production, export, import of and value addition to rosemary need to be overcome (Codex Alimentarius Commission, 2013).

In addition to fresh and dried rosemary leaves, other products such as rosemary extract and rosemary oil are in demand in the international market. The import of rosemary oil to the USA (Fig. 18.3) shows an annual growth rate of 10%, and if this is assumed to be one third of the global demand, the total global demand is estimated at 323 t p.a. (Codex Alimentarius Commission, 2013), which is a considerable amount.

18.1.4 Morphology

Rosemary (*Rosmarinus officinalis* L.) belongs to the Lamiaceae (Labiatae) family under the lamids, asterids, monocots, magnolids, and finally the angiosperm phylogeny. It is native to the Mediterranean region, and fresh and dried leaves are available commercially, with the dried form in either crushed or

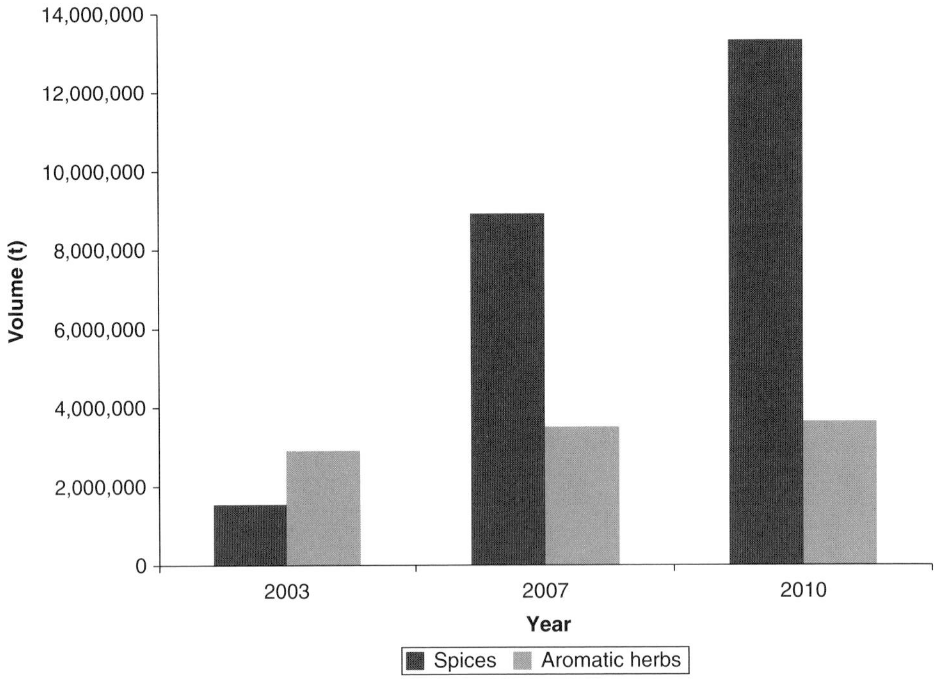

Fig. 18.2. The global export of spices and aromatic herbs based on the data from Codex Alimentarius Commission, 2013.

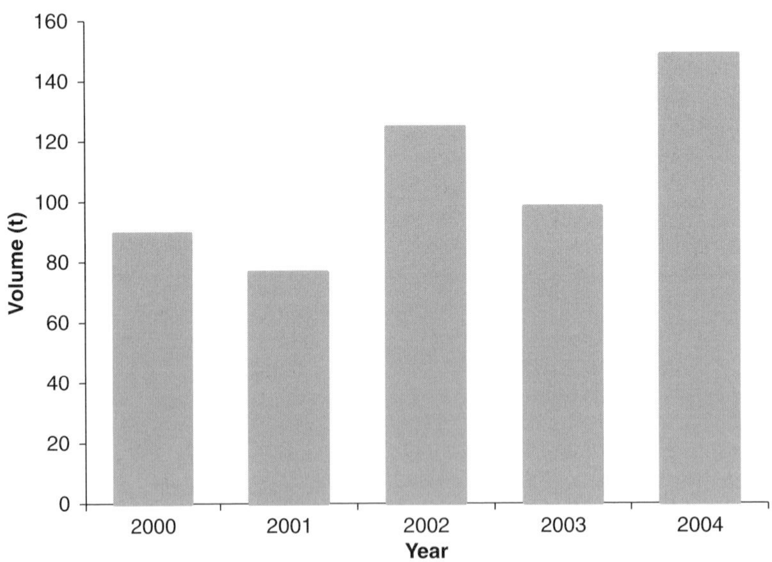

Fig. 18.3. Rosemary oil import by USA based on the data from Codex Alimentarius Commission, 2013.

powdered form. It is an aromatic herb and has a characteristic highly aromatic minty odour and slightly bitter, astringent taste.

The plant has straight stems with numerous long greyish or light brownish slender branches. It is an erect shrub that grows up to 6.6 ft (2 m) and the needle-like leaves are 0.8–1.6 in (2–3 cm) long and the width 0.05–0.14 in (1.2–3.5 mm). On the branches, the leaves are placed opposite to one another and are leathery, thick, lustrous, linear and dark green, but whitish underneath with short, dense, felted masses of woolly hairs. Illustrated in Fig. 18.4 are a rosemary branch and its different components (i.e. flower, seeds, etc.). The flowers are small and are white, pink, pale blue to deep blue or purple; the volatile essential oils are mostly found in their calyces. The calyx is 0.1–0.2 in (3–4 mm) long, green or purplish and coated with matted, tangled hairs when young, but it later develops to a 0.2–0.3 in (5–7 mm) long smooth structure.

The upper epidermis of the leaf consists of a layer of tabular polygonal to irregular cells with no stomata. The hypodermis has one to several layers of irregularly rounded to ovoid cells with thickened walls. The lower epidermis of the leaf has a lot of scattered glandular trichomes.

Fig. 18.4. Drawings of a rosemary shoot, flowers and seeds. Illustration from Köhler's Medicinal Plants (Wikimedia Commons, 2014b).

18.2 Chemistry

18.2.1 Chemical composition

Rosemary is one of the most important aromatic herbs, and its leaves are widely used as a flavouring in meats and as a perfume in soaps, detergents and household sprays. It is also extensively used in the pharmaceutical and cosmetic industries. Table 18.1 presents data on the chemical composition of rosemary. Rosemary is a good source of iron, calcium, Vitamin B_6 and carbohydrates, especially dietary fibre. It contains a decent amount of lipids and proteins, but as the amount consumed in food as a flavouring or condiment is small, rosemary is not a significant source of these nutrients. Typical products available on the market for culinary use are the fresh, dried (whole) and powdered leaves.

18.2.2 Phytochemistry

The aromatic herbs in the mint family contain a significant amount of effective antioxidant compounds (see Table 18.2), and they are used as antioxidants in food to prevent the development of rancidity and to improve the shelf life of oils and cosmetics. Rosemary is particularly effective in enhancing the stability of omega-3 rich oils.

Table 18.3 summarizes the effectiveness of rosemary and of rosemary-derived products in preventing oxidation in foods and also in preventing the production of deleterious compounds during high-temperature cooking. The table also includes mention of some of the recent findings on the health benefits of consuming rosemary and other spices and herbs.

As rosemary has been found to be a very effective antioxidant, a number of antioxidants derived from it are now available on the market (Table 18.4), and it is the main

Table 18.1. Composition of fresh and dried rosemary leaves. Adapted from data in the USDA National Nutrient Database for Standard Reference Release 28 (USDA ARS, 2015).

Component	Fresh/100 g	Dried/100 g
Carbohydrate, by difference	20.70 g	64.06 g
Energy	131 kcal	331 kcal
Protein	3.31 g	4.88 g
Total dietary fibre	14.1 g	42.6 g
Total lipid (fat)	5.86 g	15.22 g
Water	67.77 g	9.31 g
Minerals		
Calcium, Ca	317 mg	1280 mg
Iron, Fe	6.65 mg	29.25 mg
Magnesium, Mg	91 mg	220 mg
Phosphorus, P	66 mg	70 mg
Potassium, K	668 mg	955 mg
Sodium, Na	26 mg	50 mg
Zinc, Zn	0.93 mg	3.23 mg
Vitamins		
Vitamin A, (RAE)[a]	146 μg	156
Vitamin A (units)	2924 IU	3128 IU
Thiamine (B_1)	0.036 mg	0.514 mg
Riboflavin (B_2)	0.152 mg	0.428 mg
Niacin (B_3)	0.912 mg	1.000 mg
Vitamin B_6	0.336 mg	1.740 mg
Folate (B_9) (DFE)[a]	109 g	307 g
Vitamin B_{12}	0.00 μg	0.00 μg
Vitamin C (total ascorbic acid)	21.8 mg	61.2 mg
Vitamin D (D_2 + D_3)	0.0 μg	0.0 μg
Vitamin D (units)	0 IU	0 IU
Lipids		
Cholesterol	0 mg	0 mg
Fatty acids, total monounsaturated	1.160 g	3.014 g
Fatty acids, total polyunsaturated	0.901 g	2.339 g
Fatty acids, total saturated	2.838 g	7.371 g

[a]Retinol activity equivalents.
[b]Dietary folate equivalents.

and most significant raw material for the manufacture of antioxidants that are used in food and food products. Added benefits of using rosemary or extracts from rosemary are that: (i) the product is natural or from a natural source; (ii) it is 'clean label'; (iii) it has a long history of safe usage; (iv) it is non-GMO; and (v) an organic variety is generally available. All of these qualities are added benefits from rosemary in what consumers are looking for in food ingredients.

18.3 Postharvest Technology

Rosemary is harvested once or twice a year, typically 18 months after sowing, and in some instances after 24 months, then more frequently when the plant matures. Harvesting is done before the plant goes through the flowering stage or phase. The harvested leaves are processed by removing the leaves from the stems and removing the dirt and sand thoroughly by sieving. The leaves are then dried either by spreading out on a tray or hanging in bunches away from direct sunlight. Better results are typically obtained when artificial drying is used rather than sun drying, because the drying and heating are better controlled when artificial drying method is used. Drying should be done at temperatures lower than 40°C to reduce loss of flavour and to maintain the dark green colour of the leaves. After drying, the leaves

Table 18.2. Antioxidant compounds identified in rosemary and other aromatic herbs. Compiled from various sources.

Aromatic herbs in the mint family Lamiaceae	Scientific name	Antioxidant compounds
Rosemary	*Rosmarinus officinalis*	Caffeic acid, caffeoyl derivatives (rosmarinic acid), phenolic diterpenes (carnosic acid, carnosal, carnosol, epirosmanol, 12-*O*-methylcarnosic acid, rosmanol), flavonoids, camphor, caffeic acid, ursolic acid, betulinic acid, 1,8-cineole
Basil	*Ocimum basilicum*	Eugenol, citral, citronellol, linalool, myrcene, pinene, ocimene, terpineol, linalyl acetate, *trans*-ocimene, 1,8-cineole, camphor octanane, methyl eugenol, methyl chavicol, β-caryophyllene
Lavender	*Lavandula angustifolia*	Linalyl acetate, linalool, camphor, β-ocimene, 1,8-cineole, borneol, hotrienol, hexyl butyrate, α-bisabolol, caryophyllene oxide
Marjoram	*Origanum marjorana*	β-Carotene, β-sitosterol, caffeic acid, carvacrol, eugenol, hydroquinone, linalyl-acetate, myrcene, rosmarinic acid, terpinen-4-ol
Oregano	*Origanum vulgare*	Caffeic acid, *p*-coumaric acid, rosmarinic acid, caffeoyl derivatives, carvacrol, flavonoids
Sage	*Salvia officinalis*	Rosmanol, epirosmanol, phenolic acids (rosmarinic acid), phenolic diterpenes (carnosic acid), flavonoids
Thyme	*Thymus vulgaris*	Phenolic acids (gallic acid, caffeic acid, rosmarinic acid), thymol, phenolic diterpenes, flavonoids

are cleaned further (to remove any stems, foreign material, etc.), sieved and graded. The dried product is packaged and stored in tightly sealed containers. Dried rosemary is packaged and sold either in cartons or in glass or plastic containers, tightly sealed and stored at low temperatures to keep moisture, heat, oxygen and light out because these will destroy the quality of the dried herb. Dried whole rosemary leaves of good quality should contain at least 1.2% volatile oil and a maximum of 2% woody stems and a maximum of 7% ash.

Fresh clean rosemary leaves devoid of dirt and twigs and extraneous materials should have a fresh and crisp appearance, a dark green colour and a good flavour. Typically, young, fresh shoots are used for culinary purposes. These are typically marketed and sold in bunches of two or three measuring 10–12 inches or in clear cellophane sachets. They are packaged in crates when marketed in bulk. Packaged fresh rosemary leaves are stored at cool refrigerated temperatures (~5°C), when they can remain fresh for at least 2–3 weeks.

18.3.1 Processing

Drying of rosemary has been discussed above. The main thing to bear in mind is that heat, air/oxygen and light will have a deleterious effect on the quality of the dried herb and thus the choice of packaging material or packaging container is also critical. In a commercial setting, sterilization of the spice and herb is necessary to destroy pathogenic microorganisms, yeasts and moulds and other pests. There are a number of choices that can be employed for the sterilization of herbs like rosemary: irradiation, microwave sterilization and steam sterilization. The use of ethylene oxide was once popular but has now been banned in Europe and other countries. Irradiation, although not popular to consumers, can effectively kill bacteria, mould and insects while maintaining the flavour of the spice or herb. This is achieved by using γ-rays emitted from cobalt-60 or by an accelerated electron beam. A dosage of 4 to 12–15 kGy can eliminate coliforms or reduce the total plate count to below detectable levels. Microwave

Table 18.3. Selected publications on the antioxidant and antimicrobial effects of rosemary and its health benefits. Adapted from Embuscado (2015).

Spice	Summary	Reference
Rosemary and sage	Bland natural antioxidants were obtained using an extraction process and these extracts improved the flavour stability of soybean oil and potato chips.	Chang *et al.*, 1977
Rosemary	Antioxidant properties of rosemary oleoresin in turkey sausage – rosemary oleoresin (20 ppm) was comparable with a commercial blend of butylated hydroxyanisole/butylated hydroxytoluene (BHA/BHT, 200 ppm) in preventing lipid autoxidation in turkey sausage stored at 4°C as indicated by measurement of 2-thiobarbituric acid-reactive substances.	Barbut *et al.*, 1985
Rosemary extracts	Water-soluble rosemary extracts showed reduction of lipid oxidation and colour change in cooked turkey products during storage as measured by monitoring changes in thiobarbituric acid-reactive substances, hexanal production and colour of the cooked turkey samples.	Yu *et al.*, 2002
	Rosemary extracts, VivOX 20 and VivOX 4 had antioxidant and antimicrobial properties in vacuum-packed chicken frankfurters based on the Rancimat test and total plate count, and compared with a commercially available preservative Robid LI LS and a control sample without test additives.	Rižnar *et al.*, 2006
	Rosemary extracts were able to inhibit heterocyclic amines (HCAs) formation during cooking of beef patties at high temperature.	Puangsombat and Smith, 2010
Oregano and rosemary extracts	Methanol extracts from oregano and rosemary were able to retard oxidation of docosahexaenoic acid C22:6 (DHA) and eicosapentaenoic acid C20:5 (EPA) in menhaden oil. The antioxidant activity of the rosemary extract was greater than that of the oregano extract but it was sensitive to heat.	Bhale *et al.*, 2007
Marinade seasoning blends	Commercial marinades (Caribbean, Southwest and Herb) inhibited formation of heterocyclic amines (HCAs) in beef round steaks cooked at 204°C (400°F). HCAs are suspected human carcinogens formed in meats during grilling or cooking.	Smith *et al.*, 2008
Several spices	Rosemary, oregano and borage reduced formation of TBARS (thiobarbituric acid reactive substances), myoglobin oxidation, colour fading and extended shelf life in beef patties.	Sánchez-Escalante *et al.*, 2003
Spice blend containing ground rosemary	When added to hamburger meat before cooking, the spice mixture resulted in a reduction in malondialdehyde, suggesting potential health benefits for atherogenesis and carcinogenesis.	Li *et al.*, 2010

sterilization is a newer milder method that is also effective and at the same time can preserve the colour and flavour of the spice or herb. There are a number of pieces of microwave sterilization equipment that are being advertised on the Internet that promote effective sterilization while maintaining high quality of the herb or spice.

In addition to drying and sterilization, rosemary is processed using a variety of extraction methods to produce rosemary oil, a very important product in the aromatic herb

Table 18.4. Commercial antioxidants from rosemary.[a] Adapted from Embuscado (2015).

Products	Applications/usage	Manufacturer/ website
StabilEnhance rosemary extracts (powder or liquid forms):		Naturex (http://www.naturex.com)
StabilEnhance OSR – carnosic acid up to 50%	Foods with high fat content and/or sensitive to oxidation; prevent colour fading in paprika oleoresins, meat products, seasonings	
StabilEnhance WSR – rosmarinic acid up to 40%	Beverages, emulsions, foods with low fat content	
OxyBlock and XtraBlend antioxidant blends:		
OxyBlock – optimized synergistic combinations rosemary extracts and natural tocopherols, citric acid and ascorbic acid	Paprika oleoresins, seasonings, animal fats, dehydrated meats	
XtraBlend – optimized combinations of botanical extracts	Cooked meat products, raw ground meats	
Fortium® – derived from mixed tocopherols and Kemin's proprietary rosemary extracts with minimal impact on flavour, colour and odour profiles	Food products with high fat oil content such as; meat and poultry products, nuts, spices, salad dressings, mayonnaise, marine oils, lard vegetable oils, instant noodles and cereals	Kemin (http://www.kemin.com/)
Fortium® TR30 liquid (sunflower oil, rosemary extract, mixed tocopherols)		
Herbalox® seasoning - derived from natural rosemary extract, available in oil dispersible, water dispersible and water soluble forms	Beverage and energy drinks, meat and meat products, mayonnaise, sauces and dressings, breakfast cereals and energy bars	Kalsec (http://www.kalsec.com/products/antioxidants/)
Duralox® oxidation management systems – include spice extractives, organic acids, antioxidants and vitamins	Food colour stabilization and as oil-soluble antioxidants	
CO_2-extracts		Flavex (http://www.flavex.com/en/naturextrakte/home/)
Rosemary antioxidant extract (14% or 25% diterpene phenols), powder (25% diterpene phenols)	0.05–0.4% depending on application: stabilization of fatty and essential oils, carotenoids, etc. against oxidation; in the food industry (dressings, meat, sausages, snacks, etc.); food supplements; in cosmetic and pharmaceutical preparations; antioxidative, antimicrobial and anti-inflammatory properties; 0.05–0.1% in case of saturated fats, 0.2–0.4% in case of polyunsaturated oils	
Rosemary antioxidant extract, water dispersible (14% diterpene phenols) – >9% carnosic acid (calculated as diacetyl carnosic acid, 7–10% carnosic and carnosol (calculated as carnosic acid)	Extract is water dispersible with antioxidative and antimicrobial properties; used in water-based systems, in meat and food industry and in cosmetic products for retarding oxidation of fats, oils and carotenes and contributes to microbial stability	

Continued

Table 18.4. Continued

Products	Applications/usage	Manufacturer/ website
AquaROX® – water soluble natural rosemary extract; rosmarinic range 1–40%	Beverages and energy drinks, meat and meat products, mayonnaise, sauces and salad dressings, breakfast cereals, energy bars	Vitiva (http://www.vitiva.eu/VITIVA_EN, product_portfolio/rosemary_extracts/)
INOLENS® – reduced odour and bitterness of oil soluble rosemary extracts with standardized carnosic acid; carnosic acid 1–40%	Chocolate and confectionery, special meat products, mayonnaise and salad dressings, bakery products, polyunsaturated oils (PUFA, fish oil), dry pet food, high-end oils, essential oils, edible oils, flavours, seafood, etc.	
VivOX® – carnosic acid concentration 1–100%; available in powder form and in liquid form (in oils, propylene glycol or ethanol)	Fresh and processed meats, oils, fats and shortenings (vegetable and animal), snacks, mayonnaise, sauces and salad dressings, dry pet foods, carotenoids, cookies, fillings for cookies, flavours, seafood products, cereals, energy bars, nuts, milk and dairy products, fried and processed foods, pastas	

[a]Listing these rosemary products is not a sign of endorsement from the author.

market. The parts used for extraction are the stems, leaves and flowers, but rosemary oil extracted from the flowering tops produces superior oil compared with that obtained from the stems and leaves.

The chart in Fig. 18.5 shows the diagram of the processes used for extraction and the products obtained from the different processes. This covers all the extraction methods employed, not only for aromatic herbs or spices, but also other sources of aromatic compounds (e.g. citrus).

The processes are summarized as follows:

1. Maceration using a solvent– the raw material is crushed in the presence of a solvent (e.g. alcohol). The product derived from this process is known as the tincture.
2. Enfleurage method – the raw material (leaves or flower petals) is laid on a thin layer of fat. The essential oil from the leaves or flower petals transfers to the layer of fat. The leaves or petals are replaced with new ones and the process is repeated until the fat is saturated with the essential oils from the leaves or flower petals. The final product is known as the pomade.
3. Solvent extraction – solvents typically used are non-aqueous solvents such as hexane, acetone or alcohols. The resulting product is typically viscous, semi-solid or solid.
4. Expression – the material is pressed manually or by employing a machine. The pressure exerted on the raw material forces the oils to be released from the material via puncture, rasting (a process wherein the raw material is under pressure using an implement that moves back and forth to express the fluid from the material) or cutting.
5. Distillation – this is accomplished by using steam or saturated steam. It can also be accomplished using hydro-diffusion.

a. Simple steam distillation – the material is immersed directly in water and this is boiled. The vapours containing the oil are condensed and the essential oil is separated based on density and immiscibility in water.
b. Saturated steam distillation – the raw material is not in direct contact with the water. Steam is generated by boiling water and this steam is injected through the plant material placed on perforated trays. This process has several advantages over simple steam distillation. The quality of the oil constituents is preserved, the process of extraction is shorter, thus conserving energy, and pressure can be applied making the extraction process more

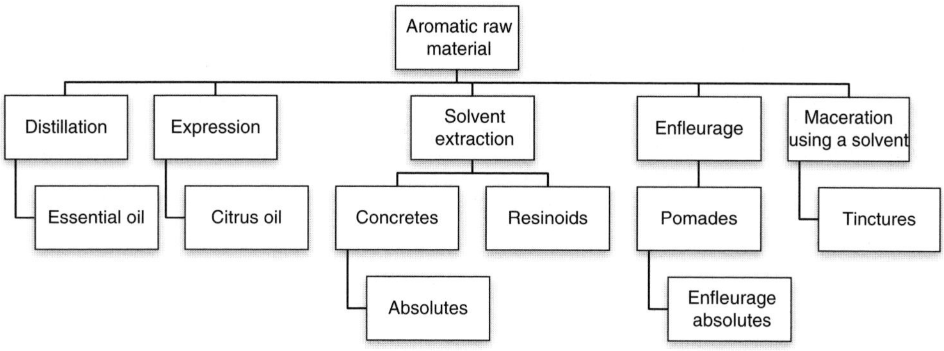

Fig. 18.5. Flow chart for the manufacture of different forms of extracts from aromatic plants such as rosemary.

efficient. This process can also be automated and controlled.
c. Hydro-diffusion – pulses of steam are sent through the plant material at a very low pressure. This is also an energy-efficient process that produces a high-quality product.

For rosemary, the highest quality of essential oils is obtained through distillation using only the flowering tops gathered when the plant is in bloom. The oil produced from this process is clear to slightly yellow thin oil with a powerful minty smell. The essential oil derived from the whole plant will have a higher camphor content and lower quality than that distilled from the flowering tops. The rosemary essential oil is stored in cool, dry area, tightly capped, and kept in the refrigerator once the container has been opened. The oil is stored in glass bottles that are preferably amber coloured. The product should not be exposed to light or heat, or come into contact with any metals.

18.3.2 Value addition

Rosemary essential oil is one of the value-added products from rosemary, and is produced through the distillation process as a clear to slightly yellow thin fluid. The essential oil is used by the food, flavouring, chemical and pharmaceutical industries in products such as flavours for foods, and flavour and perfumery in cosmetics, personal health care and for aromatherapy. The major market for rosemary oils is the USA, followed by Japan and Europe. The seven large major processing plants of rosemary essential oil are in Europe and these make up the seven largest essential oil processing companies.

In addition to essential oil, other extracts can also be derived from rosemary (i.e. extracted by alcohol such as ethanol, or water extracts). Another product is from supercritical extraction, which is typically used in dietary supplements. Rosemary leaf in powdered form is also used in dietary supplements. So depending on the end product desired, a specific type of extraction or distillation can be used to obtain a specific type of product, for example, for use as a flavouring, or for use in perfumes, in pharmaceuticals or as a dietary supplement. It is also understood that targeting a specific end product with required composition will require usage of a specific part of the plant (including specifying its age, i.e. mature versus young leaves) or harvesting the leaves or the flowers at a specific stage of its life/maturity.

18.4 Uses

18.4.1 General uses

Rosemary leaves in fresh or dry forms are used in food preparation to flavour meats and vegetables. The dry leaves are also used in crushed or powder form, primarily in food preparations. Figure 18.6 shows fresh, dry

Fig. 18.6. The different forms of rosemary leaves in the market from left to right: fresh, dried, and crushed dried leaves.

and crushed forms of rosemary leaves. Drying typically results in at least 50% loss in weight. As mentioned earlier, rosemary acts as a powerful antioxidant to preserve oils and minimize the production of harmful substances during grilling or pan frying. The products from rosemary that are so used are fresh rosemary leaves, leaf powder, rosemary essential oil, oil soluble rosemary extracts and water-soluble rosemary extracts.

Other uses of rosemary powder are in dietary supplements, either by itself or in combination with other spices and herbs. It is used as a source of powerful antioxidants, for an uplifting effect on mood, for the improvement of health and for longevity, for assisting memory and clarity of thought, and for the alleviation of muscle pain. Rosemary leaf extract is also touted as promoting optimal cognitive function and aiding in keeping a clear head for improved concentration and improved short-term memory. Rosemary has been the subject of numerous research and clinical tests that have probed into its many health benefits.

To summarize, the general uses of rosemary and rosemary products can be classified into the following categories:

1. Culinary preparations for flavour and aroma – rosemary is used in food products (meat, poultry, fish, soups, stews, vegetables) in fresh or dry forms and in non-alcoholic beverages. Fresh and dried rosemary leaves, whole or ground, are used as seasonings for soups, stews, sausages, meat, fish and poultry. Rosemary leaves are also used to extend the shelf life of oils/fats and meats because of their antioxidant and antibacterial activity.
2. Consumer products – rosemary extracts or essential oils are used in soaps, creams, deodorants, hair tonics/shampoos, household cleaners and air fresheners.
3. Cosmetics – rosemary is used in creams for acne and dermatitis and also in preparations to stimulate hair growth and to prevent dandruff. It is also used for its fragrance in the preparation of eau de cologne.
4. Pharmaceutical uses – there are many useful properties of rosemary and thus it is considered as a very important plant in the field of herbal or alternative/complementary medicine. It is employed because of its anti-inflammatory properties and is used as an astringent, expectorant, tonic and stimulant. It has been known to enhance circulation and improve food absorption through the stimulation of digestion. It is also used in oral preparations for antiseptic gargles for sore throat, inflammation of the gums and canker sores.

Rosemary is also used as a relaxant and to diminish mental fatigue and nervous exhaustion. It is also been found to be effective for asthma, bronchitis and whooping cough, as well as appropriate cure for intestinal infections and diarrhoea. It has been used since ancient times for its medicinal properties, and was traditionally used to alleviate muscle pain and spasm, improve the immune system and support the circulatory and nervous systems. Rosemary is also known to treat indigestion and improve memory, and has been recognized to have anticancer properties.
5. Other uses – rosemary is a nice decorative plant or for use as a hedge in gardens. It is also used as a good ground cover to prevent soil erosion. It is a good source of food for bees and has also used as an insect repellent.

18.4.2 Pharmacological uses

The use of rosemary is as old as humankind. A record of its usage has been traced back as early as 5000 BC. Based on historical accounts, rosemary and its constituents have a number of therapeutic properties. A number of studies have been carried out to determine the potential therapeutic effects of rosemary and products derived from the rosemary plant. Here is a list of some selected scientific publications:

1. A short-term study on the effects of rosemary on cognitive function in an elderly population was conducted and showed positive effect of the dose on cognitive performance (Pengelly *et al.*, 2012).
2. Essential oils of lavender (*Lavandula angustifolia*) and rosemary were found to have an effect on cognitive performance and subjective effects on mood (Moss *et al.*, 2003).
3. Carnosic acid from rosemary showed antimutagenic properties in bacteria and anticarcinogenic activity in various cell and animal models (Steiner *et al.*, 2001).
4. Carnosol, which is found in rosemary, exhibits antioxidant and anticarcinogenic properties (Lo *et al.*, 2002).

Although several scientific research works have been conducted on the therapeutic and medicinal effects of rosemary, work in this important area continues to proliferate, primarily due to the increasing demand for more natural solutions to counteract illnesses and diseases.

18.5 Summary

Rosemary is an evergreen, sun-loving perennial plant that is native to south of France and the Mediterranean regions. It is cultivated because of its aromatic and medicinal properties. Rosemary and products derived from rosemary are a very important group of products because of the many benefits derived from using this natural ingredient in food or due to their health benefits. Published scientific research works support the effectiveness of rosemary as an antioxidant in foods. Numerous research findings also substantiate the pharmacological benefits of rosemary and rosemary products. Rosemary extracts are widely used in the food, pharmaceutical and cosmetic industries. It is a very vital herb from the point of view of both human health and food safety, as it can be used as an effective food preservative.

According to the Codex Alimentarius Commission (2013), even if rosemary is a very small crop that is clubbed together with other aromatic herbs, it is important that the statistics on production, export, import and value addition are available for the herb because of the numerous benefits and functionality derived from it. Due to the importance of this herb, a proposal on the development of a Codex Standard for rosemary has been submitted (Codex Alimentarius Commission, 2013). According to this report:

> The objective is to develop a world-wide standard based on basic characteristics. The need to have a harmonized standard for Rosemary stems from the fact the crop is grown in developing countries in fragmented areas by marginal farmers. The marginal farmers do not have the capability to collectively organize to manage the factors which influence their output and therefore the whole food chain will be put to risk by these external factors if these risks are not recognized or mitigated by an international committee under the aegis of Codex.

The main aspects to be covered according to this proposal (Codex Alimentarius Commission, 2013), as quoted verbatim are:

> The standard entails aspects related to the properties of Rosemary in dehydrated and extract form incorporating physical parameters, presence of extraneous matters, oil content, safety and labeling in order to provide adequate product characteristics and to protect consumer's health. To supply high quality safe products, the objective of the standards are to:
>
> - Compile production, export and import figures for Rosemary and its products to overcome the current impediment in sourcing data for standardization and harmonization.
> - Establish the minimum requirements for Rosemary in its dehydrated and extract from [form] including and in additions to the quality parameters like the physical appearance, uniformity of the product, free from pest and other extraneous matter etc.
> - Define the categories to classify Rosemary in accordance with the characteristics of the herb; such as cut herbs, essential oil, fixed oil, extracts etc.
> - To monitor and strengthen the cross border phytosanitary regulations so

that the pests/microbes do not travel to other countries and cross contaminate the delicate ecosystem of marginal growers of spices and herbs.
- Include the provisions to be considered related to the uniformity of the packaged product and the packaging used.
- Include provisions for the labeling and marking of the product in accordance with the general standard for the labeling of prepackaged foods.
- Establish tolerances regarding quality and size permitted in packaged Rosemary.
- Include provisions for hygiene with reference to the recommended international Code of Practice for hygiene and general principles of food hygiene.

Consumers have increasingly favoured food products which contain natural ingredients due to concerns over the adverse health effects of synthetic raw materials, particularly some synthetic antioxidants and food additives. Rosemary and rosemary products are classified as all natural, 'clean label' and non-GMO, attractive qualities sought by consumers. As consumers' demand for natural ingredients in their food products increases, the utilization of rosemary and rosemary products will continue to expand, and scientific research works on rosemary will continue to increase. Rosemary will remain an important aromatic herb throughout the world for years to come.

References

Barbut, S., Josephson, D.B. and Maurer, A.J. (1985) Antioxidant properties of rosemary oleoresin in turkey sausage. *Journal of Food Science* 50, 1356–1359.

Bhale, S.D., Xu, Z., Prinyawiwatkul, W., King, J.M. and Godber, J.S. (2007) Oregano and rosemary extracts inhibit oxidation of long-chain n-3 fatty acids in menhaden oil. *Journal of Food Science* 72, C504–C508.

Chang, S.C., Ostric-Matijasevic, B., Hsieh, O.L. and Huang, C.L. (1977) Natural antioxidants from rosemary and sage. *Journal of Food Science* 42, 1102–1106.

Codex Alimentarius Commission (2013) *Discussion Paper for the Establishment of Codex Committee on Spices, Aromatic Herbs and Their Formulations. Joint FAO/WHO Food Standards Programme, Codex Alimentarius Commission, 36th Session, FAO Headquarters, Rome, Italy, 1–5 July 2013.* Document No. CX/CAC 13/36/10-Add.2, Food and Agriculture Organization of the United Nations, Rome. Available at: ftp://ftp.fao.org/codex/Meetings/cac/cac36/cac36_10_add2e.pdf (accessed 28 January 2016).

Embuscado, M.E. (2015) Herbs and spices as antioxidants for food preservation. In: Shahidi, F. (ed.) *Handbook of Antioxidants for Food Preservation.* Woodhead Publishing (Elsevier), Sawston, Cambridge, UK, pp. 251–284.

Li, Z., Henning, S.M., Zhang, Y., Zerlin, A., Li, L., Gao, K., Lee, R., Karp, H., Thames, G., Bowerman, S. and Heber, D. (2010) Antioxidant-rich spice added to hamburger meat during cooking results in reduced meat, plasma, and urine malondialdehyde concentrations. *The American Journal of Clinical Nutrition* 91, 1180–1184.

Lo, A.H., Liang, Y.C., Lin-Shiau, S.Y., Ho, C.T. and Lin, J.K. (2002) Carnosol, an antioxidant in rosemary, suppresses inducible nitric oxide synthase through down-regulating nuclear factor-κB in mouse macrophages. *Carcinogenesis* 23, 983–991.

Moss, M., Cook, J., Wesnes, K. and Duckett, P. (2003) Aromas of rosemary and lavender essential oils differentially affect cognition and mood in healthy adults. *International Journal of Neuroscience* 113, 15–38.

Pengelly, A., Snow, J., Mills, S.Y., Scholey, A., Wesnes, K. and Butler, L.R. (2012) Short-term study on the effects of rosemary on cognitive function in an elderly population. *Journal of Medicinal Food* 15, 10–17.

Puangsombat, K. and Smith, J.S. (2010) Inhibition of heterocyclic amine formation in beef patties by ethanolic extracts of rosemary. *Journal of Food Science* 75, T40–T47.

Rižnar, K., Čelan, S.T., Knez, Z., Škerget, M., Bauman, D. and Glaser, R. (2006) Antioxidant and antimicrobial activity of rosemary extract in chicken frankfurters. *Journal of Food Science* 71, C425–C429.

Sánchez-Escalante, A., Djenane, D., Torrescano, G., Beltrán, J.A. and Roncales, P. (2003) Antioxidant action of borage, rosemary, oregano and ascorbic acid in beef patties packaged in modified atmosphere. *Journal of Food Science* 68, 339–344.

Smith, J.S., Ameri, F. and Gadgil, P. (2008) Effect of marinades on the formation of heterocyclic amines in grilled beef steaks. *Journal of Food Science* 73, T100–T195.

Steiner, M., Priel, I., Giat, J., Levy, J., Sharoni, Y. and Danilenko, M. (2001) Carnosic acid inhibits proliferation and augments differentiation of human leukemic cells induced by 1,25-dihydroxyvitamin D3 and retinoic acid. *Nutrition and Cancer* 41, 135–144.

USDA ARS (2015) USDA National Nutrient Database for Standard Reference, Release 28. Nutrient Data Laboratory, Agricultural Research Service, US Department of Agriculture, Beltsville, Maryland. Available at: http://www.ars.usda.gov/Services/docs.htm?docid=8964 (accessed 28 January 2016). Nutrient Data Laboratory home page available at: http://www.ars.usda.gov/ba/bhnrc/ndl (accessed 28 January 2016).

Wikimedia Commons (2014a) THOR (Flowering Rosemary). Available at: https://commons.wikimedia.org/wiki/File:Rosmarinus_officinalis133095382.jpg (accessed 28 January 2016). Public domain, via Wikimedia Commons.

Wikimedia Commons (2014b) Franz Eugen Köhler's Medizinal-Pflanzen (List of Koehler Images). Available at: https://en.wikipedia.org/wiki/Rosemary#/media/File:Rosmarinus_officinalis_-_K%C3%B6hler%E2%80%93s_Medizinal-Pflanzen-258.jpg (accessed 28 January 2016). Public domain, via Wikimedia Commons.

Yu, L., Scanlin, L., Wilson, J. and Schmidt, G. (2002) Rosemary extracts as inhibitors of lipid oxidation and color change in cooked turkey products during refrigerated storage. *Journal of Food Science* 67, 682–685.

19 Sage

Ahmad Ghorbani*
Mashhad University of Medical Sciences, Mashhad, Iran

19.1 Botany

19.1.1 Introduction

Sage is a shrubby perennial herb native to the Mediterranean and Balkans regions. It is an evergreen plant in the family Lamiaceae and belongs to the genus *Salvia*, which is one of the widely spread members of the family. Its botanical name is *Salvia officinalis*. Vernacular names used in different languages are shown in Table 19.1.

Sage is a popular kitchen plant in many countries and has been long used in a variety of culinary preparations. It also has a long history of traditional use as a medicinal plant and as a beauty aid (Zargari, 1990; Bisset and Wichtl, 2001; Miura *et al.*, 2001).

19.1.2 History/origin

The words *Salvia* and sage are said to come from the Latin word *salvere* which means 'to heal' and 'to save'. This plant was called *elifagus* by the Greeks; this became *spahkos*, and later, in old English, the plant came to be known as sawge. The word *officinalis* refers to the therapeutic use of the plant – the *officina* was the traditional place in a monastery where plants and medicines were stored (Baricevic and Bartol, 2000; Stearn, 2004; Harrison, 2012; Russo *et al.*, 2013; HerbNet, 2014). Sage has been used in food and traditional medicine for centuries. In the Medieval period, the production of essential oil from sage was common and people believed that the herb could cure all ailments. In that period, Arab physicians thought that the plant extended life. Folk healers in America used sage to treat epilepsy, insomnia, seasickness, measles and intestinal worms, and in the 1920s, US medical texts recommended sage leaf poultices for swellings and sprains, and sage tea for sore throats. In Asia, traditional healers believed that sage had useful effects on gout, rheumatism, headache, oedema, night sweats, dizziness, tremors, fatigue, paralysis, chronic coughs, tonsillitis and ulcers (Zargari, 1990; HerbNet, 2014).

19.1.3 Location

Salvia is the largest genus of the family Lamiaceae and has about 900 species that are spread throughout the world (Delamare *et al.*,

*Corresponding author, e-mail: ghorbania@mums.ac.ir

© CAB International 2016. *Leafy Medicinal Herbs: Botany, Chemistry, Postharvest Technology and Uses* (eds D.C.P. Ambrose *et al.*)

Table 19.1. Vernacular names of sage in different languages.

Country	Vernacular names
Arabic	Shalbiyah, asfaqs
English	Garden sage, common sage, red sage, great sage
French	Feuillos de sauge commune, sague officinale, herbe sacrée, grande sauge, sauge
German	Salbei, grosse salbei, gebräuchliche, salbeiblätter, edelsalbei, gartensalbei
Hindi	Salvia, sefakus
Italian	Salvia, salvia domestica, salvia maggiore
Persian	Maryam goli

2007). The species *S. officinalis* is native to the western and southern Balkans, the Mediterranean and Middle Eastern regions. The plant has been naturalized and is cultivated in several regions of the world, particularly European countries (Bisset and Wichtl, 2001; PDR, 2004). It is also cultivated in North America and North Africa (Zargari, 1990; PDR, 2004).

19.1.4 Morphology

Sage is a perennial round-shaped shrub with a strongly based root system. It grows up to 60–70 cm high and wide (Fig. 19.1). The base of the stem is woody and has leafy quadrangular branches (Bisset and Wichtl, 2001; PDR, 2004). The leaves are simple, textured, evergreen, oblong or oblong–lanceolate and range in size up to 6.4 cm long by 2.5 cm wide. They are narrowed at the base, rugose on the upper side and nearly white underneath because of the presence of many short soft hairs (Clebsch and Barner, 2003; HerbNet, 2014). The leaves are also petiolate, crenate, ribbed–wrinkled, tough, tangy and aromatic (PDR, 2004). In modern cultivars, the leaves may be purple, cream, rose and yellow in the many variegated forms that are grown (Clebsch and Barner, 2003; HerbNet, 2014).

The flowers of sage are small and bloom in the summer. They are pale violet, white or pink and have 6–12-blossomed false whorls which are organized in 4–8 rows. The calyx is funnel shaped, bilabiate, glandular punctuate and 10–14 mm long. The lower lip has three segments and two thorny awned teeth, and upper lip is straight and has three thorny awned teeth. The flowers have two stamens with semicircular bent filaments (PDR, 2004).

19.2 Chemistry

19.2.1 Chemical composition

A wide range of chemical constituents with different functional groups and polarities is found in sage. In the last few decades, different solvents have been used to extract the active chemical constituents, and some of the best studied sage preparations include essential oil, alcoholic and aqueous extracts, and a butanol fraction.

More than 120 components have been identified in sage essential oil. Cineole, borneol, thujone, camphor, elemene, ledene, pinene, humulene and caryophyllene are of the major components of the oil (Langer *et al.*, 1996; Hayouni *et al.*, 2008; Badiee *et al.*, 2012). Alcoholic and aqueous extracts of sage contain phenolic compounds such as phenolic acids (e.g. rosmarinic acid) and flavonoids (e.g. luteolin-7-glucoside). Compared with the aqueous extract, the alcoholic extract has a higher content of rosmarinic acid, while the aqueous extract has a higher content of luteolin-7-glucoside (Lima *et al.*, 2007b). A number of water-soluble carbohydrates have been characterized, the most abundant of which are arabinose, galactose, glucose, mannose, rhamnose, xylose and uronic acids (Capek and Hríbalová, 2004). Although the chemical components of sage tea have not been well studied, rosmarinic acid, luteolin-7-glucoside, and volatile constituents such as 1,8-cineole, *cis*-thujone, *trans*-thujone, camphor and borneol have been identified so far. The percentages of these components differ depending on environmental parameters such as soil and climatic conditions, water availability and altitude (Russo *et al.*, 2013).

Table 19.2. Chemical compounds present in sage essential oil, tea and different extracts.

Type of sage extract	Compounds	References
Essential oil	*Allo*-aromadendren; α-amorphene; aromadendrene; (*Z*)-α-*trans*-bergamatolcetate; β-bisabolene; borneol; bornyl acetate; β-burbonene; 1-butyl acetate; cadina-1,4-diene; α-cadinene; δ-cadinene; γ-cadinene; α-cadinol; *t*-cadinol; α-calacorene; β-calacorene; calarene; *cis*-calamenene; camphene; camphor; δ-3-carene; carveol; *trans*-carvyl acetate; caryophyllene; β-caryophyllene; (*E*)-caryophyllene; caryophyllene oxide; β-caryophyllene oxide; α-cedrene; β-cedrene; cermacrene B; cineole; 1,8-cineole; α-copaene; α-cubebene; β-cubebene; cyclohexadiene; cyclohexene; *H*-cycloprop; *p*-cymene; *p*-cymene-8-ol; *n*-decane; *cis*-dihydrocarvone; 2,5-dimethylstyrene; β-elemene; δ-elemene; γ-elemene; elemol; α-fenchene; fenchyl acetate; germacrene D; α-gurjenene; β-gurjenene; *n*-hexacosane; α-humulene; humulene epoxside; isoaromadendrene epoxide; isoborneol; α-kubeben; ledene; limonene; linalool; *cis*-linalool oxide; linalyl acetate; longifolene; manool; menthone; methyl chavicol; *cis*-2-methyl-3-methylene-hep-5-ene; *trans*-2-methyl-3-methylene-hep-5-ene; α-muurolene; γ-muurolene; α-muurolol; *t*-muurolol; myrcene; β-myrcene; myrtenal; myrtenol; naphthalene; naphthalene methanol; nerol; neryl acetate; β-ocimene; *cis*-ocimene; *cis*-β-ocimene; (*E*)-β-ocimene; *trans*-β-ocimene; (*Z*)-β-ocimene; *n*-octacosane; α-phellandrene; β-phellandrene; phenanthrene; α-pinene; β-pinene; pinocamphon; *trans*-pinocamphon; sabinene; *cis*-sabinene hydrate; *trans*-sabinene-hydrate; *cis*-sabinyl acetate; *trans*-sabinyl acetate; *cis*-salven; sclareol; β-selinene; γ-selinene; terpinen-4-ol; α-terpinene; γ-terpinene; α-terpinol; terpinolene; α-terpinolene; α-terpenyl acetate; α-thujone; β-thujone; tricyclene; *n*-undecane; viridiflorol; α-ylangene	Länger *et al.*, 1996; Venskutonis, 1997; Lima *et al.*, 2004; Mitic-Culafic *et al.*, 2005; Hayouni *et al.*, 2008; El Hadri *et al.*, 2010; Badiee *et al.*, 2012; Russo *et al.*, 2013
Tea	Borneol; camphor; 1,8-cineole; luteolin-7-glucoside; rosmarinic acid; *cis*-thujone; *trans*-thujone	Lima *et al.*, 2005
Alcoholic extract of flowers, leaves and stems	*Allo*-aromadendrene; aromadendrene; (*E*)-γ-bisabolene; borneol; bornyl acetate; β-bourbonene; δ-cadinene; γ-cadinene; α-calacorene; β-calacorene; *trans*-calamenene; camphene; camphor; (*E*)-caryophyllene; caryophyllene oxide; carvacrol; 1,8-cineole; α-copaene; α-cubebene; β-cubebene; *p*-cymene; *n*-decanol; 3-decanone; dibutyl phthalate; *n*-dodecane; *n*-eicosane; ethyl hexadecanoate; eugenol; *cis*-ferruginol; *trans*-ferruginol; furfural; *n*-hexadecane; *n*-hexadecanol; isobornyl acetate; limonene; linalool; *trans*-linalool oxide; manool; methyl octadecanoate; α-muurolene; γ-muurolene; *epi*-α-muurolol; myrcene; (*E*)-β-ocimene; (*Z*)-β-ocimene; *n*-octadecanol; 3-octanol; 3-octanone; *n*-pentacosane; α-pinene; β-pinene; α-phellandrene; sabinene; terpinen-4-ol; α-terpinene; γ-terpinene; α-terpineol; terpinolene; *cis*-thujone; *trans*-thujone; *trans*-totarol; tricyclene; *n*-tridecane; viridiflorol; α-ylangene	Veličković *et al.*, 2003; Lima *et al.*, 2007a

Aqueous acetone extract	*Cis*-*p*-coumaric acid 4-*O*-(2′-*O*-β-D-apiofuranosyl)-β-D-glucopyranoside; *trans*-*p*-coumaric acid 4-*O*-(2′-*O*-β-D-apiofuranosyl)-β-D-glucopyranoside 6,8-di-*C*-β-D-glucosylapigenin (vicenin-2); 3′-*O*-β-D-glucuronide; 7-*O*-glucuronide; 7-*O*-β-D-glucuronide; 4-hydroxyacetophenone 4-*O*-(6′-*O*- β-D-apiofuranosyl)-β-D-glucopyranoside; 6-hydroxyluteolin-7-*O*-β-D-glucoside; luteolin 7-*O*-β-D-glucoside	Lu and Foo, 2000
Butanol fraction of leaf	Caffeic acid; 2,3-dihydro-2-(4′-hydroxy-3′-methoxyphenyl)-3-(hydroxymethyl)-7-methoxy-5-benzofuranpropanol 4′-*O*-β-glucopyranoside; ethyl β-D-glucopyranosyltuberonate; 6-*O*-(*E*)-feruloyl-(α and β)-glucopyranoside; 4-*O*-β-D-glucopyranosylacetophenone; homoplantaginin; 4-hydroxyacetophenone-4-*O*-β-D-apiofuranosyl-(1→6)-*O*-β-D-glucopyranoside; 4-hydroxyacetophenone 4-*O*-[5-*O*-(3,5-dimethoxy-4-hydroxybenzoyl)-β-D-apiofrunosyl]-(1→2)-β-D-glucopyranoside; *p*-hydroxybenzoic acid; (–)-hydroxyjasmonic acid; (+)-1-hydroxypinoresinol-1-β-D-glucoside; icariside F2; (–)-isolariciresinol-3α-*O*-β-D-glucopyranoside; luteolin-7-*O*-β-glucopyranoside; rosmarinic acid; 1-*O*-(2,3,4-trihydroxy-3-methyl)-butyl-6-*O*-feruloyl-β-D-glucopyranoside	Wang *et al.*, 1998, 2000
Water-soluble carbohydrates	Arabinose; fucose; galactose; glucose; mannose; 3-*O*-methyl-galactose; rhamnose; uronic acids; xylose	Capek and Hríbalová, 2004

19.2.2 Phytochemistry

A comparison of the constituents of the flowers, leaves and stems of sage shows that the flowers contain the highest amounts of 1,8-cineole and α-pinene. Linalool is the most abundant component in the stem extract. In the leaves, camphene, limonene, *cis*-thujone, *trans*-thujone, camphor, bornyl acetate, and α-humulene are the most abundant chemical constituents (Veličković *et al.*, 2003).

Table 19.2 lists the chemical compounds present in sage tea, essential oil and various extracts.

19.3 Postharvest Technology

19.3.1 Processing

Drying and grinding

The harvested plant material is washed to clean off soil and debris. The material is then sliced into small pieces and stored in a room with adequate ventilation at ambient temperature. It should be protected from microbial fermentation, which causes the degradation of metabolites, and should also be stored in the shade to reduce ultraviolet ray-induced chemical reactions (Seidel, 2006).

The dried plant material can be powdered using a hammer mill or blender. Grinding improves subsequent extraction because it increases the surface area and increases the penetration of solvent into the cells. The powdered sample should be stored in a cool (4°C) and dry place; storage for long time may decompose some of the active compounds (Seidel, 2006).

Extraction

MACERATED EXTRACT. Maceration is a simple, but still widely used, method for initial and bulk extraction. The sage leaf powder is suspended in a suitable volume of solvent (3–5 ml/g powder) and incubated for 48–72 h at 22–37°C. The solvent is chosen based on its selectivity for the compounds to be extracted. Non-polar solvents such as *n*-hexane and dichloromethane are used to extract the lipophilic constituents of sage (e.g. fatty acids and some terpenoids). Solvents with medium polarity, such as acetone and ethyl acetate, can be used to extract constituents with intermediate polarity (e.g. flavonoids). To extract more polar constituents (e.g. glycosides and tannins), a more polar solvent such as ethanol (50–70% diluted with water) is used. After the extraction, the solvent is passed through a filter to separate the residual sage material from the solvent. The resulting extract can be dried over a water bath; the resulting extract is kept at −20°C until use (Miura *et al.*, 2001; Seidel, 2006; Ghorbani *et al.*, 2014).

SOXHLET EXTRACT. In the Soxhlet extraction method, the sage leaf powder is placed in a cellulose covering and inserted into the extraction chamber of the Soxhlet apparatus. A suitable solvent (as described above) is then added to the collecting flask of the apparatus and heat is applied for approximately 48 h. This procedure is less solvent- and time-consuming than maceration and is suitable for extracting a small amount of plant material. However, because the extract is constantly heated, this method may lead to the degradation of thermolabile ingredients (Seidel, 2006; Mortazavian *et al.*, 2012).

PERCOLATED EXTRACT. A percolator is a conical or cylindrical container that has a lid at the top and a perforated plate at the bottom. The fragmented sage leaves are soaked in a solvent in the container and the percolate (extract) is allowed to slowly flow out of the bottom. This technique is suitable for large-scale extraction, although it does needs large volumes of solvent and may damage thermolabile metabolites (Seidel, 2006).

Essential oil

Volatile constituents from sage (e.g. phenylpropanoids and some terpenoids) can be extracted using the steam distillation method. The fresh or dried leaves are covered with the solvent in a round-bottomed flask connected to a condenser. Upon heating, the vapours (essential oil and water) condense and the distillate is collected. The aqueous phase of the collected liquid is recirculated into the flask and the volatile oil is separated into a container. The white–yellow oil can be dried over anhydrous sodium sulfate and kept at 4°C. Optimal conditions for essential oil extraction are a solvent-to-sage ratio of 6:1, leaf particle size of 1 mm, temperature 40°C, and 55–75% ethanol (by weight) (Ahmadi and Mirza, 1999; Seidel, 2006; Durling *et al.*, 2007; El Hadri *et al.*, 2010). The main components of sage essential oil are listed in Table 19.2.

Tablets/capsules

Homogenous crude powder of the leaves or prepared extract can be encapsulated or compressed into a tablet of the desired dosage.

19.3.2 Value addition

Sage can be used to prepare a variety of cosmetic products. Suggested uses include as a deep cleansing mask or a facial steam for oily skin, and for enhancing the colour of dark and grey hair, increasing hair shine, cleaning teeth and treating dandruff, as well as using as refreshing and deodorizing footbath. Commercially, the leaves of sage can be used to flavour vinegar, meat, soups, cheeses and sauces (Zargari, 1990; HerbNet, 2014).

19.4 Uses

19.4.1 General uses

Historically, sage has been used in the preparation of many foods because of its flavouring and seasoning properties. In Italy, sage leaves are fried and eaten with potatoes, gnocchi and veal dishes. In the UK, cottage cheese is mixed with chopped fresh sage to spread on bread, and sage honey is used on bread and muffins. Sage tea is made from dried or fresh leaves and it is suggested that it improves the digestion (Zargari, 1990; Bisset and Wichtl, 2001; Miura *et al.*, 2001; HerbNet, 2014).

19.4.2 Pharmacological uses

Sage has been long used in traditional medicine to treat different illnesses (Zargari, 1990). In recent decades, the plant has been a subject of intensive study for its pharmacological effects, and experimental and clinical studies have revealed several biological activities, including hypoglycaemic, hypolipidaemic, anti-inflammatory, antimicrobial, antioxidant, antitumor and immunomodulatory effects.

Hypoglycaemic and hypolipidaemic effects

Recent studies have confirmed the beneficial effects of natural compounds on glycaemic status and serum lipids in diabetes (Ghorbani, 2013a,b). Different types of sage extracts are able to decrease the blood glucose of normal and diabetic animals. The plant decreases the levels of triglyceride, cholesterol, urea, uric acid, creatinine and the liver enzymes alanine aminotransferase (ALT) and aspartate aminotransferase (AST) in diabetic animals. The mechanisms proposed for the antidiabetic actions of sage involve: (i) increased plasma insulin; (ii) decreased insulin resistance through the stimulation of peroxisome proliferator-activated receptor gamma (PPAR-γ); and (iii) increased hepatocyte responses to insulin and the inhibition of gluconeogenesis (Baricevic and Bartol, 2000; Alarcon-Aguilar *et al.*, 2002; Eidi *et al.*, 2005; Lima *et al.*, 2006; Eidi and Eidi, 2009; Mueller and Jungbauer, 2009; Shafiee-Nick *et al.*, 2012; Ghorbani *et al.*, 2013). In a randomized controlled clinical trial on 67 hyperlipidaemic patients, sage leaf extract lowered the blood levels of triglyceride, total cholesterol, low-density lipoprotein (LDL, bad cholesterol) and low-density lipoprotein (VLDL, also bad cholesterol) without any adverse effects on the blood levels of AST, ALT and creatinine. Therefore, sage may be effective and safe in the management of hyperlipidaemia (Kianbakht *et al.*, 2011).

Anti-inflammatory effects

Experimental studies have suggested that the extract of sage leaf has an anti-inflammatory effect, which confirms the traditional use of this plant for the alleviation of wounds and pain (Qnais *et al.*, 2010; Pra *et al.*, 2011; Rodrigues *et al.*, 2012). Between different extracts, the chloroform extract has more anti-inflammatory effect than the methanol extract, which has low anti-inflammatory activity, while the essential oil is inactive (Baricevic *et al.*, 2001). The carnosol and ursolic acid/oleanolic acid contained in sage extract contribute to its anti-inflammatory and antinociceptive effects (Rodrigues *et al.*, 2012), and it has been demonstrated that the anti-inflammatory effect of ursolic acid is twofold greater than that of the non-steroidal anti-inflammatory (NSAID) drug indomethacin (Baricevic *et al.*, 2001). The results of a clinical trial showed that the risk of post-operative infection in patients undergoing tonsillectomy and adenoidectomy was similar between the group receiving sage and the group treated with benzydamine hydrochloride, another NSAID (Lalićević and Djordjević, 2004).

Antimicrobial effects

The antimicrobial effect of sage was demonstrated decades ago. In recent years, medicinal plants with antimicrobial effects have gained special attention as some microbes have acquired resistance to antibiotics. The antimicrobial activities of various sage extracts are summarized in Table 19.3.

Sage essential oil demonstrates potent bactericidal and bacteriostatic effects against Gram-positive bacteria (Carta *et al.*, 1996; Mitic-Culafic *et al.*, 2005; Bozin *et al.*, 2007; Delamare *et al.*, 2007; Badiee *et al.*, 2012). This antimicrobial effect is attributed to several active ingredients in the plant oil. For example thujone, camphor and 1,8-cineole inhibit the growth of *Aeromonas hydrophila*, *A. sobria*, *Bacillus megatherium*, *B. subtilis*, *B. cereus* and *Klebsiella oxytoca* (Delamare *et al.*, 2007). Also, oleanolic acid and ursolic acid have antimicrobial activity against multidrug-resistant bacteria such as methicillin-resistant *Staphylococcus aureus*, penicillin-resistant *Streptococcus pneumonia* and vancomycin-resistant enterococci. Ursolic acid has a stronger activity than ampicillin on *Enterococcus faecium* and multidrug-resistant bacteria (Horiuchi *et al.*, 2007b). Carnosol and carnosic acid are two other antimicrobial compounds isolated from sage leaves; these compounds potentiate the antimicrobial effect of aminoglycosides against methicillin-resistant *S. aureus*.

It seems that Gram-positive bacteria are more sensitive than the Gram-negative bacteria to the antimicrobial action of sage essential oil. The transport of sage constituents through the cell wall may be the major process that limits their antibacterial action on Gram-negative bacteria (Mitic-Culafic *et al.*, 2005). The essential oil of sage also shows good antifungal activity against *Botrytis cinerea* (a chrysanthemum pathogen) and against *Candida* species such as *C. albicans*, *C. parapsilosis*, *C. krusei* and *C. glabrata* (Carta *et al.*, 1996; Badiee *et al.*, 2012). An ethanol extract of sage may show a higher antibacterial effect than its essential oil (Velickovic *et al.*, 2003). Also, leaf extract has a stronger antimicrobial activity than extracts from the flower and stem (Velickovic *et al.*, 2003). Safficinolide and sageone are two diterpenoids isolated from sage that have antiviral activity against vesicular stomatitis virus (Tada *et al.*, 1994). The hydroalcoholic extract of sage is also reported as a potential antimalarial agent (Akkawi *et al.*, 2012).

Antioxidant effects

Evidence from several studies has confirmed that sage has potent antioxidant properties. Evaluation of antioxidant activity of phenolic compounds from sage shows that rosmarinic acid and luteolin-7-*O*-β-glucopyranoside are potent antioxidant agents. The superoxide scavenging effect of rosmarinic acid derivatives is 15–20 times stronger than that of Trolox, a synthetic water-soluble vitamin E that is used as a standard in antioxidant assays. The antioxidative activity of phenolic compounds is related to their conjugated rings and hydroxyl groups (Wang *et al.*, 1998; Lu and Foo, 2001). Sage also contains 12-*O*-methyl carnosol and a large number of diverse catechol compounds that are very effective

Table 19.3. Antimicrobial effects of different extracts of sage: (+) good antimicrobial effect; (×) weak antimicrobial effect; (–) no antimicrobial effect; (*) controversy about antimicrobial effect.

Sage extract	Type of organism (bacterium, virus, fungus)	Reference
Essential oil	**Gram-positive bacteria:** *Bacillus cereus* (+), *B. megatherium* (+), *B. subtilis* (+), *Enterococcus faecalis* (+), *Listeria monocytogenes* (+), *Staphylococcus aureus* (–), *S. epidermidis* (+), *S. faecalis* (×) **Gram-negative bacteria:** *Aeromonas hydrophila* (+), *A. sobria* (+), *Escherichia coli* (+), *Klebsiella oxytoca* (+), *K. pneumonia* (+), *Pseudomonas aeruginosa* (*), *P. morgani* (+), *Salmonella anatum* (+), *S. enteritidis* (+), *S. typhi* (+), *Shigella sonei* (+) **Fungi:** *Botrytis cinerea* (+), *Candida albicans* (+), *C. glabrata* (+), *C. krusei* (+), *C. parapsilosis* (+)	Carta *et al.*, 1996; Mitic-Culafic *et al.*, 2005; Bozin *et al.*, 2007; Delamare *et al.*, 2007; Hayouni *et al.*, 2008; Badiee *et al.*, 2012
Ethanolic extract	**Gram-positive bacteria:** *Bacillus subtilis* (+), *Sarcina lutea* (+), *S. aureus* (+) **Gram-negative bacteria:** *E. coli* (×), *P. aeruginosa* (×); *S. enteritidis* (×) **Fungi:** *Aspergillus niger* (+), *C. albicans* (×), *Saccharomyces cerevisiae* (×)	Veličković *et al.*, 2003
Carnosol and carnosic acid isolated from sage	**Gram-positive bacteria:** *E. faecalis* (+), *E. faecium* (+), *S. aureus* (+) **Gram-negative bacteria:** *P. aeruginosa* (–), *Serratia marcescens* (–)	Horiuchi *et al.*, 2007a
Oleanolic acid and ursolic acid isolated from sage	**Gram-positive bacteria:** *E. faecalis* (+), *E. faecium* (+), *S. aureus* (+), *S. pneumoniae* (+) **Gram-negative bacteria:** *E. coli* (–), *P. aeruginosa* (–), *Serratia marcescens* (–)	Horiuchi *et al.*, 2007b
Safficinolide and sageone isolated from sage	Vesicular stomatitis virus (+)	Tada *et al.*, 1994

scavengers of superoxide anions. The radical scavenging effect of carnosol is comparable with that of α-tocopherol (Miura *et al.*, 2002; Bors *et al.*, 2004). According to the results of a study by Cuvelier *et al.* (1996), the most active antioxidant constituents of sage are carnosol, rosmarinic acid and carnosic acid, followed by caffeic acid, rosmanol, rosmadial, genkwanin and cirsimaritin.

Antitumour effects

Both sage essential oil and the aqueous extract of sage show growth-inhibitory and pro-apoptotic effects on cell lines derived from colorectal cancer (HCT-116, HCT15, CO115), breast cancer (MCF-7), melanoma (A375, M14, A2058) and squamous human cell carcinoma of the oral cavity (UMSCC1) (Xavier *et al.*, 2009; El Hadri *et al.*, 2010; Sertel *et al.*, 2011; Russo *et al.*, 2013). The potential antitumor activity of sage is due to its anti-angiogenic, anti-migratory and anti-proliferative effects (Keshavarz *et al.*, 2010, 2011). These effects may be related to the presence of *trans*-caryophyllene, α-humulene, thujone, camphor, rosmarinic acid and ursolic acid in sage (Jedinak *et al.*, 2006; Xavier *et al.*, 2009; El Hadri *et al.*, 2010; Russo *et al.*, 2013). *In vitro* studies have reported that ursolic acid inhibits angiogenesis, proteases and invasion by tumour cells (Jedinak *et al.*, 2006).

Immunomodulatory effects

Some studies have reported that the water-soluble polysaccharides of sage have immunomodulatory activity (Capek *et al.*, 2003; Capek and Hríbalová *et al.*, 2004; Capek, 2009). However, it seems that this effect is mediated only by the aqueous extract, because other work has demonstrated that sage essential oil does not show any immunomodulatory activity (Carrasco *et al.*, 2009).

Effects on the nervous system

The ethanolic extract of sage interacts with the muscarinic and nicotinic cholinergic systems and potentiates memory retention (Eidi *et al.*, 2006). Also, some constituents of sage function as benzodiazepine receptor-active components (Kavvadias *et al.*, 2003). Clinical studies have shown that the cholinesterase-inhibiting property of sage improves mood and cognitive performance in healthy young participants (Kennedy *et al.*, 2006). Furthermore, the administration of a sage hydroalcoholic extract (60 drops/day) to patients with Alzheimer's disease improves cognitive functions (Akhondzadeh *et al.*, 2003). The hydroalcoholic extract of sage may also have beneficial effects on vincristine-induced peripheral neuropathic pain (Abad *et al.*, 2011).

Dosage and side effects

Different types of sage preparations (e.g. essential oil, alcoholic extract and distillate) are used in traditional medicine to treat various diseases. The recommended doses for adults are shown below (PDR, 2004; Mills and Bone, 2005).

- 0.1–0.3 g of the essential oil;
- 2.5–7.5 g of the tincture;
- 3–12 g/day of dried leaf or by infusion;
- 3–12 ml/day of a 1:1 liquid extract; and
- 2–4.5 ml/day of a 1:2 liquid extract.

These proposed doses and mode of administrations are yet to be approved in clinical studies.

Allergic reactions, tachycardia, vertigo, hot flushes and convulsions may occur on prolonged use or following overdose of the ethanolic extract and volatile oil of sage (this corresponds to more than 15 g of the leaves). After taking high dose of sage, patients may experience vomiting, salivation, epileptic spasms, cyanosis and tongue swallowing (Bisset and Wichtl, 2001; PDR, 2004; Mills and Bone, 2005). The convulsive effect of sage oil is because of its direct effect on the central nervous system at doses above 0.5 g/kg. Thujone, camphor and terpene ketones are the most toxic compounds of sage. When consumed orally, the LD_{50} of sage oil is 2.6 g/kg in rats (Mill and Bone, 2005), and data from animal studies have also demonstrated that sage enhances CCl_4-induced hepatotoxicity. Therefore, sage preparations may have a negative effect on the safety of drugs metabolized by the phase I drug-detoxifying enzymes (Lima *et al.*, 2007a).

Because of the potential toxicity of thujone, the consumption of sage may be associated with the risk of inducing harmful effects on the fetus. Therefore, the use of essential oil and other types of sage extracts is contraindicated in pregnancy and lactation (Bisset and Wichtl, 2001; PDR, 2004; Mills and Bone, 2005).

19.5 Summary

Sage is a perennial evergreen shrub in the Lamiaceae family and now grows in several regions of the world, but especially in Mediterranean countries and Europe. For centuries, the plant has been used in the preparation of many foods because of its flavouring and seasoning properties. Sage leaves also have a long history of use in traditional medicine and cosmetic products. In experimental studies, extracts of sage have shown a number of pharmacological activities, such as hypoglycaemic, hypolipidaemic, anti-inflammatory, antimicrobial, antioxidant, antitumor and immunomodulatory effects. Several chemical compounds are found in sage, including fatty acids, steroids, waxes, alkaloids, phenolics, terpenoids, glycoside derivatives and polyacetylenes. Cineole, borneol, thujone, camphor, elemene, ledene, pinene, humulene and caryophyllene are some of the major ingredients present in its essential oil. Despite the promising results that have been obtained from experimental studies, there are not enough well-designed clinical trials that have investigated the healing effects of sage. Also, although this plant is commonly used as a traditional medicinal plant, it should not be consumed in large amounts or over prolonged periods.

Fig. 19.1. Cultivated sage.

References

Abad, A.N.A., Nouri, M.H.K. and Tavakkoli, F. (2011) Effect of *Salvia officinalis* hydroalcoholic extract on vincristine-induced neuropathy in mice. *Chinese Journal of Natural Medicines* 9, 354-358.

Ahmadi, L. and Mirza, M. (1999) A study of chemical composition of essential oil from *Salvia officinalis* L. during different growth stages. *Journal of Water and Soil Science – Isfahan University of Technology* 3, 93–100.

Akhondzadeh, S., Noroozian, M., Mohammadi, M., Ohadinia, S., Jamshidi, A.H. and Khani, M. (2003) *Salvia officinalis* extract in the treatment of patients with mild to moderate Alzheimer's disease: a double blind, randomized and placebo-controlled trial. *Journal of Clinical Pharmacy and Therapeutics* 28, 53–59.

Akkawi, M., Sharif, A.A., Salem, K., Saleh, A. and Aburemeleh, Q. (2012) Wild sage (*Salvia officinalis* L.) as a potential anti-malarial drug. *Malaria Journal* 11, P3.

Alarcon-Aguilar, F.J., Roman-Ramos, R., Flores-Saenz, J.L. and Aguirre-Garcia, F. (2002) Investigation on the hypoglycaemic effects of extracts of four Mexican medicinal plants in normal and alloxan-diabetic mice. *Phytotherapy Research* 16, 383–386.

Badiee, P., Nasirzadeh, A.R. and Motaffaf, M. (2012) Comparison of *Salvia officinalis* L. essential oil and antifungal agents against *Candida* species. *Journal of Pharmaceutical Technology and Drug Research* 1, 7.

Baricevic, D. and Bartol, T. (2000) The biological/pharmacological activity of the *Salvia* genus. In: Kintzios, S.E. (ed.) *Sage – the genus* Salvia. *Harwood Academic Publishers*, Amsterdam, pp. 143–184.

Baricevic, D., Sosa, S., Della Loggia, R., Tubaro, A., Simonovska, B., Krasna, A. and Zupancic, A. (2001) Topical anti-inflammatory activity of *Salvia officinalis* L. leaves: the relevance of ursolic acid. *Journal of Ethnopharmacology* 75, 125–132.

Bisset, N.G. and Wichtl, M. (2001) *Herbal Drugs and Phytopharmaceuticals: A Handbook for Practice on a Scientific Basis with Reference to German Commission E Monographs*, 2nd edn. CRC Press, Boca Raton, Florida, pp. 440–443.

Bors, W., Michel, C., Stettmaier, K., Lu, Y. and Foo, L.Y. (2004) Antioxidant mechanisms of polyphenolic caffeic acid oligomers, constituents of *Salvia officinalis*. *Biological Research* 37, 301–311.

Bozin, B., Mimica-Dukic, N., Samojlik, I. and Jovin, E. (2007) Antimicrobial and antioxidant properties of rosemary and sage (*Rosmarinus officinalis* L. and *Salvia officinalis* L., Lamiaceae) essential oils. *Journal of Agricultural and Food Chemistry* 55, 7879–7885.

Capek, P. (2009) A water soluble glucomannan isolated from an immunomodulatory active polysaccharide of *Salvia officinalis* L. *Carbohydrate Polymers* 75, 356–359.

Capek, P. and Hríbalová, V. (2004) Water-soluble polysaccharides from *Salvia officinalis* L. possessing immunomodulatory activity. *Phytochemistry* 65, 1983–1992.

Capek, P., Hríbalová, V., Švandová, E., Ebringerová, A., Sasinková, V. and Masarová, J. (2003) Characterization of immunomodulatory polysaccharides from *Salvia officinalis* L. *International Journal of Biological Macromolecules* 33, 113–119.
Carrasco, F.R., Schmidt, G., Romero, A.L., Sartoretto, J.L., Caparroz-Assef, S.M., Bersani-Amado, C.A. and Cuman, R.K. (2009) Immunomodulatory activity of *Zingiber officinale* Roscoe, *Salvia officinalis* L. and *Syzygium aromaticum* L. essential oils: evidence for humor- and cell-mediated responses. *Journal of Pharmacy and Pharmacology* 61, 961–967.
Carta, C., Moretti, M.D.L. and Peana, A.T. (1996) Activity of the oil of *Salvia officinalis* L. against *Botrytis cinerea*. *Journal of Essential Oil Research* 8, 399–404.
Clebsch, B. and Barner, C.D. (2003) *The New Book of Salvias: Sages for Every Garden*. Timber Press, Portland, Oregon.
Cuvelier, M.E., Richard, H. and Berset, C. (1996) Antioxidative activity and phenolic composition of pilot-plant and commercial extracts of sage and rosemary. *Journal of the American Oil Chemists' Society* 73, 645–652.
Delamare, A.P.L., Moschen-Pistorello, I.T., Artico, L., Atti-Serafini, L. and Echeverrigaray, S. (2007) Antibacterial activity of the essential oils of *Salvia officinalis* L. and *Salvia triloba* L. cultivated in south Brazil. *Food Chemistry* 100, 603–608.
Durling, N.E., Catchpole, O.J., Grey, J.B., Webby, R.F., Mitchell, K.A. and Yeap Fu, L. (2007) Extraction of phenolics and essential oil from dried sage (*Salvia officinalis* L.) using ethanol–water mixtures. *Food Chemistry* 101, 1417–1424.
Eidi, A. and Eidi, M. (2009) Antidiabetic effects of sage (*Salvia officinalis L.*) leaves in normal and streptozotocin-induced diabetic rats. *Diabetes Metabolic Syndrome: Clinical Research and Reviews* 3, 40–44.
Eidi, M., Eidi, A. and Zamanizadeh, H. (2005) Effect of *Salvia officinalis L.* leaves on serum glucose and insulin in healthy and streptozotocin-induced diabetic rats. *Journal of Ethnopharmacology* 100, 310–333.
Eidi, M., Eidi, A. and Bahar, M. (2006) Effects of *Salvia officinalis* L. (sage) leaves on memory retention and its interaction with the cholinergic system in rats. *Nutrition* 22, 321–326.
El Hadri, A., Goméz del Río, M.Á., Sanz, J., González Coloma, A., Idaomar, M., Ribas Ozonas, B., Benedí González, J. and Sánchez Reus, M.I. (2010) Cytotoxic activity of α-humulene and *trans*-caryophyllene from *Salvia officinalis* in animal and human tumor cells. *Anales de la Real Academia Nacional de Farmacia* 76, 343–356.
Ghorbani, A. (2013a) Phytotherapy for diabetic dyslipidemia: evidence from clinical trials. *Clinical Lipidology* 8, 311–319.
Ghorbani, A. (2013b) Best herbs for managing diabetes: a review of clinical studies. *Brazilian Journal of Pharmaceutical Sciences* 49, 413–422.
Ghorbani, A., Shafiee-Nick, R., Rakhshandeh, H. and Borji, A. (2013) Antihyperlipidemic effect of a polyherbal mixture in streptozotocin-induced diabetic rats. *Journal of Lipids* 2013, Article ID 67579.
Ghorbani, A., Hadjzadeh, M.R., Rajaei, Z. and Zendehbad, S.B. (2014) Effects of fenugreek seeds on adipogenesis and lipolysis in normal and diabetic rat. *Pakistan Journal of Biological Sciences* 17, 523–528.
Harrison, L. (2012) *RHS Latin for Gardeners*. Mitchell Beazley (Octopus Publising Group), London.
Hayouni, E.A., Chraief, I., Abedrabba, M., Bouix, M., Leveau, J.Y., Mohammed, H. and Hamdi, M. (2008) Tunisian *Salvia officinalis* L. and *Schinus molle* L. essential oils: their chemical compositions and their preservative effects against *Salmonella* inoculated in minced beef meat. *International Journal of Food Microbiology* 125, 242–251.
HerbNet (2014) *Sage*. From HerbNet, The Herb Growing & Marketing Network, Silver Spring, Pennsylvania Available at: http://www.herbnet.com/sage.pdf (accessed 2 July 2014).
Horiuchi, K., Shiota, S., Kuroda, T., Hatano, T., Yoshida, T. and Tsuchiya, T. (2007a) Potentiation of antimicrobial activity of aminoglycosides by carnosol from *Salvia officinalis*. *Biological and Pharmaceutical Bulletin (Tokyo)* 30, 287–290.
Horiuchi, K., Shiota, S., Hatano, T., Yoshida, T., Kuroda, T. and Tsuchiya, T. (2007b) Antimicrobial activity of oleanolic acid from *Salvia officinalis* and related compounds on vancomycin-resistant enterococci. *Biological and Pharmaceutical Bulletin (Tokyo)* 30, 1147–1149.
Jedinak, A., Muckova, M., Kostalova, D., Maliar, T. and Masterova, I. (2006) Antiprotease and antimetastatic activity of ursolic acid isolated from *Salvia officinalis*. *Zeitschrift fur Naturforschung C* 61, 777–782.

Kavvadias, D., Monschein, V., Sand, P., Riederer, P. and Schreier, P. (2003) Constituents of sage (*Salvia officinalis*) with *in vitro* affinity to human brain benzodiazepine receptor. *Planta Medica* 69, 113–117.

Kennedy, D.O., Pace, S., Haskell, C., Okello, E.J., Milne, A. and Scholey, A.B. (2006) Effects of cholinesterase inhibiting sage (*Salvia officinalis*) on mood, anxiety and performance on a psychological stressor battery. *Neuropsychopharmacology* 31, 845–852.

Keshavarz, M., Mostafaie, A., Mansouri, K., Bidmeshkipour, A., Motlagh, H.R.M. and Parvaneh, S. (2010) *In vitro* and *ex vivo* antiangiogenic activity of *Salvia officinalis*. *Phytotherapy Research* 24, 1526–1531.

Keshavarz, M., Bidmeshkipour, A., Mostafaie, A., Mansouri, K. and Mohammadi-Motlagh, H.R. (2011) Antitumor activity of *Salvia officinalis* is due to its anti-angiogenic, anti-migratory and anti-proliferative effects. *Cell Journal* 12, 477–482.

Kianbakht, S., Abasi, B., Perham, M. and Dabaghian, H.F. (2011) Antihyperlipidemic effects of *Salvia officinalis* L. leaf extract in patients with hyperlipidemia: a randomized double-blind placebo-controlled clinical trial. *Phytotherapy Research* 25, 1849–1853.

Lalićević, S. and Djordjević, I. (2004) Comparison of benzydamine hydrochloride and *Salvia officinalis* as an adjuvant local treatment to systemic nonsteroidal anti-inflammatory drug in controlling pain after tonsillectomy, adenoidectomy, or both: an open-label, single-blind, randomized clinical trial. *Current Therapeutic Research* 65, 360–372.

Länger, R., Mechtler, C. and Jurenitsch, J. (1996) Composition of the essential oils of commercial samples of *Salvia officinalis* L. and *S. fruticosa* Miller: a comparison of oils obtained by extraction and steam distillation. *Phytochemical Analysis* 7, 289–293.

Lima, C.F., Carvalho, F., Fernandes, E., Bastos, M.L., Santos-Gomes, P.C., Fernandes-Ferreira, M. and Pereira-Wilson, C. (2004) Evaluation of toxic/protective effects of the essential oil of *Salvia officinalis* on freshly isolated rat hepatocytes. *Toxicology in Vitro* 18, 457–465.

Lima, C.F., Andrade, P.B., Seabra, R.M., Fernandes-Ferreira, M. and Pereira-Wilson, C. (2005) The drinking of a *Salvia officinalis* infusion improves liver antioxidant status in mice and rats. *Journal of Ethnopharmacology* 97, 383–389.

Lima, C.F., Azevedo, M.F., Araujo, R., Fernandes-Ferreira, M. and Pereira-Wilson, C. (2006) Metformin-like effect of *Salvia officinalis* (common sage): is it useful in diabetes prevention? *British Journal of Nutrition* 96, 326–333.

Lima, C.F., Fernandes-Ferreira, M. and Pereira-Wilson, C. (2007a) Drinking of *Salvia officinalis* tea increases CCl_4-induced hepatotoxicity in mice. *Food and Chemical Toxicology* 45, 456–464.

Lima, C.F., Valentao, P.C.R., Andrade, P.B., Seabra, R.M., Fernandes-Ferreira, M. and Pereira-Wilson, C. (2007b) Water and methanolic extracts of *Salvia officinalis* protect HepG2 cells from *t*-BHP induced oxidative damage. *Chemico-Biological Interactions* 167, 107–115.

Lu, Y. and Foo, L.Y. (2000) Flavonoid and phenolic glycosides from *Salvia officinalis*. *Phytochemistry* 55, 263–267.

Lu, Y. and Foo, L.Y. (2001) Antioxidant activities of polyphenols from sage (*Salvia officinalis*). *Food Chemistry* 75, 197–202.

Mills, S. and Bone, K. (2005) *The Essential Guide to Herbal Safety*. Elsevier, St Louis, Missouri, pp. 558–559.

Mitic-Culafic, D., Vukovic-Gacic, B., Knezevic-Vukcevic, J., Stankovic, S. and Simic, D. (2005) Comparative study on the antibacterial activity of volatiles from sage (*Salvia officinalis* L.). *Archives of Biological Sciences* 57, 173–178.

Miura, K., Kikuzaki, H. and Nakatani, N. (2001) Apiananeterpenoids from *Salvia officinalis*. *Phytochemistry* 58, 1171–1175.

Miura, K., Kikuzaki, H. and Nakatani, N. (2002) Antioxidant activity of chemical components from sage (*Salvia officinalis* L.) and thyme (*Thymus vulgaris* L.) measured by the oil stability index method. *Journal of Agricultural and Food Chemistry* 50, 1845–1851.

Mortazavian, S.M., Ghorbani, A. and Hesari, T.G. (2012) Effect of hydro-alcoholic extract of *Viola tricolor* and its fractions on proliferation of uterine cervix carcinoma cells. *Iranian Journal of Obstetrics, Gynecology and Infertility* 15, 9–16.

Mueller, M. and Jungbauer, A. (2009) Culinary plants, herbs and spices – a rich source of PPARc ligands. *Food Chemistry* 117, 660–667.

PDR (2004) *PDR [Physicians Desk Reference] for Herbal Medicines*, 3rd edn. Thomson PDR, Montvale, New Jersey.

Pra, V.D., Bisol, L.B., Detoni, S., Denti, M. and Grando, J. (2011) Anti-inflammatory activity of fractionated extracts of *Salvia officinalis*. *Journal of Applied Pharmaceutical Science* 1, 67–71.

Qnais, E.Y., Abu-Dieyeh, M., Abdulla, F.A. and Abdalla, S.S. (2010) The antinociceptive and anti-inflammatory effects of *Salvia officinalis* leaf aqueous and butanol extracts. *Pharmaceutical Biology* 48, 1149–1156.
Rodrigues, M.R., Kanazawa, L.K., das Neves, T.L., da Silva, C.F., Horst, H., Pizzolatti, M.G., Santos, A.R., Baggio, C.H. and Werner, M.F. (2012) Antinociceptive and anti-inflammatory potential of extract and isolated compounds from the leaves of *Salvia officinalis* in mice. *Journal of Ethnopharmacology* 139, 519–526.
Russo, A., Formisano, C., Rigano, D., Senatore, F., Delfine, S., Cardile, V., Rosselli, S. and Bruno, M. (2013) Chemical composition and anticancer activity of essential oils of Mediterranean sage (*Salvia officinalis* L.) grown in different environmental conditions. *Food and Chemical Toxicology* 55, 42–47.
Seidel, V. (2006) Initial and bulk extraction. In: Sarker, S.D., Latif, Z. and Gray, A.I. (eds) *Natural Product Isolation*, 2nd edn. Humana Press, Totowa, New Jersey, pp. 27–46.
Sertel, S., Eichhorn, T., Plinkert, P.K. and Efferth, T. (2011) Anticancer activity of *Salvia officinalis* essential oil against HNSCC cell line (UMSCC1). *HNO* 59, 1203–1208.
Shafiee-Nick, R., Ghorbani, A., Vafaee, F. and Rakhshandeh, H. (2012) Chronic administration of a combination of six herbs inhibits the progression of hyperglycemia and decreases serum lipids and aspartate amino transferase activity in diabetic rats. *Advances in Pharmacological Sciences* 2012, 789–796.
Stearn, W.T. (2004) *Botanical Latin*. Timber Press, Bath, UK.
Tada, M., Okuno, K., Chiba, K., Ohnishi, E. and Yoshii, T. (1994) Antiviral diterpenes from *Salvia officinalis*. *Phytochemistry* 35, 539–541.
Veličković, D.T., Ranđelović, N.V., Ristić, M.S., Veličković, A.S. and Šmelcerović, A.A. (2003) Chemical constituents and antimicrobial activity of the ethanol extracts obtained from the flower, leaf and stem of *Salvia officinalis* L. *Journal of the Serbian Chemical Society* 68, 17–24.
Venskutonis, P.R. (1997) Effect of drying on the volatile constituents of thyme (*Thymus vulgaris* L.) and sage (*Salvia officinalis* L.). *Food Chemistry* 59, 219–227.
Wang, M., Li, J., Rangarajan, M., Shao, Y., LaVoie, E.J., Huang, T.-C. and Ho, C.-T. (1998) Antioxidative phenolic compounds from sage (*Salvia officinalis*). *Journal of Agricultural and Food Chemistry* 46, 4869–4873.
Wang, M., Kikuzaki, H., Zhu, N., Sang, S., Nakatani, N. and Ho, C.-T. (2000) Isolation and structural elucidation of two new glycosides from sage (*Salvia officinalis* L.). *Journal of Agricultural and Food Chemistry* 48, 235–238.
Xavier, C.R.R., Lima, C.F., Fernandes-Ferreira, M. and Pereira-Wilson, C. (2009) *Salvia fruticosa*, *Salvia officinalis*, and rosmarinic acid induce apoptosis and inhibit proliferation of human colorectal cell lines: the role in MAPK/ERK pathway. *Nutrition and Cancer* 61, 564–571.
Zargari, A. (1990) *Medicinal Plants, Vol. 4*, 4th edn,. Tehran University Press, Tehran, pp. 59–64.

20 Senna

Kuntal Das*
Krupanidhi College of Pharmacy, Bangalore, India

20.1 Botany

20.1.1 Introduction

Senna belongs to the genus *Senna*, a large group of around 350 species of trees, shrubs, vines and herbs with numerous species growing in the South American rainforests and tropics formerly included in the genus *Cassia*, and belonging to the family Fabaceae. Many of the species have been used medicinally and these tropical plants have a range of therapeutic activities. Various cassias have been known since the 9th or 10th centuries as purgatives and laxatives. One of the most important species of *Senna* is *S. alexandrina* Mill. (syn. *C. acutifolia* Delile, the name formerly used for Alexandrian or Egyptian senna, *C. angustifolia* Vahl, the name formerly used for Indian senna, *C. senna* L. and *C. alexandrina* (Garsault) Thell.). This is a shrubby species (Fig. 20.1), which grows abundantly in many countries. It is a traditional medicinal plant used in the Ayurvedic system of medicine and was domesticated in India from the Arabian region (Yemen). Senna was introduced into European medicine by the Arabs in the 9th century. The Arabian physicians Serapion and Mesue were the first to utilize the therapeutic properties of senna for the treatment of constipation. Thereafter, it became a well-known drug in the Unani system of medicine, and eventually, it was listed as a drug by the various Pharmacopoeias of the world (Arya, 2003). Senna has been used as a laxative since ancient times in Asia, especially by the South-east Asians.

Indian senna is collected as the dried leaf and leaflets of what is now known as *S. alexandrina*, and is also known as Tinnevelly (or Tinnervelly) senna (derived from the name of the city Tinnevelly in Tamil Nadu). The pods and leaves are the main useful parts of the senna plant, and both are used as over-the-counter pharmaceutical preparations (Chaubey and Kapoor, 2001). The plant is cultivated in dry lands, especially in southern and western India. There is some controversy about the origin of senna and its cultivation, with reports in the literature indicating that there are two varieties of senna: the first found in Egypt and Sudan (in the Nile river basin), and the second widely cultivated in India (Balasanka *et al.*, 2013). Although these are now treated as a

*Corresponding author, e-mail: drkkdsd@gmail.com

© CAB International 2016. *Leafy Medicinal Herbs: Botany, Chemistry, Postharvest Technology and Uses* (eds D.C.P. Ambrose *et al.*)

Fig. 20.1. Senna plants in field.

Table 20.1 Vernacular names of senna.

Language	Name
Indian languages	
Bengali	Sanna-makki, sonpat
Gujarati	Mindhiaval, sonamukhi
Hindi	Hindisana, sanaya
Kannada	Nelavarike
Malayalam	Adapatiyan, chinnukki, nilavaka
Marathi	Sonamukhi
Oriya	Sunamukhi
Sanskrit	Markandika, swarnpatri, bhumiari
Tamil	Nilavirai
Telegu	Nela tangedu
Other languages	
Chinese	Fan-hsieh-yeh
French	Casse
Latin (medicinal name)	Folium Cassiae
Spanish	Sen de España
Thai	Ma kham khaek

single species, *S. alexandrina*, they differ greatly in their morphological and anatomical characteristics.

The medicinal importance of the senna is largely due to its sennoside content, which is used to cure intestinal disorders, jaundice, anaemia, typhoid and dermal diseases, and stimulates intestinal peristalsis (Majid, 2010). Depending on the dosage used, senna has two type of effects: (i) as a laxative, which acts by loosening the bowel contents and enhancing evacuation; and (ii) as a purgative, which acts to cause the evacuation of liquid diarrhoeic matter. Hence, senna should be recommended only for use in the short term, and its use depends on the condition of the patient; it can be given in a large dose, but only as per the physician's direction.

Senna has a number of vernacular names in India, and these are listed in Table 20.1, along with some of the names by which it is known in the rest of the world.

20.1.2 History/origin

Traditionally, senna was used in 'love sachets' in the Middle East. In China, it is known as fan-hsieh-yeh, which means 'foreign country laxative herb'. Among the different senna varieties, *C. acutifolia* (now *S. alexandrina*) is one of the oldest varieties of herbal medicines. By 1640, senna was cultivated and utilized in England for its cathartic activity and included officially in the *British Pharmacopoeia*. After this, and due to commercial demand, the cultivation of senna started in different countries, and Indian senna is now cultivated commercially in several Indian states and in China (where it is cultivated on a mass scale in Guangdong, Guangxi and Yunnan) and Pakistan (Hoffmann, 1990; Lust, 1990; Balasanka *et al.*, 2013). When the leaves have reached maturity, they are sun dried and used without further processing.

20.1.3 Location

Senna is native to Yemen and is widely cultivated in different parts of India, especially in the dry areas of Haryana, Rajasthan, Gujarat, Maharashtra, Madhya Pradesh, Andhra Pradesh and Tamil Nadu, with commercial plantations in certain coastal parts of Gujarat, especially in the Bhuj region, in Rajasthan, in southern regions such as Tamil Nadu (Tinnevelly and Ramnathpuram), western regions such as Karnataka (Mysore) and northern India (Jammu) (Gupta *et al.*, 1977). Senna germinates easily from seed and grows well on light well-drained sandy loam and lateritic soil of pH 7–8.5, when it produces high biomass yields. Bright sunlight gives optimum growth.

20.1.4 Morphology

Senna is a shrub of less than 1 m in height. It has many ascending branches. The leaves are compound pinnate, petiolate, about 10 cm long and bear 5–8 pairs of leaflets each on a small stalk. The stem is erect, smooth, and pale green, with long, spreading branches. The asymmetric flowers are small and yellow in colour without nectar. They have five uniform and similar petals. The pods are oblong, about 2 inches long and contain 6–7 seeds. (Kirtikar and Basu, 1999) (Fig. 20.2). The leaflets are pale yellowish green and have a slight odour and a mucilaginous taste; they are about 2.5–6 cm long and 7–15 mm wide, lanceolate, slightly asymmetric at the base and acute at the apex with a sharp spine. Both of the leaf surfaces are smooth.

20.2 Chemistry

20.2.1 Chemical composition/ phytochemistry

The leaves of Indian senna contain anthranoid compounds, of which dianthrones make up 75–80% and anthrones 20–25% (Werner and Merz, 2007). The leaves also contain flavonol and anthraquinone glycosides (1–3%). Among the flavonols, kaempferol, kaempferin and isorhamnetin are important compounds (Srivastava *et al.*, 1983). Of the anthraquinones, senna contains rhein (cassic acid or 1,8-dihydroxyanthraquinone-3-carboxylic acid) and emodin (also called aloe-emodin, as it is found in the latex of the aloe plant). These two anthraquinones are components of the glycosides sennoside A and B (rhein dianthrones), which play a major role in the laxative activity of senna (Bala *et al.*, 2001), and also of sennoside C and D; all four of these compounds are described as rhein aloe-emodin heterodianthrones (WHO, 1999) (Fig. 20.3).

The leaves, pods and roots of Indian senna contain chrysophenol, imodin and aloe-imodin (Srivastava *et al.*, 1983). Apart from these compounds, the leaves also contain β-sitosterol (0.33%) (Rastogi and Mehrotra, 1993), 10–12% mineral matter, 7–10% mucilage, about 8% polyol, mannitol, sodium potassium tartrate, salicylic acid and a small amount of resinous compounds. Other compounds that have been found in the leaves include a new anthraquinone glycoside, emodin-8-*O*-sophoroside and the known glycosides kaempferol 3-*O*-β-glucoside and isorhamnetin-3-O-β-glucoside, tinnevellin glycoside, isorhamnetin-3-*O*-β-gentiobioside, apigenin-6,8-di-*C*-glycoside, emodin-8-*O*-β-D-glucopyranoside, D-3-*O*-methylinositol and sucrose ((Kinjo *et al.*, 1994; Singh *et al.*, 1995; Wu *et al.*, 2007). The flowers contain crisofenic acid. Two new naphthalene glycosides have been isolated from the seeds (Wang *et al.*, 2007), as well as an oleanen-type triterpenoid glycoside from butanolic extracts of the seeds (Khan and Srivastava, 2009).

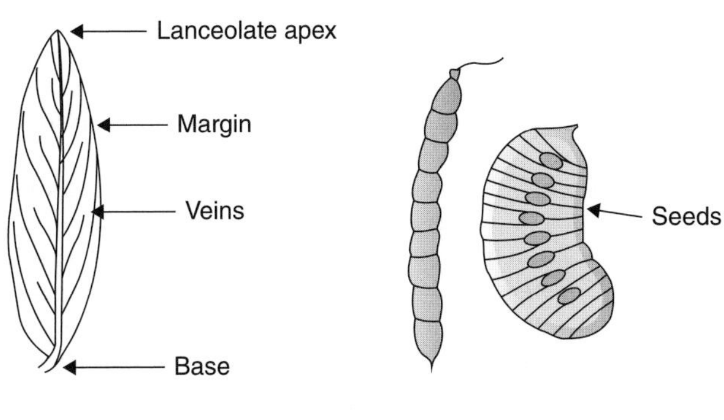

Senna leaf **Senna pods**

Fig. 20.2. Senna leaf morphology and senna pod.

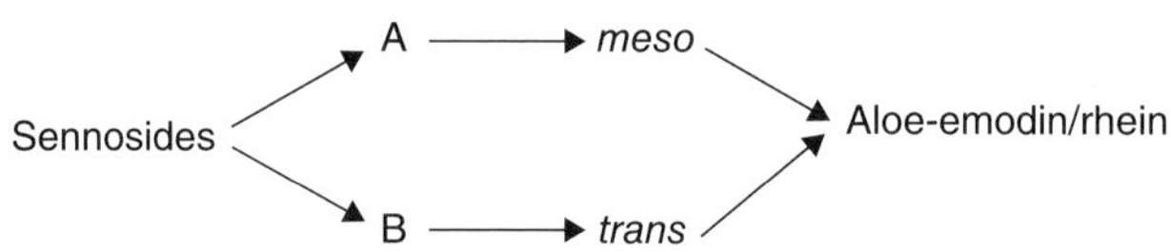

Interrelationship between sennosides with rhein component

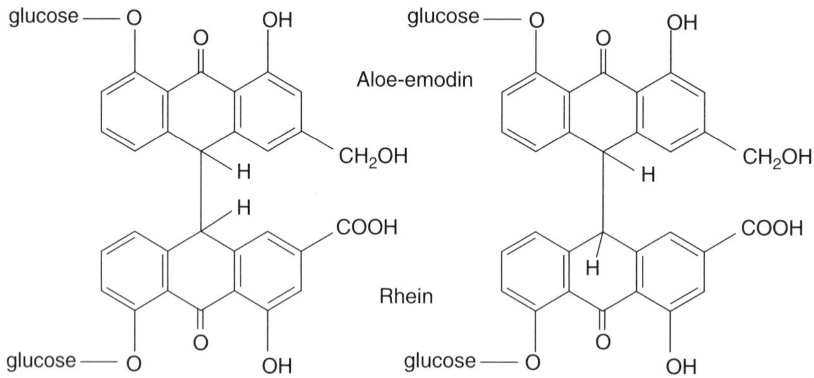

Sennoside-A (*meso* form) Sennoside-B (*trans* form)

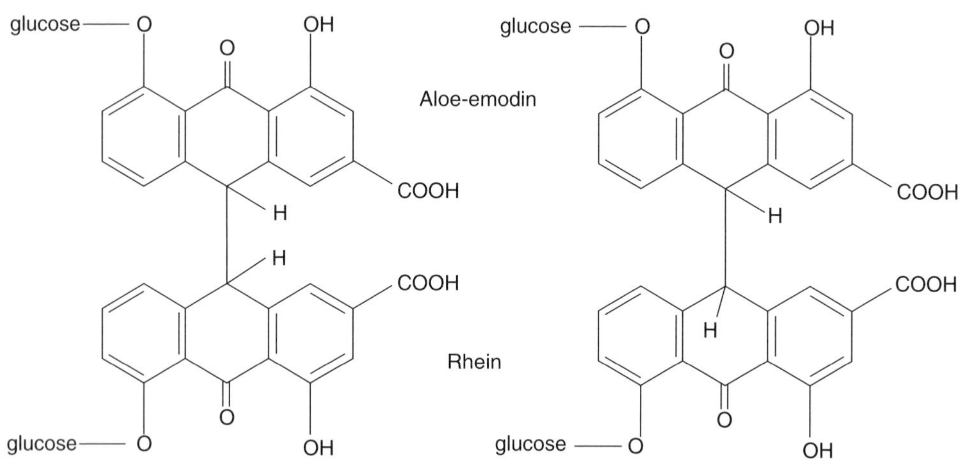

Sennoside-C (*meso* form) Sennoside-D (*trans* form)

Fig. 20.3. Chemical structure of the sennosides of senna.

20.3 Postharvest Technology

The plants produce foliage containing high amounts of sennosides 50–90 days from sowing (Upadhyay *et al.*, 2011). A good crop of senna can give 15 q/ha of dry leaves and 7 q/ha of pods under irrigated and good horticultural management conditions (Farooqi and Sreeramu, 2010). The plant is harvested three times over the cultivation period in the season, with the first picking in March and the others before May; there are further pickings in October and December (Pareek *et al.*, 1983). The flowers contain 2.6% of sennosides, whereas the pods (shells) contain 3–5% and the foliage 2.5–4% (Srivastava *et al.*, 1980; Atal and Kapoor, 1982; Pareek *et al.*, 1983).

20.3.1 Processing

The harvested leaf material is spread in a thin layer in an open field to reduce the moisture content and then dried in well-ventilated sheds for a week until the colour of the dried leaves is a yellowish green. If the leaves turn black or brown that indicates deterioration of their quality. Generally, senna leaves are dried under sunlight or in shade at a normal room temperature.

Studies on drying senna were carried out in a forced flow type mechanical dryer at different air temperatures. Based on the sennoside concentration, it was shown that 45°C was the optimum temperature for drying (Ambrose and Naik, 2013). The dried leaves retained their greenish colour (Fig. 20.4).

The dried leaves are stored in light-proof containers at room temperature to minimize the degradation of active constituents (i.e. sennosides) and protect them from moisture. It has been reported that black polyethylene bags, aluminium foil bags and transparent polyethylene bags are suitable for the storage of dried senna leaves (European Pharmacopoeia, 2005; Upadhyay *et al.*, 2011).

There are various extraction methods described to isolate different constituents from senna leaves. Generally, alcoholic and aqueous solutions are used for the basic extraction of Indian senna leaves using the Soxhlet, reflux and maceration methods. Traditionally, the sennosides are extracted in the form of calcium sennosides by using an organic solvent such as benzene and with methanol (4–6 h), acidification with hydrochloric acid (pH 3.2), and the addition of anhydrous calcium chloride (dissolved in denatured spirit), which precipitates calcium sennosides in the presence of ammonia solution (pH 8) (Kokate, 2005). Mehta and Laddha (2009) described an efficient method for the isolation of rhein from senna leaves using mixed water and alcohol solvents in which the hydrolysis of the sennosides and extraction of the hydrolysis products (free anthraquinones) is carried out. A microwave-assisted extraction technique has been developed for the extraction of calcium sennosides from senna leaflets using benzene and methanol as solvents; this method reduced the amount of solvent consumed by the extraction process and the time needed for the extraction (Srikanth *et al.*, 2011; Shukla *et al.*, 2013).

Fig. 20.4. Senna leaves dried in forced flow dryer.

20.3.2 Value addition

The form of senna that is available in the market is calcium sennoside. Senna products are commonly used in different kinds of herbal formulations that are available in various pharmaceutical dosage forms, such as tablets, capsules, powder, granules, liquid, paste and suppositories:

- Tablets or capsules: enteric coated or uncoated tablets are available that contain 40–90 mg of calcium sennosides equivalent to 7.5–18 mg of hydroxyanthracene glycosides as sennosides B per tablet.
- Powders: senna leaf powder is available either alone prepared directly from the raw dried leaves or admixed with some other powder, such as isabgol (from seed husks of psyllium, *Plantago ovata*) or cascara (bark from *Rhamnus purshiana*), to give a synergistic laxative effect. Senna leaf powder contains calcium sennosides

equivalent to 5–30 mg hydroxyanthracene glycosides as sennosides B per dose.

- Liquids and granules: senna is also available in the form of syrup or fluid with or without alcohol and containing 5–30 mg of hydroxyl anthracene glycosides as sennosides B per dose.
- Pastes: another form of senna is as the thick paste form of the crude extract, which is used for curing different skin ailments.
- Suppositories: the crude extract of senna residue is used in the form of suppositories, which are packed in aluminium wrappers.

Senna is also available in the form of herbal teas and chocolate.

20.4 Uses

20.4.1 General uses

Home-made preparations of senna, such as decoctions, powders, syrups, infusions and confections are prepared from senna leaf powder. In cosmetic preparations, senna leaves are blended with henna leaf and are used as black hair dye (Chopra *et al.*, 1958). Powdered leaves are mixed with vinegar and made into a plaster that is applied locally in certain skin diseases. The sennosides that are present in the senna leaves are available when they are used as health teas (Kojima *et al.*, 2001). The paste of the ground, dried root is used in Ayurvedic medicine as a treatment for ringworm and snakebites. Senna leaves are also used as one of the ingredients of Nilaavarai churnam, a product used for treating stomach distention, hiccups, vomiting and biliousness (Atal and Kapoor, 1982).

Apart from its laxative action, several other therapeutic applications of senna have been reported, such as in the treatment of liver disease, splenetic extension, hepatitis, anaemia, abdominal pain, leprosy, foul-smelling breath, bronchitis and tumours, etc. (Kirtikar and Basu, 1999). Senna leaf when combined with other aromatic herbs is useful in the treatment of flatulence or colic. Senna also helps in weight loss and is used as a blood purifier. The infusion of the leaf extract is used as an anti-anaemic and for the treatment of dysentery and fevers.

20.4.2 Pharmacological uses

Senna helps to loosen the bowels and increases the peristaltic movement of the intestine. This happens via several mechanisms. Sennosides, the active constituents, are converted by bacteria in the colon into another substance, i.e. the anthrone rhein, which has two beneficial effects, stimulation of colon activity and increased fluid secretion. The laxative effect is further promoted by the inhibition of water and electrolyte absorption from the large intestine, which increases the volume and pressure of the intestinal contents. As a result, the stimulation of colon motility increases, so producing propulsive contractions. The action on the colon is mediated by the stimulation of endogenous prostaglandin formation and secretion (de Witte, 1993).

Indian senna purifies the blood and restores the metabolic imbalance that occurs as a result of indigestion. The herb has purgative and cathartic (Grieve, 1931), antipyretic, laxative, vermifuge and diuretic properties. Indian senna also has a marked hepatoprotective activity (Ilavarasan *et al.*, 2001). It has further been reported that due to the presence of sennosides in the leaf, senna inhibits bovine serum monoamine oxidase enzyme activity and acts against skin diseases like scabies and itching (Al-Masry, 1975).

The powder made from crushing leaves and fruit is helpful in treating constipation and indigestion (Lust, 1990). Other pharmacological uses of senna have also been considered: as an astringent, cathartic, depurative, anthelmintic, cholagogue, expectorant and febrifuge; and for leucoderma, jaundice and typhoid fever (Varier *et al.*, 1994). A clinical study showed that Indian senna extract can be used as an enema after abdominal operations, and regulates the disordered function of the gastrointestinal tract after abdominal operations (Wang *et al.*, 1998). The early literature gave the standard dose of Indian senna powder as 1–2 g before bedtime, with the onset of bowel movement about 8–10 h later (Agra *et al.*, 1998).

Senna is also used as an anthelmintic treatment and a mild liver stimulant. Other uses include the treatment of tumours, foul breath, bronchitis and leprosy (Pullaiah, 2002). Due to the presence of rhein glycoside in the leaf, it shows antiviral, antioxidant and anti-inflammatory activities and marked reduction of osteoarthritis, gout and rheumatoid arthritis (Mehta and Laddha, 2009). A recent study demonstrated that the administration an aqueous Soxhlet extract of Indian senna leaf arrests spermatogenesis in male rats (Dhanapal *et al.*, 2012).

Sennosides are strong purgatives and they should be taken with care and in the proper dosage, especially in pregnant, menstruating or post-partum women. It cannot be used to treat inflammation of the gastrointestinal tract because it causes irritation (Mengs, 1986). The irritant effect upon the intestinal membrane may cause griping, pain or nausea, along with liquid stools or diarrhoea in the case of an overdose (Gruenwald *et al.*, 1998).

20.5 Summary

Senna is a well-known traditional drug in the Unani, Ayurvedic and allopathic (homeopathic) systems of medicine and also as a home remedy. The plant was introduced into Tamil Nadu in the 18th century, and it is grown there as an annual crop; hence, it has the alternative name of Tinnevelly senna. The versatile medicinal properties of senna have increased demand so that it is now a commercial crop. The dried leaves and pods contain sennosides, which are used for their laxative properties, so a major part of the crop is exported in the form of leaves, pods and sennoside concentrates. The pharmaceutical industries in India use senna for the manufacture of various herbal formulations, and Indian senna is an important medicinal and economic crop for use in novel herbal formulations.

References

Agra, Y., Sacristan, A. and Gonzalez, M. (1998) Efficacy of senna versus lactulose in terminal cancer patients treated with opioids. *Journal of Pain Symptom Management* 15, 1–7.

Al-Masry, H. (1975) *Al Aqrabazin wa Almostahdarat Alsaydalaniah (Pharmacology and Pharmacetics Preparations)*, 1st edn. Dar Al-Qualm, Kuwait.

Ambrose, D.C.P. and Naik, R. (2013) Mechanical drying of senna leaves (*Cassia angustifolia*). *Current Agriculture Research Journal* 1, 65–68.

Arya, R. (2003) Yield of *Cassia angustifolia* in combination with different tree species in a silvi herbal trial under hot arid conditions in India. *Bioresource Technology* 86, 165–169.

Atal, C.K. and Kapoor, B.M. (eds) (1982) *Cultivation and Utilization of Medicinal Plants*. Regional Research Laboratory, Jammu Tawi, India.

Bala, S., Uniyal, G.C., Dubey, T. and Singh, S.P. (2001) Improved method for analysis of sennosides in *Cassia angustifolia* by HPLC. *Phytochemical Analysis* 12, 277–280.

Balasankar, D., Vanilarasu, K., Selva Preetha, P., Rajeswari, S., Umadevi, M. and Bhowmik, D. (2013) Senna – a medical miracle plant. *Journal of Medicinal Plants Studies* 1(3), 41–47.

Chaubey, M. and Kapoor, V.P. (2001) Structure of a galactomannan from the seeds of *Cassia angustifolia* Vahl. *Carbohydrate Research* 332, 439–444.

Chopra, R.N., Chopra, I.C., Handa, K.L. and Kapur, D. (1958) *Chopra's Indigenous Drugs of India*. U.N. Dhur, Calcutta [Kolkata], India.

de Witte, P. (1993) Metabolism and pharmacokinetics of anthranoids. *Pharmacology* 47(Suppl. 1), 86–97.

Dhanapal, R., Babitha, J., Kandeepan, S. and Murugaian, P. (2012) Testicular antifertility action of *Cassia angustifolia* in male albino rats. Presented to: *Proceedings of the National Seminar on Current Perspectives in Biological Sciences (NSOCPIBS – 2012) 11 and 12 October. Advanced BioTech* 12(07, Supplement), 59–62.

European Pharmacopoeia (2005) Senna leaf. Sennae folium. In: *European Pharmacopoeia 5.0*, 5th edn. EDQM (European Directorate for the Quality of Medicines and HealthCare), Council of Europe, Strasbourg, France, pp. 2402–2403.

Farooqi, A.A. and Sreeramu, B.S. (2010) Senna. In: Farooqi, A.A. and Sreeramu, B.S. *Cultivation of Medicinal and Aromatic plants*, rev. edn. Universities Press, Hyderabad, India, pp. 294–301.

Grieve, M. (1931) *A Modern Herbal, Volume II.* Originally published by Harcourt, Brace & Company, New York, in 1931 and edited by C.F. Leyel. Republished in 1971 by Dover Publications, New York and indexed by M. Marshall in 1981–1982. Also available at: https://www.botanical.com/botanical/mgmh/s/senna-42.html (accessed 1 February 2016).

Gruenwald, J., Brendler, T. and Jaenicke, C. (1998) Senna. In: Gruenwald, J. *et al.* (eds) *PDR [Physician's Desk Reference] for Herbal Medicines*, 1st edn. Medical Economics Company, Montvale, New Jersey, pp. 722–724.

Gupta, R., Modi, J.N. and Mehta, K.G. (1977) Studies on cultivation of senna (*Cassia angustifolia* Vahl.) in north Gujarat. *South Indian Horticulture* 25, 26–29.

Hoffmann, D. (1990) *The New Holistic Herbal*, 2nd edn. Element Books, Shaftesbury, UK.

Ilavarasan, R., Mohideen, S., Vijay, L.M. and Manonmani, G. (2001) Hepatoprotective effect of *Cassia angustifolia* Vahl. *Indian Journal of Pharmaceutical Sciences* 63, 504–507.

Khan, N.A. and Srivastava, A. (2009) Antifungal activity of bioactive triterpenoid saponin from the seeds of *Cassia angustifolia. Natural Product Research* 23, 1128–1133.

Kinjo, J., Ikeda, T., Watanabe, K. and Nohara, T. (1994) An anthraquinone glycoside from *Cassia angustifolia* leaves. *Phytochemistry* 37, 1685–1687.

Kirtikar, K.R. and Basu, B.D. (eds) (1999) Senna. In: Kirtikar, K.R. and Basu, B.D. (eds) *Indian Medicinal Plants, Volume II.* International Book Distributors, Dehra Dun, India, pp. 876–877.

Kojima, T., Kishi, M., Sekita, S. and Satake, M. (2001) Origin of sennosides in health teas including malva leaves. *Shokuhin Eiseigaku Zasshi (Food Hygiene and Safety Science)* 42, 202–205.

Kokate, C.K. (2005) *Practical Pharmacognosy*, 14th edn. Nirali Prakashan, Pune, India.

Lust, J. (1990) *The Herb Book.* Bantam, London.

Majid, U. (2010) Effects of agrochemical pollution on growth, structure and some physiochemical aspects of *Cassia angustifolia* Vahl. PhD thesis, Jamia Hamdard, New Delhi.

Mehta, N. and Laddha, K.S. (2009) A modified method of isolation of rhein from senna. *Indian Journal of Pharmaceutical Sciences* 71, 128–129.

Mengs, U. (1986) Reproductive toxicological investigations with sennosides. *Arzneimittelforschung* 36, 1355–1358.

Pareek, S.K., Srivastava, V.K., Maheswari, M.L., Mandal, S. and Gupta, R. (1983) Investigation in agronomic parameters of senna (*Cassia angustifolia* Vahl) as grown in north-western India. *International Journal of Tropical Agriculture* 1, 139–144.

Pullaiah, T. (2002) Senna. In: Pullaiah, T. *Medicinal Plants in India.* Regency Publications, New Delhi, pp.137–139.

Rastogi, R.P. and Mehrotra, B.N. (1993) *Compendium of Indian Medicinal Plants, Vol. 2.* Publications and Information Directorate, New Delhi.

Shukla, A., Gupta, R., Sharma, P. and Jain, A.P. (2013) Comparative study of microwave assisted with conventional extraction of calcium sennosides from senna leaf. *Research Journal of Pharmaceutical, Biological and Chemical Sciences* 4, 103–109.

Singh, M., Chaudhuri, P.K. and Sharma, R.P. (1995) Constituents of the leaves of *Cassia angustifolia. Fitoterapia* 66, 284–286.

Srikanth, S., Sandeep Kumar, V., Shireesh Kiran, R., Prasad, M.V.V. and Krishna Mohan, G. (2011) Microwave assisted extraction of calcium sennosides from senna leaflets. *Research Journal of Pharmaceutical, Biological and Chemical Sciences* 2, 137–145.

Srivastava, V.K., Gupta, R. and Maheshwari, M.L. (1980) Photocontrol of anthracene compounds formation in senna (*Cassia acutifolia* Vahl) leaves. *Indian Journal of Experimental Biology* 18, 1318–1319.

Srivastava, V.K., Maheshwari, M.L. and Mandal, S. (1983) A rapid high performance liquid chromatography method for analysis of sennoside in senna. *Indian Journal of Pharmaceutical Sciences* 45, 230–233.

Upadhyay, A., Chandel, Y., Nayak, P.S. and Khan, N.A. (2011) Sennoside contents in senna (*Cassia angustifolia* Vahl) as influenced by date of leaf picking, packaging material and storage period. *Journal of Stored Products and Postharvest Research* 2, 97–103.

Varier, P.S., Warrier, P.K., Nambiar, V.P.K. and Ramankutty, C. (1994) Senna. In: Varier *et al. Indian Medicinal Plants – A Compendium, Volume 2.* Orient Longman Publication, Hyderabad, India, p. 31.

Wallis, T.E. (2004) *Textbook of Pharmacognosy*, 5th edn. CBS Publishers, New Delhi.

Wang, M., Yan, S. and Wang, J. (1998) Clinical and experimental study on using *Cassia angustifolia* extract as enema after abdominal operation. Zhongguo Zhong Xi Yi Jie He Za Zhi *(Chinese Journal of Integrated Traditional and Western Medicine)* 18, 540–542.

Wang, Z.J., Wu, Q.P., Tang, L.Y., Fu, M.H., He, Y., Gong, Q.F. and Huang, L.Q. (2007) Two new glycosides from the genus of *Cassia. Chinese Chemical Letters* 18, 1218–1220.

Werner, C. and Merz, B. (2007) *Assessment Report on* Cassia senna *L. and* Cassia angustifolia *Vahl, Folium.* European Medicines Agency (EMEA), London. Available at: http://www.ema.europa.eu/docs/en_GB/document_library/Herbal_-_HMPC_assessment_report/2009/12/WC500018219.pdf (accessed 1 February 2016).

WHO (1999) *WHO Monographs on Selected Medicinal Plants – Volume 1.* World Health Organization, Geneva, Switzerland.

Wu, Q.P., Wang, Z.J., Fu, M.H., Tang, L.Y., He, Y. and Fang, J. (2007) Chemical constituents from the leaves of *Cassia angustifolia. Zhong Yao Cai (Journal of Chinese Medicinal Materials)* 30, 1250–1252.

21 Spinach

Periyasamy Suganya[1]* and A. Sangamithra[2]
[1]*Avinashilingam Institute for Home Science and Higher Education for Women, Coimbatore, Tamil Nadu, India;* [2]*Kongu Engineering College, Perundurai, Erode, Tamil Nadu, India*

21.1 Botany

21.1.1 Introduction

Spinach (*Spinacia oleracea* L.), an annual plant, was originally placed in the family Chenopodiaceae, but this family was combined with the family Amaranthaceae in 2003, within the order Caryophyllales. The plant has a significant role as herb, although it is often consumed as a leafy vegetable. The leaves and stems are tender and delicate (Fig. 21.1), and can be eaten either fresh or cooked, or processed into different forms (Rubatzky *et al.*, 1997). Spinach is abundant in core nutrients and many phytochemicals. It is an excellent source of vitamins A, C and K, and of minerals such as calcium, potassium, magnesium, manganese and iron. It is also high in oxalates (Kawazu *et al.*, 2003), which have been reported to affect iron metabolism and kidney health (see Section 21.2.2). Spinach contains an ample amount of fibre and is low in calories. Carotenoids, including β-carotene, zeaxanthin and lutein, and phenolic compounds are the most important phytochemicals present in spinach. It is thought that cancer, heart disease and problems related to ageing can be prevented by these phytochemicals. The crop also has a hypoglycaemic effect, which has been used in the treatment of urinary calculi and lung inflammation. The seeds of spinach are used as a laxative and for treating breathing difficulties and liver inflammation. Canning, modified atmospheric packaging, freezing, drying and minimal processing are important processing and preservation techniques for spinach.

21.1.2 History

Spinach is sometimes known as the 'Queen of the garden patch' and can be available all year round when the climate is suitable. The origin of the crop is believed to have been from Persia, but spinach is also a native herb of south-west Asia. It had spread into the Mediterranean by the 5th century and thereafter to India and China. The diffusion of spinach into the Mediterranean area was almost certainly the result of Arab ingenuity, as it was they who introduced it into cultivation in Europe in the 12th century, where it was greatly appreciated for its medicinal properties. The Italians promoted spinach as a leafy vegetable that was used in many dishes. Spinach was originally cultivated as a

*Corresponding author, e-mail: suganya.abe@gmail.com

© CAB International 2016. *Leafy Medicinal Herbs: Botany, Chemistry, Postharvest Technology and Uses* (eds D.C.P. Ambrose *et al.*)

medicinal plant some 2000 years ago in Iran, where the ancient Iranians appreciated its properties (Splittstoesser, 1990). The crop is now widely grown in regions with a relatively moderate temperature and is considered to be a 'super food'.

Vernacular names of spinach in various countries/languages are listed in Table 21.1.

21.1.3 Location

Spinach is produced worldwide for its edible leaves. The global commercial production of spinach was 21,662,608 million t in the year 2012. China stands first in spinach production, accounting for about 19,500,000 million t p.a. The next second largest producer is the USA, with production at around 354,050 million t, followed by Japan, Turkey and Indonesia. The USA is the world's largest exporter of spinach, followed by Spain and Italy. Fig. 21.2 depicts the global production of spinach from 1961 to 2012 (FAOSTAT, 2014).

Spinach is a cold hardy plant and is especially popular in the autumn season, when the days are short and cool. It grows in

Fig. 21.1. Fresh spinach.

Table 21.1. Vernacular names of spinach by country or language. From SciNameFinder, 2013.

Country/ language	Vernacular name	Country/ language	Vernacular name
Bulgarian	Spanak	India	Pinni, palak
Chinese	Bocai	Italian	Spinacio
Danish	Spinat	Japanese	Horenso
Dutch	Spinazie	Korean	Sigeumchi
English	Spinach	Norwegian	Spinat
French	Épinard	Polish	Szpinak
German	Spinat	Portuguese	Espinafre
Greek	Spanaci	Spanish	Espinaca
Icelandic	Spínat	Swedish	Spenat

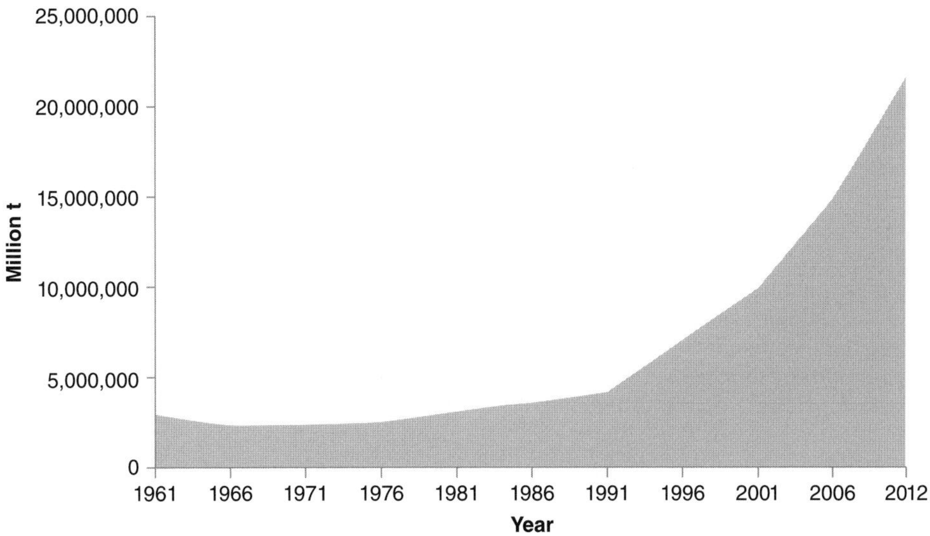

Fig. 21.2. Global production of spinach from 1961 to 2012.

well-drained, fertile soil rich in organic matter. It can tolerate slightly alkaline soils but is highly sensitive to acid soils. The preferred pH is in the range of 6.5–7.5. Young plants can tolerate temperatures as low as −9°C.

21.1.4 Morphology/types

Spinach is a dioecious, annual herb with a height of up to 150 cm and a long taproot. The leaves are simple, ovate to triangular at the base, crinkled or fat, and are arranged on stems in a spiral manner. The leaf size varies from 2 to 30 cm long and 1 to 15 cm wide. The base of the plant holds the larger leaves and the flowering stem the smaller leaves. The petiole has round to sharply pointed basal lobes and is around 6–12 cm long. In male plants, the inflorescence is spike-like and extends up to 10 cm in length. The yellow–green flowers are small, with a diameter of 3–4 mm; they are usually unisexual and rarely bisexual, and they produce small, dry, lumpy fruit clusters that contain the seeds.

Spinach grows well in the cool temperatures of the spring and autumn seasons. Curly (or savoy), semi-curly and flat leaf spinach are the main varieties that are cultivated all over the world. Sometimes, the use of curly and flat leaf spinach can extend the growing season into summer and winter. Curly leaf spinach is more cold resistant than the other varieties. It has deeply wrinkled leaves that make the cleaning process difficult, and provide a characteristic crisp texture when they are eaten raw. The dark fresh green colour makes curly leaf spinach best for fresh salads and sandwiches. Semi-curly spinach varieties have straighter leaves that are not wrinkled. The disease and bolt resistance of semi-curly spinach makes it a good choice for home growing. Semi-curly spinach can also be cooked for longer periods without changes to its texture and shape, but it is also tasty, crisp and delicate enough to be a salad green. The smooth leaves of flat leaf spinach make the cleaning process easier and make it suitable for further processing. Baby spinach is a type of flat leaf spinach that has been picked in the early stages of growth, when the leaves are particularly small, tender and sweet.

Several different varieties of spinach were formerly classified on the basis of the characteristics of the fruit and leaf, leading to the definition of *S. oleracea* var. *oleracea* (also given as *S. oleracea* var. *spinosa* (Moench) Celak and *S. spinosa* Moench) as the major variety, which had prickly fruits, and *S. inermis* (Moench) and *S. glabra* (Mill) as varieties with non-prickly, round fruits. However, these former varieties are all now regarded as one species, *Spinacia oleracea*. Other cultivars are grouped according to the colour and texture of the leaf, i.e. pale or dark green, and leaf texture, i.e. smooth, semi-smooth or crumpled. Asian type cultivars have thin and smooth leaves, long and purple red base petioles, and often bear prickly fruits, are fast growing and have quick bolting characteristics. American and European cultivars have thin, dark green and smooth to crumpled leaves, and the slow-bolting summer type has a green or pink petiole. In general, the Japanese prefer dark green leaves whereas the Chinese prefer pale green leaves.

21.2 Chemistry

Spinach is known for its rich nutritive value (it has abundant macronutrients and micronutrients) and for its medicinal properties, which include its abilities to increase vitality, restore energy and improve blood quality. The plant is plentiful in phytochemicals that decrease cholesterol levels, prevent cell damage and also play a vital role with the prevention cancer, diabetes, cardiovascular disease and hypertension.

21.2.1 Chemical (nutritional) composition

Data on the nutritional composition of fresh spinach are given in Table 21.2, based on measurements made on 15 cultivars grown in different areas of Iran. These data indicate that spinach contains high amounts of vitamin A, vitamin E and several vital antioxidants,

Table. 21.2. Nutritional content of fresh spinach from 15 cultivars grown in Iran. From Barzegar *et al.*, 2007.

Component	Content/100 g fresh weight	Component	Content/100 g fresh weight
Major components		**Vitamins**	
Ash	1.63–2.46 g	Vitamin A	9377 IU
Crude fibre	1.80–2.22 g	Thiamine (B_1)	0.078 mg
Moisture	88.2–91.1 g	Riboflavin (B_2)	0.189 mg
Oxalate	0.53–1.17 g	Niacin (B_3)	0.724 mg
Protein	1.62–3.19 g	Pantothenic acid (B_5)	0.065 mg
Total fat	0.14–0.36 g	Pyridoxine (B_6)	0.195 mg
Total phenolics	55.2–103.9 mg TAE[a]	Folate (B_9)	194 µg
Minerals		Vitamin E	2.03 mg
Copper	0.074–0.443 mg	**Fatty acids**	
Iron	2.61–8.01 mg	Erucic	0.55–5.43%
Magnesium	67.4–108.3 mg	Hexadecatrienoic	1.85–6.10%
Phosphorus	40.4–90.9 mg	Linoleic	13.59–22.50%
Potassium	595.8–1106.6 mg	Linolenic	31.60–44.04%
Sodium	34.0–117.7 mg	Oleic	6.68–16.51%
Zinc	0.963–2.606 mg	Palmitic	17.27–24.58%
		Palmitoleic	0.85–11.00%
		Stearic	0.81–3.17%

[a]Tannic acid equivalents

and also of several minerals, other vitamins and unsaturated fatty acids. Spinach contains omega-3 fatty acids, which prevent inflammatory reactions. It is also rich in vitamin C (ascorbic acid), which is a strong biological antioxidant and is involved in the synthesis of neurotransmitters, steroid hormones and collagen, the absorption of iron and calcium, and the conversion of cholesterol to bile acids. Vitamin C also assists in the healing of wounds and burns, the prevention of blood clotting and strengthening the walls of the capillaries (Combs, 1998). Spinach contains a considerable quantity of soluble dietary fibre, which is recommended for cholesterol control and weight reduction programmes – and a cup of raw spinach contains only around 7 kcal of energy.

Figure 21.3 depicts the amino acid content of spinach.

21.2.2 Phytochemistry

Spinach is rich in powerful antioxidant and health-promoting phytochemicals, including carotenoids such as lutein, β-carotene and zeaxanthin, and secondary metabolites such as phenolic/polyphenolic compounds, which include the flavonoids and compounds such as chlorogenic acid.

Carotenoids are fat-soluble vitamins, they have a high absorption capacity by the human body. Carotenoids are released from foods during cutting, chopping, blanching and cooking, which increases their bioavailability. Zeaxanthin and lutein are xanthophylls, which are yellow pigments, although at high concentrations they may appear orange or red. These two compounds are important for vision, and lutein protects against atherosclerosis (a build-up of fatty deposits in the arteries), which leads to heart attacks. Fresh spinach contains 5626 µg β-carotene/100 g and 12,198 µg/100 g lutein plus zeaxanthin (USDA, 2015).

Fresh spinach has a total phenolic content of about 1629–4835 mg of chlorogenic acid equivalents/kg of fresh weight (Gil *et al.*, 1999). Spring-grown spinach contains a greater amount of total phenols than autumn-grown spinach (Howard *et al.*, 2002). Spinach contains large amounts of flavonoids, 807–2241 mg/kg in freshly cut leaves. The

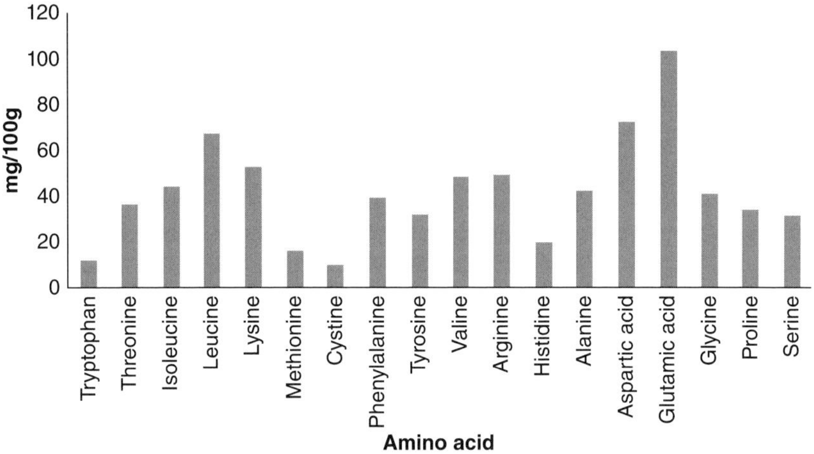

Fig. 21.3. Amino acid content in raw spinach.

flavonoids exhibit antioxidant, antimutagenic, anti-inflammatory, and anticarcinogenic properties (Lomnitski *et al.*, 2000a,b, Edenharder *et al.*, 2001, Nyska *et al.*, 2001). According to the USDA flavonoid database, the main flavonoids in spinach are kaempferol (6.38 mg/100 g fresh weight, FW), quercetin (3.97 mg/100 g FW), luteolin (0.74 mg/100 g FW) and myricetin (0.35 mg/100 g FW) (USDA, 2013). Quercetin has anti-inflammatory properties and plays a vital role in reducing the risk of cancer and heart diseases. Glucuronides and acylated diglycosides and triglycosides of methylated and methylene-dioxy derivatives of 6-oxygenated flavonols are some of the unusual flavonoid compounds reported in spinach (Pandjaitan *et al.*, 2005), which include the following (Shahidi and Naczk, 2003):

- patuletin (quercetagetin 6-methyl ether);
- spinacetin (quercetagetin 6,3′-dimethyl ether);
- spinatoside (3,6-dimethylquercetagetin 4′-*O*-glucuronide);
- jaceidin 4′-glucuronide;
- patuletin 3-gentobioside;
- patuletin 3-glycosyl(1→6)-[apiosyl)(1→2)]-glucoside;
- spinacetin 3-glycosyl(1→6)-[apiosyl(1→2)]-glucoside;
- patuletin 3-(2″-feruloylglycosyl)(1→6)-[apiosyl(1→2)]-glucoside;
- spinacetin 3-(2″-feruloylglycosyl)(1→6)-[apiosyl(1→2)]-glucoside;
- spinacetin 3-(2″-*p*-coumaroylglycosyl)(1→6)-[apiosyl(1→2)]-glucoside; and
- spinacetin 3-gentiobioside.

Figure 21.4 shows the structures of some of these spinach flavonoids.

Anti-nutritional factors

Anti-nutritional factors are natural compounds that interfere with the absorption of nutrients. They benefit plants by helping to prevent pests from eating them. They include phytate, oxalate and trypsin inhibitor. Oxalates occur naturally in plants and animals, mostly as calcium salts, and consequently they are also consumed in a variety of different foods. They are controversial compounds among nutritionists and health experts. Fresh spinach contains high amounts of oxalates, on average around 645 mg/100 g. The oxalates help to prevent the absorption of calcium and other minerals. They form insoluble salts with calcium, and the calcium in spinach is one of the least bioavailable of calcium sources. Oxalate binds to both iron and calcium, and also magnesium, to form oxalate salts, such as calcium oxalate and magnesium oxalate, thus preventing these minerals from being absorbed from the small intestine. Calcium oxalate is less soluble than the other oxalate salts and so tends to form crystals instead of dissolving and staying in the intestine; furthermore, it can form kidney stones.

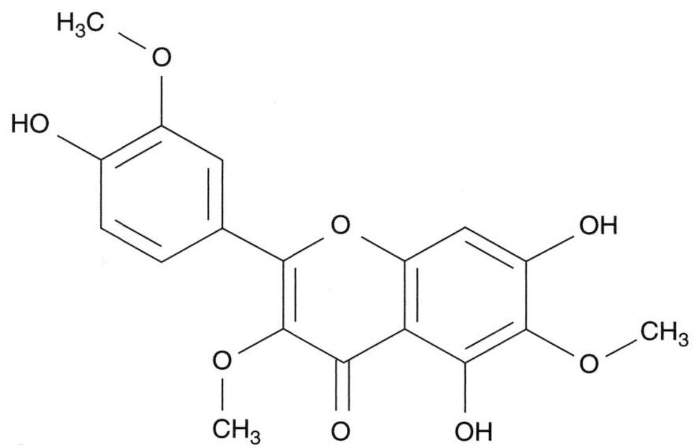

Jaceidin

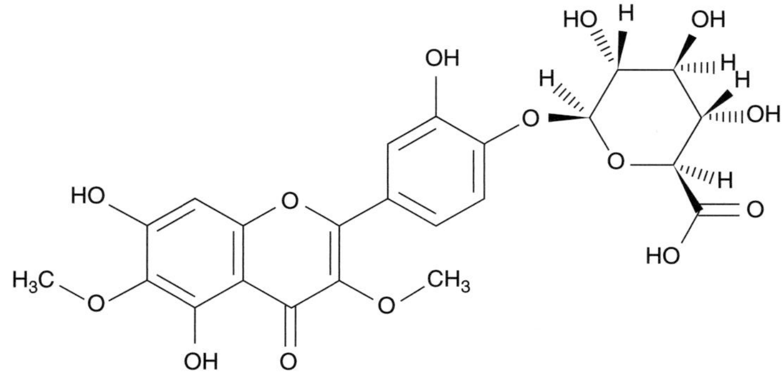

Spinatoside

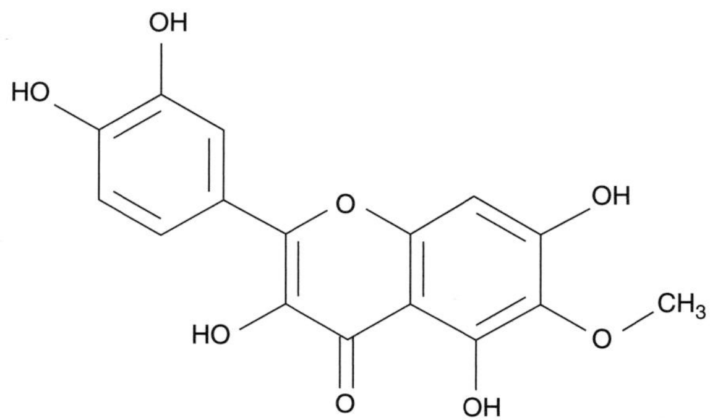

Patuletin

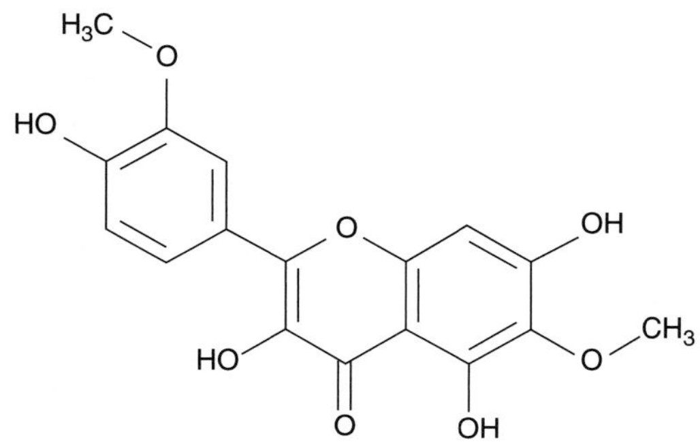

Spinacetin

Fig. 21.4. Chemical structures of some flavonoids found in spinach.

Spinach also contains purines, organic compounds which on breakdown are converted into uric acid. High levels of uric acid in the body increases the risk of developing kidney stones. Hence, patients with or prone to kidney stones are recommended to avoid spinach and other high oxalate foods. The cooking of spinach reduces oxalate content from 5–15%. Moreover, during blanching, spinach takes up calcium from water, and the consequent increase in calcium content has a direct influence on the ratio between the soluble and insoluble forms of oxalate (Bengtsson, 1969).

21.3 Postharvest Processing

Spinach leaves are harvested after the plants attain maturity, which requires 37–45 days. The leaves are detached from the plant at the base of their petioles simply by cutting. They should be removed when green in colour within a week of the formation of the full leaves, and it is advised that yellow leaves are avoided during harvesting. A single plant should have a minimum of six leaves of 5–7 inches (12–15 cm) long. The outer leaves are cut at 5–10 day intervals.

21.3.1 Preharvest and postharvest factors affecting quality

Preharvest factors such as climate and light conditions affect the quality of the leaves, but photosynthesis may be supported in light-stored leaves, and light storage may also increase ascorbic acid synthesis (Toledo *et al.*, 2003). Postharvest management for spinach focuses mainly on temperature control and water loss. Leafy vegetables generally have relatively high respiration rates and a large surface to volume ratio, resulting in a relatively short storage time. Temperature, both in terms of total and average temperature and the extremes that occurred during the growth period, may influence leaf chemical composition (Lefsrud *et al.*, 2005). Because of the high respiration rate (40–70 ml CO_2/kg/h at 10°C), and sensitivity to ethylene and elevated CO_2 concentrations, spinach leaves need to be cooled immediately or very soon after harvest. Table 21.3 shows effect of different temperatures on leaf respiration rate. Storability can be improved by lowering the temperature, increasing humidity and modifying the surrounding atmosphere. A lower temperature decreases metabolic rates, and thereby slows down deterioration. However, a very low temperature may cause chilling or freezing injury to the produce (Wills *et al.*, 1981). Storage also has an impact on the bioactive components of spinach, with ascorbic acid quickly decreasing, whereas carotenoids and flavonoids appear to be more stable (Kalt, 2005).

21.3.2 Preprocessing steps

The preprocessing steps for spinach include cleaning, pre-cooling, pretreatment and storage. The harvested spinach leaves are placed on a perforated shaking platform to facilitate the removal of soil and debris. The leaves are then dipped into cold water at about 10°C, to lower field heat, remove impurities left on the leaves and also to inhibit microbial activity. A manual inspection is then carried out to remove any unwanted materials. The leaves are packed into nylon mesh bags and dipped into another container of circulating water at 10°C for a final rinse. The bags are then spun to remove water. Following this, the spinach is mechanically packed into bags and stored at 2°C; it is now ready for shipment to retail outlets (Hodges *et al.*, 2000).

Fresh spinach can be stored for about 10–14 days under refrigerated conditions.

Table 21.3. Variation in respiration rate of harvested spinach leaves with temperature. Adapted from Suslow and Cantwell, 2013.

Temperature (°C)	Respiration rate (ml CO_2/kg/h)
0	9–11
5	17–29
10	41–69
15	67–111
20	86–143

The most suitable temperature for storage is 5–10°C. Spinach loses up to 80% of its ascorbic acid in 50 h at room temperature, probably due to wilting. Treatment with 0.1 or 1.0 μl/l 1-MCP (1-methylcyclopropene) inhibits ethylene sensitivity, and can be successfully used to retain the quality and chlorophyll content of leaves at 23°C for 6 days (Grozeff *et al.*, 2010). Maintenance of the appropriate relative humidity and atmosphere during storage improves shelf life and reduces the losses of bioactive compounds.

21.3.3 Processing

Implementation of an appropriate postharvest technology can reduce postharvest losses and enhance quality of the product. Postharvest losses of fresh vegetables are estimated at 20–50% in tropical countries. Thus, processing is an essential step in reducing postharvest losses and improving the eating quality of spinach; this can include the use of techniques such as freezing, canning, dehydration and minimal processing.

Blanching

Catalase and peroxidase enzymes are a major cause of the deterioration of spinach. Peroxidase, for example, reacts with hydrogen peroxide, to form phototoxic free radicals which cause the breakdown and loss of vitamins. These enzymes are inactivated by blanching, but the process must be done properly so as to preserve chlorophyll, which otherwise is broken down to pheophytin, and imparts a brown colour to over-blanched spinach. Olayinka *et al.* (2012) found that blanching samples of Nigerian spinach (or tete, *Celosia argentea*) for 10 min leached out 49–51% of its antioxidant contents into the boiling water used, and that only around 36% remained after 20 min. Other reports have indicated that blanching for up to 20 min seriously affects the antioxidant content of vegetables (Hunter and Fletcher, 2002). At above 95°C, the blanching process would decompose the antioxidants altogether. The optimum period for blanching spinach leaves in boiling water is 2 min, and microwave blanching reduces the loss of vitamin C by 18%.

Freezing

Freezing is one the most commonly used methods to maintain the high quality of fresh produce. In this method, the temperature is lowered to freezing point in order to retain nutrients and to increase the shelf life of produce by preventing (or slowing) microbial growth. The fresh spinach is washed thoroughly and woody stems are cut off. It is then blanched in boiling water for 2 min and immediately cooled down using cold water. The excess moisture is drained off and the spinach is packed into airtight freezer containers or bags, labelled and placed in a freezer.

The freezing-point of spinach is about −0.7°C, and spinach stored at a controlled freezing point had superior quality to that of cold stored spinach (Shen *et al.*, 2012). Glutamic acid is the dominant amino acid in fresh spinach, but this is reduced in frozen spinach. The main folate (vitamin B_9) derivative in spinach is 5-methyltetrahydrofolate. Folate content decreased significantly (by 25%) during the washing of spinach due to leaching (Delchier *et al.*, 2013).

Minimal processing

Minimally processed products have been defined as ‘any fruit or vegetable, or any combination thereof, which has been physically altered from its original form, but has remained in its fresh condition’ (IFPA, 1999). Minimal processing can provide fresh and quality products via a hurdle approach. It provides consumer with a new class of products for which convenience, quality and a relatively long shelf life are desirable (Siriphanich, 1993). In order to minimally process freshly cut spinach, the leaves are initially sorted and, after removal of the leaf stalks, are washed in cold running water. After draining, they are immersed in a mixture of ascorbic acid (0.5%) and citric acid (0.5%) at a temperature of 6°C for 10 min. The leaves are then chopped into fragments, packaged and stored for around 12 days while maintained at 4°C.

The shelf life of minimally processed baby spinach can be extended by up to 7 days when it is stored at 7°C. However, the complications associated with baby spinach are the development of strong off-odours and decay associated with softening of the tissue (Medina *et al.*, 2012). The fresh cut spinach is packed in mono-oriented polypropylene (PP) or low-density polyethylene (LDPE) bags. The type and permeability of the film affected off-odour development but did not influence visual sensory attributes or chlorophyll retention of the spinach (Piagentini *et al.*, 2002).

Drying

The consumption period for spinach can be extended by drying. The high moisture content of spinach is similar to that of other green leafy vegetables. Apart from improving the storability of the product at ambient temperatures, drying also considerably reduces weight and volume, and minimizes the cost of packaging, storage and transportation (Baysal *et al.*, 2003). After thorough washing and trimming, the spinach is loaded on to trays and subsequently on to carts for dehydration. The drying temperature should be maintained at 80°C. The dehydration time is about 4 h. After dehydration, the moisture content of the dehydrated spinach should be below 6.5%.

MICROWAVE DRYING. In microwave drying, the power and temperature used are the most important factors. Microwave drying of spinach in a 750 W oven requires only 350 s to produce the best quality dried spinach with respect to colour and vitamin C content. The energy requirement for this is 0.12 kWh (Ozkan *et al.*, 2007). A higher microwave power and temperature gives a shorter drying time, and the drying rate increases remarkably with power output and temperature of the oven. The microwave drying process may induce a colour shift towards the darker region. Dadali *et al.* (2007) observed a decreasing trend in the values of the parameters L (lightness) and b (yellow/blue), and an increasing trend in a (red/green) and the total colour change (ΔE) during microwave drying. Microwave drying of spinach using 430 kV/m for 7 h removed 80.1% of the total moisture. The total chlorophyll content of microwave-dried spinach is noticeably higher than that of oven-dried spinach. Similarly, the ascorbic acid content is almost three times higher than that from oven drying after 6 weeks of storage (Bajgai and Hashinaga, 2001).

Fig. 21.5. Freeze-dried spinach.

OTHER DRYING METHODS. Spinach can also be dried using freeze-drying (Fig. 21.5), but the drying time (45 h) is much longer than that for controlled low-temperature vacuum dehydration (6 h). However, in controlled low-temperature vacuum dehydration, product collapse is observed as a result of the evaporation of moisture (King *et al.*, 2001). The activation energy for thin-layer convective drying of spinach was found to be 34.35 kJ/mol (Doymaz, 2009).

Modified atmosphere packaging

Modified atmosphere packaging (MAP) has been developed to ensure that packaged food products stay fresh and attractive for a longer period by using defined gas mixtures. The objective is to create inside the packaging an appropriately balanced gas composition that will allow a decline in the physiological activity of the product that will delay deterioration. The effectiveness of MAP depends on factors such as freshness, degree of product processing, appropriate composition of the gas mixture used, general product properties, microbiological quality, barrier potential of the packaging material used and its dependence on the temperature and on the intensity of product respiration (Hertog *et al.*, 2004; Fonseca *et al.*, 2006).

Young, tender spinach leaves 8–10 cm long stored in an atmosphere of 5% O_2 + 10% CO_2 maintained their initial appearance for up to 12 days at 7.5°C, but the carotenoid concentration was lower than that in leaves that had been stored in air. The chlorophyll content was highest after storage in 5% O_2 + 20% CO_2, but the leaves showed rapid deterioration in visual quality. Storage at 10°C in 2% O_2 + 3% or 10% CO_2 controlled the enzyme activity of the leaves and contributed to the preservation of ascorbic acid and the activity of L-ascorbate peroxidase activity. Storage at 2°C prevented a reduction in ascorbic acid content and L-ascorbate peroxidase activity. Ascorbic acid content, L-ascorbate peroxidase activity and yellowing decreased rapidly when storage was at 25°C (Mizukami *et al.*, 2003). There was an increase in leaf pH after 1 day of storage in 20% O_2, and a reduction of pH with high contents of CO_2 which may be due to the dissolution of CO_2 and the formation of carbonic acid (Babic and Watada, 1996; Bieganska-Marecik *et al.*, 2007). High concentrations of CO_2 could damage the tips of leaves, but it was later reported that storage in 10–40% CO_2 + 10% O_2 retarded yellowing (Gross *et al.*, 2004).

The development of off-odour in baby spinach stored by MAP was due to the build-up of volatiles in the headspace, which, in turn, lead to changes in volatile generation. It is thought that the accumulation of non-respiratory volatiles may have negative effects in MAP (Toivonen and Deell, 2001). The occurrence of off-odours was significantly increased by the headspace composition in terms of low O_2 combined with CO_2, and their development can be controlled by moderate O_2 and CO_2 levels. However, these conditions may reduce shelf life by accelerating senescence (Tudela *et al.*, 2013). The shelf life of spinach can be prolonged by more than 7 days at 7°C by the using micro-perforated films of high CO_2 permeability.

Canning

Canning is a method of preserving food in which the food contents are processed and sealed in an airtight container. The spinach leaves are first thoroughly washed and then visually sorted to remove defective leaves, weeds and rubbish. They are then blanched in water at 77°C for about 6 min. While still hot, the blanched leaves are loosely packed into cans and hot 2% brine is added at an initial temperature of 60°C. The temperature is then raised; the process takes about 95 min at 116°C.

21.3.4 Value addition

Extraction of alkaloids

The medicinal properties of spinach flavonoids are remarkable enough to have urged researchers to create specialized spinach extracts. Spinach leaves contain chlorophyll 'a', chlorophyll 'b' and β-carotene as the major pigments, and also smaller amounts of other pigments such as xanthophylls, which are oxidized versions of carotenes and pheophytins. The presence of flavonols (as exemplified by quercetin, kaempferol and myricetin), and flavones (as exemplified by apigenin and luteolin) were investigated in raw spinach extract.

Fresh spinach contains total polyphenols of about 270 and 390 mg/kg as tannic acid and catechin equivalents, respectively. Flavonoids were effectively extracted in the temperature range 50–130°C with water as solvent and in the range 50–150°C with ethanol as solvent (Howard *et al.*, 2002). Under aqueous extraction, the highest extraction efficiency (19.7%) was obtained for β-carotene. By electroplasmolysis treatment at 60 V/cm for 8 s, the extraction efficiencies for chlorophyll 'a', chlorophyll 'b' and total chlorophyll were 14.9, 12.6 and 13.7%, respectively (Toprak *et al.*, 2011).

21.4 Uses

21.4.1 General uses

Spinach can be eaten raw in form of salad or cooked with other ingredients. It may be cooked by boiling, sautéing or creaming, or with a brine solution. To achieve the best flavour, sautéing and creaming require additional ingredients. Spinach provides more nutrients than any other food and is distinguished by its high iron content and that of other phytonutrients. It is an excellent source of vitamins A, B_2, B_6, B_9 (folate), C, E, K, potassium, magnesium, manganese and calcium. Dietary magnesium is required for energy metabolism, maintaining muscle, nerve and heart function, a healthy immune system and blood pressure. Spinach is also a good source of protein, dietary fibre and omega-3 fatty acids. It has several bioactive components, e.g. flavonoids, which act as antioxidants and so protect the body from free radicals and other oxidants. It also exhibits antiproliferative, anti-inflammatory, anticancer and anti-ageing properties (Lomnitski *et al.*, 2003).

21.4.2 Medicinal uses

Antioxidant benefits

Most of the spinach flavonoid and carotenoid nutrients act as anti-inflammatory and anticancer agents. A study in mice showed that oral administration of glyconutrients from spinach hindered the destruction of DNA and growth of cancer and tumour cells. A decrease in 56.1% in solid tumour volume was observed in mice after feeding these nutrients for a period of 2 weeks. The spinach nutrients decreased the power of tumours to supply themselves with blood, apparently without any side effects. Markers of cell proliferation were drastically reduced (Maeda *et al.*, 2008). Neoxanthin, a carotenoid from spinach, inhibited the multiplication of prostate cancer cells (Asai *et al.*, 2004). From the studies on laboratory animals, it was found that spinach extracts have the ability to slow down the growth of stomach cancer and skin cancer cells.

Thirteen compounds were separated from spinach that had antioxidant and antimutagenic properties against the dietary carcinogen 2-amino-3-methylimidazo[4,5-*f*] quinoline. Glutathione, an extremely important spinach antioxidant, protects DNA from oxidation, enhances the immune system, detoxifies carcinogens, assists liver health and decreases inflammation. α-Lipoic acid is also an essential antioxidant synthesized by both plants and animals. It is necessary for cellular energy production, helps to neutralize the damage caused by free radicals, lowers blood cholesterol and helps to prevent chronic diseases. Another endogenous antioxidant rich in spinach is coenzyme Q_{10} or ubiquinone, a crucial component in energy metabolism which helps in protecting the heart and skeletal muscles (Edenharder *et al.*, 2001; Joseph *et al.*, 2002).

Other medicinal uses

The consumption of spinach reduces the risk of atherosclerosis and high blood pressure. The vitamin K in spinach prevents the activation of osteoclasts, the cells that break down bones, and thereby helps in maintaining bone health. Spinach also contains other bone-supportive nutrients such as calcium and magnesium. A regular intake of carotenoids like lutein and zeaxanthin protects against age-related vision problems, such as macular degeneration and cataract formation, and high serum levels of lutein and zeaxanthin reduce the risk of coronary heart disease. Similarly, high plasma levels of carotenoids were associated with better vascular health and lower cardiovascular disease risk (Liao *et al.*, 1999). Betaine (trimethylglycine), a lesser known compound found in spinach, prevents cardiovascular disease by lowering the levels of homocysteine, a non-protein amino acid that is associated with the development of heart disease (Joseph *et al.*, 2002).

21.5 Summary

Spinach, from the Amaranthaceae family, is an edible flowering plant native to west and south-western Asia. It is considered to be one of the world's healthiest vegetables because of its nutrient richness. Spinach contains numerous health-promoting phytonutrients such as the carotenoids β-carotene, lutein and zeaxanthin, and flavonoids, which provide powerful antioxidant protection, as well as many vitamins and minerals. A low concentration of ethylene accelerates the yellowing of leaves during storage, as spinach is highly sensitive to ethylene; this yellowness can be delayed by exposing the leaves to atmospheres containing 7–10% O_2 and 5–10% CO_2. Blanching is the principal process used to process spinach, and acts by inactivating its enzymes. Freezing, dehydration and minimal processing of spinach are also practised. Research interest has also been diverted to the extraction of phytochemicals from spinach because of their great medicinal and nutritional value.

References

Aktas, E.T. and Yildiz, H. (2011) Effects of electroplasmolysis treatment on chlorophyll and carotenoid extraction yield from spinach and tomato. *Journal of Food Engineering* 106, 339–346.

Asai, A., Terasaki, M. and Nagao, A. (2004) An epoxide–furanoid rearrangement of spinach neoxanthin occurs in the gastrointestinal tract of mice and *in vitro*: formation and cytostatic activity of neochrome stereoisomers. *The Journal of Nutrition* 134, 2237–2243.

Babic, I. and Watada, A. (1996) Microbial populations of fresh-cut spinach leaves affected by controlled atmospheres. *Postharvest Biology and Technology* 9, 187–193.

Bajgai, T.R. and Hashinaga, F. (2001) Drying of spinach with a high electric field. *Drying Technology* 19, 2331–2341.

Barzegar, M., Erfani, F., Jabbari, A. and Hassandokht, M.R. (2007) Chemical composition of 15 spinach (*Spinacea* [sic] *oleracea* L.) cultivars grown in Iran. *Italian Journal of Food Science* 19, 309–318.

Baysal, T., Icier, F., Ersus, S. and Yıldız, H. (2003) Effects of microwave and infrared drying on the quality of carrot and garlic. *European Food Research and Technology* 218, 68–73.

Bengtsson, B.L. (1969) Effect of blanching on mineral and oxalate content of spinach. *International Journal of Food Science and Technology* 4, 141–145.

Bieganska-Marecik, R., Radziejewska-Kubzdela, E. and Czapski, J. (2007) Application of modified atmosphere packaging to extend shelf-life of minimally processed spinach. *Polish Journal of Food and Nutrition Sciences* 57, 13–17.

Combs, C. (1998) *The Vitamins: Fundamental Aspects in Nutrition and Health.* Academic Press, San Diego, California.

Dadali, G., Demirhan, E. and Özbek, B. (2007) Color change kinetics of spinach undergoing microwave drying. *Drying Technology* 25, 1713–1723.

Delchier, N., Ringling, C., Le Grandois, J., Aoudé-Werner, D., Galland, R., Georgé, S., Rychlik, M. and Renard, C.M. (2013) Effects of industrial processing on folate content in green vegetables. *Food Chemistry* 139, 815–824.

Doymaz, I. (2009) Thin-layer drying of spinach leaves in a convective dryer. *Journal of Food Process Engineering* 32, 112–125.

Edenharder, R., Keller, G., Platt, K.L. and Unger, K.K. (2001) Isolation and characterization of structurally novel antimutagenic flavonoids from spinach (*Spinacia oleracea*). *Journal of Agricultural and Food Chemistry* 49, 2767–2773.

FAOSTAT (2014) Food and, Agriculture Organization of the United Nations. Rome, Rome. Available at: http://faostat.fao.org/ (accessed 2 February 2016).

Fonseca, M.D.O., Leal, N., Cenci, S., Cecon, P. and Smith, R. (2006) Postharvest controlled atmosphere storage of 'Sunrise Solo' and 'Golden' pawpaws. *Revista Brasileira de Armazenamento* 31, 154–161.

Gil, M.I., Ferreres, F. and Tomas-Barberan, F.A. (1999) Effect of postharvest storage and processing on the antioxidant constituents (flavonoids and vitamin C) of fresh-cut spinach. *Journal of Agricultural and Food Chemistry* 47, 2213–2217.

Gross, K.C., Wang, C.Y. and Saltveit, M. (2004) *The Commercial Storage of Fruits, Vegetables, and Florist and Nursery Stocks.* Agriculture Handbook 66, Agricultural Research Service, US Department of Agriculture, Beltsville, Maryland.

Grozeff, G.G., Micieli, M.E., Gómez, F., Fernández, L., Guiamet, J.J., Chaves, A.R. and Bartoli, C.G. (2010) 1-Methyl cyclopropene extends postharvest life of spinach leaves. *Postharvest Biology and Technology* 55, 182–185.

Hertog, M.L., Nicholson, S.E. and Jeffery, P.B. (2004) The effect of modified atmospheres on the rate of firmness change of 'Hayward' kiwifruit. *Postharvest Biology and Technology* 31, 251–261.

Hodges, D.M., Forney, C.F. and Wismer, W. (2000) Processing line effects on storage attributes of fresh-cut spinach leaves. *HortScience* 35, 1308–1311.

Howard, L., Pandjaitan, N., Morelock, T. and Gil, M. (2002) Antioxidant capacity and phenolic content of spinach as affected by genetics and growing season. *Journal of Agricultural and Food Chemistry* 50, 5891–5896.

Hunter, K.J. and Fletcher, J.M. (2002) The antioxidant activity and composition of fresh, frozen, jarred and canned vegetables. *Innovative Food Science and Emerging Technologies* 3, 399–406.

IFPA (1999) *Fresh-Cut Produce Handling Guidelines*, 3rd edn. International Fresh-Cut Produce Association, Alexandria, Virginia.

Joseph, J.A., Nadeau, D. and Underwood, A. (2002). *The Color Code: A Revolutionary Eating Plan for Optimum Health.* Hyperion Books, New York.

Kalt, W. (2005) Effects of production and processing factors on major fruit and vegetable antioxidants. *Journal of Food Science* 70, R11–R19.

Kawazu, Y., Okimura, M., Ishii, T. and Yui, S. (2003) Varietal and seasonal differences in oxalate content of spinach. *Scientia Horticulturae* 97, 203–210.

King, V.A.E., Liu, C.-F. and Liu, Y.-J. (2001) Chlorophyll stability in spinach dehydrated by freeze-drying and controlled low-temperature vacuum dehydration. *Food Research International* 34, 167–175.

Lefsrud, M.G., Kopsell, D.A., Kopsell, D.E. and Curran-Celentano, J. (2005) Air temperature affects biomass and carotenoid pigment accumulation in kale and spinach grown in a controlled environment. *HortScience* 40, 2026–2030.

Liao, Y., Mcgee, D.L., Cooper, R.S. and Sutkowski, M.B.E. (1999) How generalizable are coronary risk prediction models? Comparison of Framingham and two national cohorts. *American Heart Journal* 137, 837–845.

Lomnitski, L., Foley, J.F., Grossman, S., Shaul, V.B., Maronpot, R.R., Moomaw, C.R., Carbonatto, M. and Nyska, A. (2000a) Effects of apocynin and natural antioxidant from spinach on inducible nitric oxide synthase and cyclooxygenase-2 induction in lipopolysaccharide-induced hepatic injury in rat. *Pharmacology and Toxicology* 87, 18–25.

Lomnitski, L., Nyska, A., Ben-Shaul, V., Maronpot, R.R., Haseman, J.K., Harrus, T.L., Bergman, M. and Grossman, S. (2000b) Effects of antioxidants apocynin and the natural water-soluble antioxidant from spinach on cellular damage induced by lipopolysaccaride in the rat. *Toxicologic Pathology* 28, 580–587.

Lomnitski, L., Bergman, M., Nyska, A., Ben-Shaul, V. and Grossman, S. (2003) Composition, efficacy, and safety of spinach extracts. *Nutrition and Cancer* 46, 222–231.

Maeda, N., Kokai, Y., Ohtani, S., Sahara, H., Kumamoto-Yonezawa, Y., Kuriyama, I., Hada, T., Sato, N., Yoshida, H. and Mizushina, Y. (2008) Anti-tumor effect of orally administered spinach glycolipid fraction on implanted cancer cells colon-26 in mice. *Lipids* 43, 741–748.

Medina, M.S., Tudela, J.A., Marín, A., Allende, A. and Gil, M.I. (2012) Short postharvest storage under low relative humidity improves quality and shelf life of minimally processed baby spinach (*Spinacia oleracea* L.). *Postharvest Biology and Technology* 67, 1–9.

Mizukami, Y., Saito, T. and Shiga, T. (2003) Enzyme activities related to ascorbic acid in spinach leaves during storage. *Japanese Society of Food Science and Technology* 50, 1–6.

Nyska, A., Lomnitski, L., Spalding, J., Dunson, D.B., Goldsworthy, T.L., Grossman, S., Bergman, M. and Boorman, G. (2001) Topical and oral administration of the natural water-soluble antioxidant from spinach reduces the multiplicity of papillomas in the Tg.AC mouse model. *Toxicology Letters* 122, 33–44.

Olayinka, O.O., Kareem, A.M., Aryo, I.B., Omotugba, S.K. and Oyebanji, A.O. (2012) Antioxidant contents (vitamin C) of raw and blanched different fresh vegetable samples. *Food and Nutrition Sciences* 3, 18–21.

Ozkan, I.A., Akbudak, B. and Akbudak, N. (2007) Microwave drying characteristics of spinach. *Journal of Food Engineering,* 78, 577–583.

Pandjaitan, N., Howard, L., Morelock, T. and Gil, M. (2005) Antioxidant capacity and phenolic content of spinach as affected by genetics and maturation. *Journal of Agricultural and Food Chemistry* 53, 8618–8623.

Piagentini, A., Güemes, D. and Pirovani, M. (2002) Sensory characteristics of fresh-cut spinach preserved by combined factors methodology. *Journal of Food Science* 67, 1544–1549.

Rubatzky, V.E., Yamaguchi, M., Rubatzky, V. and Yamaguchi, M. (1997) *World Vegetables: Principles, Production and Nutritive Values.* Chapman & Hall, New York.

SciNameFinder (2013) SciNameFinder™, compiled by A. Møller, Danish Food Informatics, Roskilde, Denmark. Available at: http://sciname.info/ (accessed 2 February 2016).

Shahidi, F. and Naczk, M. (2003) *Phenolics in Food and Nutraceuticals.* CRC Press, Boca Raton, Florida.

Shen, J., Hu, K.-Y., Liu, X.-H. and Wang, R.-L. (2012) Quality of spinach during controlled freezing point storage. *Food Science and Technology* 2012(9). Abstract available at: http://en.cnki.com.cn/Article_en/CJFDTOTAL-SSPJ201209020.htm (accessed 2 February 2016).

Siriphanich, J. (1993) Minimal processing of tropical fruits. In: Champ, B.R., Highley, E. and Johnson, G.I. (eds) *Proceedings of International Conference on Postharvest Handling of Tropical Fruits, July 19–23, Chiangmai, Thailand.* ACIAR Proceedings No. 50, Australian Centre for International Agricultural Research, Canberra, pp. 127–137.

Splittstoesser, W.E. (1990) *Vegetable Growing Handbook – Organic and Traditional Methods,* 3rd edn. Chapman & Hall, New York.

Suslow, T.V. and Cantwell, M. (2013) *Spinach: Recommendations for Maintaining Postharvest Quality.* Produce Facts, University of California, Davis, California. Available at: http://postharvest.ucdavis.edu/pfvegetable/Spinach/ (accessed 2 February 2016).

Toivonen, P. and Deell, J.R. (2001). Chlorophyll fluorescence, fermentation product accumulation, and quality of stored broccoli in modified atmosphere packages and subsequent air storage. *Postharvest Biology and Technology* 23, 61–69.

Toledo, M.E.A., Ueda, Y., Imahori, Y. and Ayaki, M. (2003) L-ascorbic acid metabolism in spinach (*Spinacia oleracea* L.) during postharvest storage in light and dark. *Postharvest Biology and Technology* 28, 47–57.

Tudela, J.A., Marín, A., Garrido, Y., Cantwell, M., Medina-Martínez, M.S. and Gil, M.I. (2013) Off-odour development in modified atmosphere packaged baby spinach is an unresolved problem. *Postharvest Biology and Technology* 75, 75–85.

USDA (2013) USDA Database for the Flavonoid Content of Selected Foods, Release 3.1. Agricultural Research Service, US Department of Agriculture, Beltsville, Maryland. Available at: http://www.ars.usda.gov/News/docs.htm?docid=6231 (accessed 2 February 2016).

USDA (2015) USDA National Nutrient Database for Standard Reference Release 28, Agricultural Research Service, US Department of Agriculture, Beltsville, Maryland. Available at: http://ndb.nal.usda.gov/ (accessed 4 February 2016).

Wills, R.H., Lee, T., Graham, D., Mcglasson, W. and Hall, E. (1981). *Postharvest. An Introduction to the Physiology and Handling of Fruit and Vegetables,* 2nd edn. New South Wales University Press, Kensington, New South Wales, Australia.

22 Stevia

Ramanathan Parimalavalli[1]* and S. Radhai Sri[2]
[1]*Periyar University, Salem, Tamil Nadu, India;* [2]*PSG College of Arts and Science, Coimbatore, Tamil Nadu, India*

22.1 Botany

22.1.1 Introduction

Stevia rebaudiana (Bertoni) is a herb belonging to the Asteraceae family. It is known by different names, which include sweet herb, sweet leaf, honey leaf, candy leaf and honey yerba (Carakostas *et al.*, 2008), the sweet herb of Paraguay (Kennelley, 2002), or just plain stevia. Among the 150 species of *Stevia*, *S. rebaudiana* is the sweetest, and has been described as a 'wonder plant' or the 'sweetness of the future' owing to its calorie-free natural sweetness (Senthil, 2002). Stevia leaves have been used as a sweetener for many hundreds of years by indigenous South American peoples, who use it in the local green tea as well as in other drinks. Stevia is a tropical herb – a tender perennial shrubby plant, which lives for 3 to 5 years. The leaves of stevia are in growing demand as a natural sweetener, and the health benefits of the herb have created an interest among researchers. It has been reported to have various nutritional and medicinal properties, such as antimicrobial, antiviral, antifungal, antihypertensive, anti-hyperglycaemic and antitumour effects (Gupta *et al.*, 2013).

22.1.2 History/origin

Stevia is native to north-eastern Paraguay and grows in sandy soil near streams (Lewis, 2003). It was first established in Japan in 1968 (Sumida, 1968).

22.1.3 Location

The native occurrence of stevia is reported to be between 22–24° S and 53–56° W in Paraguay and Brazil (Robinson, 1980; Sumida, 1980). The crop has also been introduced to other countries, including Korea, Mexico, the USA, Indonesia, Tanzania and Canada (Fors, 1995), and it is cultivated as a cash crop in number of countries. The seeds are highly heterogeneous and vary in their content of stevioside (the main component responsible for the sweetness of the plant), so it is better to start commercial cultures using rooted cuttings from selected plants that contain high level of associated bitterness, indicating a high stevioside content (Shu and Wang, 1988; Xie and Ouyang, 1998). There is no intensive mechanized production of stevia due to difficulties in the propagation of the crop by seed (Savitha *et al.*, 2004).

*Corresponding author, e-mail: parimala1996@gmail.com

© CAB International 2016. *Leafy Medicinal Herbs: Botany, Chemistry, Postharvest Technology and Uses* (eds D.C.P. Ambrose *et al.*)

22.1.4 Morphology

Stevia grows to a height of 65 cm, and up to 180 cm in fertile soil. The leaves are sessile and alternate, lanceolate to oblanceolate, and serrated above the middle (Fig. 22.1). The trichome structures on the leaf surface are of two distinct sizes, one large (4–5 μm) and one small (2.5 μm) (Shaffert and Chetobar, 1994).

22.2 Chemistry

22.2.1 Chemical composition/ phytochemistry

The main natural component of stevia leaves that is responsible for their sweetness is the glycoside known as stevioside. Stevioside is intensely sweet and is present in amounts constituting up to 13% of the leaves (Brandle *et al.*, 2002).

The phytochemicals present in stevia have been analysed by techniques including various types of chromatography, electrophoresis, magnetic resonance spectrometry, spectroscopy and enzymatic determination. As many as 100 phytochemicals from stevia have been explored, and the plant is rich in terpenes and flavonoids. Among the glycosides that have been identified, stevioside is considered the sweetest, comprising some 6–18% of leaf weight. Tests have found it to be up to 300 times sweeter than ordinary sugar (sucrose). The other main chemical compounds in stevia include: apigenin, β-sitosterol, caffeic acid, campesterol, diterpene glycosides, gibberellic acid and steviol (Taylor, 2005).

Fig. 22.1. A stevia shoot.

Oddone (1999) reported that stevioside (St) traditionally makes up the majority of the sweetener as used. Stevioside is also responsible for the somewhat bitter aftertaste which is sometimes described as a liquorice taste. Rebaudioside A is usually present as 30–40% of the total sweetener and this compound has the sweetest taste; it has been assessed as 180–400 times sweeter than sugar, but has no aftertaste. The ratio of rebaudioside A to stevioside is the accepted measure of sweetness quality – and the more rebaudioside A, the better. If the two compounds are present in equal quantities, or there is more rebaudioside A than stevioside, it appears that the aftertaste is eliminated.

The minor glycosides present are 30–80 times sweeter than sugar, with the sweet principles present in stevia being dulcoside A (50–120× sweeter than sugar), rebaudioside A (250–450×), rebaudioside B (300–350×), rebaudioside C (50–120×), rebaudioside D (250–450×), rebaudioside E (150–300×) and steviolbioside (100–125×) (Ikan *et al.*, 1997).

22.3 Postharvest Technology

Harvesting should be carried out at the flower bud appearance stage as the stevioside content of the leaves falls when flowering commences. The first harvest of the crop can be done 4–5 months after planting, and subsequent harvests once every 3 months for up to 3 years after planting (Kalpana and Khan, 2008).

22.3.1 Processing

Drying

Drying is an important aspect in the processing of stevia leaves, and if it is not done properly, there can be deterioration of the leaves as a result of their high initial moisture content. It is advised that harvested leaves are completely dried at the earliest opportunity as fresh leaves have a limited shelf life. Samsudin and Ab. Aziz (2013) conducted drying studies in a pilot scale dryer at different temperatures. With respect to the colour of the leaves, and their nutrient content, drying at 50°C was found to be the best. Nutrient analysis of the dried leaves gave 16 g protein, 65 g carbohydrate, 2805 mg potassium, 620 mg calcium, 268 mg phosphorus and 16 mg vitamin C per 100 g.

In another experiment, stevia leaves were dried at temperatures ranging from 30 to 80°C. The drying process was found to take place only in the falling rate period, and the Midilli–Kucuk model was found to give the best fit to the drying curve (Lemus-Mondaca *et al.*, 2014).

Extraction

Extracting the sweeteners from green or dried leaves can readily be done using boiling water, with extraction efficiencies of up to 98% achievable (Strauss, 1995). In contrast, the extraction procedure for steviol glycosides is a tedious process involving conventional and purification techniques. It involves dissolving the glycosides by soaking the leaves in warm water, precipitating and dissolving, evaporating the solution to obtain a concentrated solution, purification followed by drying in a spray dryer and crystallizing to produce white sugar (Rank and Midmore, 2006).

Megeji *et al.* (2005), as part of a study designed to assess the economics of growing *S. rebaudiana* as a crop for the extraction of stevioside, developed a laboratory-scale process that was able to produce 63% pure stevioside. This involved aqueous extraction of dried and powdered leaves, followed by filtration and crystallization. The stevioside content of the crop (leaves and stems) was monitored at two harvesting dates a year over 2 years by analysing a methanolic extract prepared as follows: dried and powdered material was cold extracted with methanol, and after 24 h, the solvent was decanted and the process repeated four times. Megeji *et al.* (2005) concluded that the crop could be economic as grown under the conditions prevailing at Palampur in Himachal Pradesh, India.

In another experiment, an aqueous stevia extract was prepared by mixing 25 g of dried leaves with 500 ml of water at room temperature and heating at 100°C until the volume was reduced to a quarter of the original. The resulting mixture was filtered through double lined muslin cloth.

A solid–liquid extraction for the determination of rebaudioside A in stevia was carried out by Asrul *et al.* (2013). The separation of rebaudioside A was achieved by extracting the leaf powder with solvents such as petroleum ether, methanol or diethyl ether, followed by purification to obtain the bioactive compound. Based on the extraction studies, methanol was found to be the best solvent.

The conventional methods of isolation of steviosides involve long extraction and purification procedures, so the optimization of product yields is challenging. Rao *et al.* (2012) improvised a new process of stevioside extraction in which the air-dried leaves were defatted, powdered and extracted using a pressurized hot water extractor (PHWE), followed by purification and concentration of the sweet glycosides through ultra- and nano-membrane filtration. This procedure gave high (98.2%) purity steviosides. This process established a 'green' method for the isolation of high-quality steviol glycosides, with an improved final yield of 10.1–11% from the crude leaf extract and with improved organoleptic and biological (antioxidant) activity.

22.3.2 Value addition

Both untreated stevia leaves and the herbal green powder prepared from them taste sweet. The quality of the leaf is evaluated by its sweetness. Although stevia preparations are sweeter than sugar, they have a slightly bitter aftertaste (Savitha *et al.*, 2004). One teaspoonful of stevia powder, is claimed to have a sweetening value equal to that of one cup of sugar (Fors 1995), and it has been reported that stevia leaf extracts are 152 times sweeter than 3% sucrose and 97 times sweeter than 10% sucrose (Cardello *et al.*, 1999). It has also been said that 50 g of stevia leaf can replace 1 kg of cane sugar (Barathi, 2003). A tasty flavour remains in the mouth for ½–1 h after the use of stevia as a sweetener (Ramana Rao, 2004).

The functional and sensory properties of stevia are due to the presence of glycosides, and because these are natural products, stevia can be viewed as superior to other high-potency sweeteners (Starratt and Gijzen, 2004). The palatability of both foods and drinks can be enhanced by using stevia as a sweetener (Ikan *et al.*, 1997). Unlike aspartame, a synthetic product, stevia sweeteners are heat stable to 200°C, are acid stable and do not ferment, so they can be used in the preparation of baked/cooked foods. There are no calories present in stevia and hence stevia products can be used as a 'healthy' replacement for sugar; for the same reason, they can also be used as a natural sweetener by diabetic and obese persons.

Stevia leaves can be used as flavour enhancers or sweeteners in a wide variety of foods and beverages, such as vegetables, coffee, apple sauce and hot cereal (Bonvie *et al.*, 1998). The leaf extract can be used to sweeten fruit juices, beverages, baked products, chocolates, Indian sweets, delicacies, etc. (Lewis, 2003). Stevia blends well with citrus fruit flavours such as lemon and cranberry. Stevioside, in the form of the pure compound or as stevia leaf extracts, has been widely used as food additive (Kinghorn, 1991). Leaf powder made from stevia can be used in most typical Indian dishes, such as chakkara pongal, payasam, ravaa ladoo, sauces, jams, juices, pickles, tea, coffee and even herbal tea (Barathi, 2003). In addition, stevia products can be used as tabletop sweeteners for tea and coffee, and in weight-watchers diets, pastries, pies, ice cream, yoghurts, sherbets, jellies and desserts, chewing gum, candies and other confectionery, diabetic diets, etc.

22.4 Uses

22.4.1 General uses

Natural plant origin intense sweeteners such as stevioside, glycyrrhizin and thaumatin used at very low levels in foods are generally not metabolized by the body and are excreted unchanged (George *et al.*, 2006). Sawate *et al.* (2007) noted that the level of incorporation of stevioside increased in a naturally carbonated cocoa beverage, the carbohydrate content decreased to an energy value of 7.04 kcal as a result of the replacement of sugar by stevioside. When stevioside is used to replace 40–50% of the sucrose in beverage production, it can improve the taste of beverages, making their sweetness cooling and refreshing in comparison with the thick and oily sweetness of sugar. Formulations of coffee-flavoured jelly can be made that contain a simple stevia extract that substitutes for 20% of the sugar that is commonly used in this food. Counties such as Japan use stevia both as a sweetener in soft drinks and for table use.

Geuns (1998) stated that the claimed intake per person per day in Europe of fresh powdered leaves is 2.4 g dry powder, which is equivalent to 400 mg stevioside. The acceptable daily intake of powdered stevia is 5 g. Among other countries, stevia and stevioside are permitted as food additives in Brazil, Korea and Japan, and stevia is allowed as a dietary supplement in the USA.

22.4.2 Pharmacological uses

The fresh stevia leaf is delicious and nutritious, a fantastic no calorie sweetener and an external and internal medicine. Many reports have stated that the products made from whole leaf concentrate have extraordinary health benefits (see http://www.stevia.com/Stevia_Article_List.aspx for some examples). Stevia is perhaps the only sweetener in the world that does not have any negative effects on health. This fact now seems to have been proven by the 150 odd scientific studies done on it that have gone on to validate the remarkable impact of stevia on health and its therapeutic effects on the body – any past controversy over its use seems now to have abated.

The beneficial effects of stevia extract include its activities as an antioxidant, and in the treatment of high blood pressure and hypertension. The leaf extract has antihypertensive effects and increases the level of insulin. Recent studies have indicated that stevioside and rebaudioside A can boost insulin production by the pancreatic cells and increase glucose tolerance as well (Alexander, 2007). Research conducted by Dutch and Japanese scientists in 2003 has shown that continuous intake of stevia induces the β-cells of the pancreas to produce more insulin. This, in turn, reduces dependence on oral as well as injectable insulin and controls diabetes. The use of stevia as sugar is an ideal substitution for people suffering from diabetes and or obesity linked to the overuse of cane sugar.

A number of other studies have come to similar conclusions to those outlined above, and some of these are mentioned here. Stevia leaf extract has therapeutic effects because of its ready absorption in the body (Selzer *et al.*, 2005). Melis and Salnatl (1991) and Melis (1995) reported that stevia decreased blood pressure, and increased diuretic and natriuretic effects in rats, and Melis (1996) noted that oral stevioside could be effective for treating hypertension. Maitree *et al.* (1993) suggested that the leaf of stevia could be used as a sweetener by diabetic and overweight people in their diets. Jeppesen *et al.* (2000) reported that stevioside aids in the secretion of insulin in the pancreas, and that it could play a potential role as an anti-hyperglycaemic agent in the treatment of type II diabetes. Stevia is the best non-calorie-containing alternative sweetener to cane sugar for obese and diabetic people and for those having dental problems as a result of over-consumption of ordinary sugar (Barathi, 2003; Geuns *et al.*, 2003).

Yet other studies have suggested hypotensive and hypoglycaemic effects of stevioside when administered in high dosage, and stevia effectively regulates blood sugar in people with diabetes and hypoglycaemia by bringing it towards more normal levels. In normal adult humans, stevia increases glucose tolerance (www.denutrtion.com). Savitha *et al.* (2004) reported on the effect of consumption of stevia incorporated in selected recipes as a substitute for sugar on the blood sugar and blood pressure levels in selected patients; they showed that stevia products influenced blood sugar, insulin, blood pressure, sodium excretion in the urine, and the lipid profile and weight of the subject and concluded that stevia poses a potential hypoglycaemic and anti-hypertensive effect. Ferri *et al.* (2006) described a double-blind study of 25 hospitalized patients, whose mean blood sugar dropped 35.2% some 6–8 h after ingesting stevia. Hsieh *et al.* (2003) reported that oral stevioside significantly decreased systolic blood pressure and diastolic blood pressure of patients compared with placebo in a 2 year randomized study. In a similar study, Chan *et al.* (2000) demonstrated that the systolic and diastolic blood pressure of the group taking oral stevioside decreased significantly, and that the effect persisted over the whole year of the trial, while blood biochemistry parameters, including lipid and glucose, showed no significant changes.

Sugar-rich products have become a way of life and over the years have permeated all levels of the food processing industry. However, sugar is responsible for tooth decay and in highly concentrated forms reduces the effects of other herbal elements. The naturally occurring stevia plant is 250–300 times sweeter than sugar and possesses antibacterial properties that remove plaque from and improve the overall condition of the teeth. The remarkable characteristics of stevia make it ideal for use in toothpastes and dental powders – in addition to its use as alternative substitute for sugar in food products.

Stevia is considered to be a safe natural food product and has been used for centuries by the natives of South America. There has never been a report of an adverse reaction linked to its use (Bonni *et al.*, 1996). Neither have any ill effects on the body been reported from the use of stevia, including side effects of any kind (Krishnamurthy, 2001). According to the Herb Research Foundation, numerous scientists and millions of consumers throughout the world, especially in Japan, regard stevia as a safe and intensely sweet natural product that could make it popular for use as a non-calorific sweetener (Nithyadevi 2003). Neither does stevioside appear to present a toxicity risk when consumed in low amounts (Lewis, 2003). Extensive research on stevia seems, therefore, to have proven that its leaves and extracts are safe.

22.5 Summary

Stevia is a 'friendly' plant which is beneficial as a natural sweetener. The soft green leaves and its woody stems are the useful parts of this herb. A moderate intake of stevia is not believed to be harmful, and extensive reviews of human and animal data indicate that it is safe, and that it tastes 20–30 times sweeter than ordinary dry sugar. Stevia is valuable for both its sweetening and its medicinal properties, and is extensively used in different parts of the world as a substitute for cane sugar. The clinical evidence for the medicinal properties of this plant need to be further explored in the future.

References

Alexander, C. (2007) Stevia hits the sweet spot, naturally. *Food Engineering and Ingredients* 32(3), 632–633.

Asrul, A., Shazani, S. and Shaha, R.K. (2013) Optimization of Rebaudioside A extraction from *Stevia rebaudiana* (Bertoni) and quantification by high performance liquid chromatography analysis. *Journal of Tropical Resources and Sustainable Science* 1(1), 62–70.

Barathi, N. (2003) Stevia – the calorie free natural sweetener. *Natural Product Radiance* 2(3), 120–122.

Bonvie, L., Bonvie, B. and Gates, D. (1996) *The Stevia Story: A Tale of Incredible Sweetness.* B.E.D. Publications, Atlanta, Georgia.

Bonvie, L., Bonvie, B. and Gates, D. (1998) Stevia: the natural sweetener that frightens NutraSweet. *Earth Island Journal* 13(1), 26–27.

Brandle, J.E., Starratt, A.N. and Gijzen, M. (2002) *Stevia rebaudiana*: its biological, chemical and agricultural properties. *Canadian Journal of Plant Science* 78, 527–536.

Carakostas, M.C., Curry, L.L., Boileau, A.C. and Brusick, D.J. (2008) Overview: the history, technical function and safety of rebaudioside A, a naturally occurring steviol glycoside, for use in food and beverages. *Food Chemistry and Toxicology* 46, S1–S10.

Cardello, H.M.A.B., Silva, M.A.P.A. and Damasio, M.H. (1999) Measurement of the relative sweetness of stevia extract, aspartame and cyclamate/saccharin blend as compared to sucrose at different concentrations. *Plant Foods for Human Nutrition* 54, 119–130.

Chan, P., Tomlinson, B. and Chen, Y.J. (2000) A double-blind placebo controlled study of the effectiveness and tolerability of oral stevioside in human hypertension. *British Journal of Clinical Pharmacology* 50, 215–220.

Ferri, L.A., Alves-Do-Prado, W., Yamada, S.S., Gazola, S., Batista, M.R. and Bazotte, R.B. (2006) Investigation of the antihypertensive effect of oral crude stevioside in patients with mild essential hypertension. *Phytotherapy Research* 20, 732–736.

Fors, A. (1995) A new character in the sweetener scenario. *Sugar Journal* 58(2), 30.

Geuns, J.M.C. (1998) *Stevia rebaudiana* Bertoni plants and dried leaves as novel food. Final version 21.9.1998 with addendum. Cited in Scientific Committee on Food (1999) Opinion on *Stevia rebaudiana* Bertoni Plants and Leaves (expressed on 17 June 1999). Document No. CS/NF/STEV/3 Final 17 June 1999, European Commission, Brussels; available at: http://www.food.gov.uk/sites/default/files/multimedia/pdfs/stevioside.pdf (accessed 3 February 2016).

Geuns, J.M.C., Augustijns, P., Mols, R., Buyse, J.G. and Driessen, B. (2003) Metabolism of stevioside in pigs and intestinal absorption characteristics of stevioside, rebaudioside A and steviol. *Food and Chemistry Toxicology* 41, 1599–1607.

George, V., Arora, S., Sharma, V., Wadhwa, B.K., Sharma, G.S. and Singh, A.K. (2006) Sweetener blends and their applications: a review. *Indian Journal of Dairy Science* 59, 131–134.

Gupta, E., Purwar, S., Sundaram, S. and Rai, G.K. (2013) Nutritional and therapeutic values of *Stevia rebaudiana*: a review. *Journal of Medicinal Plants Research* 7(46), 3343–3353.

Hsieh, M.-H., Chan, P., Sue, Y.-M., Liu, J.-C., Liang, T.H., Huang, T.-Y., Tomlinson, B., Chow, M.S., Kao, P.-F. and Chen, Y.-J. (2003) Efficacy and tolerability of oral stevioside in patients with mild essential hypertension: a two-year, randomized, placebo-controlled study. *Clinical Therapeutics* 25, 2797–808.

Ikan, R., Weinstein, V., Milner, Y., Bravdo, B., Shoseyov, O., Segal, D., Altman, A. and Chet, I. (1997) Natural glycosides as potential odorants and flavorants. *Acta Horticulturae* 344, 17–28.

Jeppesen, P.B., Gregersen, S., Poulsen, C.R. and Herma, K. (2000) Stevioside acts directly on pancreatic beta cells to secrete insulin actions independent of cyclic adenosine monophosphate and adenosine triphosphate-sensitive K+-channel activity. *Metabolism* 49, 208–214.

Kalpana, R. and Khan, M.D.K. (2008) Post harvest management of stevia leaves: a review. *Journal of Food Science and Technology* 45, 391–397.

Katayama, O., Sumida, T., Hayashi, H. and Mitsuhashi, H. (1976) The practical application of *Stevia* and research and development data. ISU Company, Japan. [English translation]. Cited by Brandle *et al.* (2002).

Kennelley, E.J. (2002) Sweet and non-sweet constituents of *Stevia rebaudiana* (Bertoni) Bertoni. In: Kinghorn, A.D. (ed.) Stevia*: The Genus* Stevia. Medicinal and Aromatic Plants – Industrial Profiles 19, Taylor & Francis, London and New York, pp. 68–85.

Kinghorn, A.D. (1991) Less common high potency sweeteners. *Journal of Natural Products* 53, 190–195.

Krishnamurthy, K. (2001) Stevia – the herb sweeter than sugar. *Kissan World* 28(10), 31–37.

Lemus-Mondaca, R., Vega-Gálvez, A., Moraga, N.O. and Astudillo, S. (2014) Dehydration of *Stevia rebaudiana* Bertoni leaves: kinetics, modeling and energy features. *Journal of Food Processing and Preservation* 39, 508–520.

Lewis, W.H. (2003) *Medical Botany: Plants Affecting Human Health*, 2nd edn. Wiley, Hoboken, New Jersey.

Maitree, S., Usanee, V., Umnat, M. and Duang, B. (1993) Mutagenicity and human chromosomal effect of stevioside, a sweetener from *Stevia rebaudiana* Bertoni. *Environmental Health Perspectives* 101, 53–56.

Megeji, N.W., Kumar, J.K., Virendra Singh, Kaul, V.K. and Ahuja, P.K. (2005) Introducing *Stevia rebaudiana*, a natural zero-calorie sweetener. *Current Science* 88, 801–804.

Melis, M.S. (1995) Chronic administration of aqueous extract of *Stevia rebaudiana* in rats, renal effects. *Journal of Ethnopharmacology* 47, 129–134.

Melis, M.S. (1996) A crude extract of *Stevia rebaudiana* increases the renal plasma flow of normal and hypertensive rats. *Brazilian Journal of Medical Biological Research* 29, 669–675.

Melis, M.S. and Salnatl, A.R. (1991) Effect of calcium and verapamil on renal function of rats during treatment with stevioside. *Journal of Ethnopharmacology* 33, 257–262.

Nithyadevi, A. (2003) Stevia – the natural sweetener. *Kissan World* 30(1), 58–59.

Oddone, B. (1999) How to Grow Stevia. Guarani Botanicals, Inc., Pawcatuck, Connecticut.

Ramana Rao, K.V.A. (2004) Sweetener that lowers blood sugar – Stevia. *Sanjivani Medical Times* (2004), 6–8.

Rank, A.H. and Midmore, J.D. (2006). An intense natural sweetener – laying the ground work for a new rural industry; May 2006. RIRDC Publication No 06/020, RIRDC Project No UCQ-17A, Rural Industries Research and Development Corporation, Barton and Kingston, Australian Capital Territory, Australia.

Rao, A.B., Prasad, E., Roopa, G., Sridhar, S. and Ravikumar, Y.V.L. (2012) Simple extraction and membrane purification process in isolation of steviosides with improved organoleptic activity. *Advances in Bioscience and Biotechnology* 3, 327–335.

Samsudin, A. and Ab. Aziz, I. (2013) Drying of stevia leaves using laboratory and pilot scale dryers. *Journal of Tropical Agricultural and Food Science* 41(1), 137–147.

Savitha, S.M., Sheela, K., Saran, S., Shanker, A.G. and Parama, R. (2004) *Stevia rebaudiana* – a functional component for food industry. *Journal of Human Ecology* 15, 261–264.

Sawate, A.R., Abdulraheem, Kshirsagar, R.B. and Patil, B.M. (2007) Effect of hydrocolloids and stevioside on the quality attributes of ready to serve natural carbonated cocoa beverage. *Indian Journal of Nutrition and Dietetics* 44, 324–331.

Selzer, J., Vasile, A. and Cornelia, D. (2005) All natural flavor enhancers for green tea beverages and dental hygiene product. *United States Patent Application.* Publication (10) Pub. No.: US

2005/0152997 A1, Jul. 14, 2005. Available at: http://patentimages.storage.googleapis.com/pdfs/US20050152997.pdf (accessed 3 February 2016).

Senthil, S. (2002) *Stevia* a natural sweetening plant. *Kissan World* 29(11), 60.

Shaffert, E.E. and Chetobar, A.A. (1994) Structure, topography and ontogeny of *Stevia rebaudiana*. *Botanicheskii Zhurnal* 79, 38–48.

Shu, S. and Wang, W. (1988) Variation in quantitative characters in *Stevia*. *Acta Agronomica Sinica* 14(2), 167–173.

Starratt, A.N. and Gijzen, M. (2004) Stevia rebaudiana – *Its Biological, Chemical and Agricultural Properties*. Southern Crop Protection and Food Research Centre, Agriculture and Agri-Food Canada, London, Ontario, Canada. Agriculture and Agri-food Canada. Southern Crop Protection and Food Research Centre, London.

Strauss, S. (1995) The perfect sweetener. *MIT Technology Review* 98(8), 18–20.

Sumida, T. (1968) Reports on *Stevia rebaudiana* Bertoni M, introduced from Brazil as a new sweetness resource in Japan. *Miscellaneous Publication, Hokkaido National Agricultural Experiment Station* 2, 69–83.

Sumida, T. (1980) Studies on *Stevia rebaudiana* Bertoni M, introduced from Brazil as a new possible crop for sweetness resources in Japan. *Journal of the Central Agricultural Experiment Station* 31, 1–71.

Taylor, L. (2005) *Stevia rebaudiana*: introduction. In: Taylor, L. *The Healing Power of Rainforest Herbs: A Guide to Understanding and Using Herbal Medicinals*. Square One Publishers, New York, pp. 1–4.

Xie, S. and Ouyang, X. (1998) The growth and differentiation of callus cultures of *Stevia rebaudiana* in relation to the stevioside accumulation. *Journal of Tropical and Subtropical Botany* 6, 8–14.

23 Thyme

Rashid Abdullah Al-Yahyai[1]* and Darach Lupton[2]
[1]*Sultan Qaboos University, Muscat, Oman;*
[2]*Oman Botanic Garden, Muscat, Oman*

23.1 Botany

23.1.1 Introduction

Thymus vulgaris (L.), popularly known as 'common thyme', is a popular culinary and medicinal member of the Lamiaceae family (Fig. 23.1). Its main centre of natural distribution is the Mediterranean (Spain, Italy, France, Greece, Egypt, Lebanon and Turkey) (Gonçalves *et al.*, 2011). The herb is grown throughout many parts of the world and has naturalized across many regions (Ozcan and Chalchat, 2004). The common English name for *T. vulgaris* is thyme, which refers to both the plant and the dried leaves. In Spanish, it is called *tomillo*, in Portuguese *tomilho*, in French *thym*, in Hindi *Ajavāyana kē phūla* (अजवायनकेफूल), in simplified Chinese *Shèxiāng cǎo* (麝香草), in German *Thymian*, in Russian *tim'yan* (тимьян) and in Arabic *Zatar* (زعتر) – in Arabic *Thymus* and *Zataria* (a thyme-like plant also in the Lamiaceae family) are given the same name.

Three varieties are typically grown for use, broad leaved, narrow leaved and variegated. The narrow-leaved variety, which has smaller, greyer leaves, is more fragrant than the broad-leaved variety, and is commonly called winter or German thyme. The fragrant lemon thyme is lemon scented, and its leaves are broader than those of ordinary garden (or common) thyme, and are not folded at the margins; it is classified as a variety of *T. serpyllum* (or as the species *T. citrodorus*). Its habit is trailing and it is typically smaller than garden thyme, and it is evergreen, although is considered to be not as tough as common thyme. Silver thyme is another variety and is thought to be the hardiest of all; it is also considered by some to have the best flavour (Grieve, 1971).

The vernacular word 'thyme' has been used as both the scientific and the commercially used name for *T. vulgaris*, which has led to misunderstanding and abuse. For example, more than 50 plants are named and used as 'thyme' in Turkey, and though most of them belong to the genus *Thymus*, many belong to other plant genera, e.g. *Origanum* and *Thymbra* (Özgüven and Tansi, 1998). Adding to this confusion are the large number of vernacular names for *T. vulgaris*; black thyme, common thyme, English thyme, French thyme, garden thyme, German thyme and winter thyme, to name but a few. In addition, there are also large numbers of *T. vulgaris* cultivars available commercially.

*Corresponding author, e-mail: alyahyai@squ.edu.om

© CAB International 2016. *Leafy Medicinal Herbs: Botany, Chemistry, Postharvest Technology and Uses* (eds D.C.P. Ambrose *et al.*)

Fig. 23.1. Upper portion of thyme plant showing typical whorled flowers, square stems and opposite leaves.

For example, in 2014, the Royal Horticultural Society in the UK listed 31 *T. vulgaris* cultivars available for purchase in Britain alone (RHS, 2014).

23.1.2 History/origin

The word *Thymus* may originate from the Greek word 'thyo', meaning perfume (Stahl-Biskup and Saez, 2002). Alternatively, it may have come from the Greek word 'thumus', meaning strength, as in ancient and medieval times the plant was thought to be a great source of strength and bravery.

The cleansing properties of thyme were known to the ancient Greeks, and Pliny tells us that, when burnt, 'it puts to flight all venomous creatures'. Lady Northcote (in *The Herb Garden*) says that among the Greeks, thyme denoted graceful elegance, and that 'to smell of Thyme' was an expression of honour, bestowed upon those of venerable character. To the ancient Greeks, thyme came to denote elegance and style; their use of thyme was frequent – medicinally, in massage and bath oils, as incense in the temples and as an aphrodisiac. The Romans associated thyme with courage and bathed in water prepared with thyme to ready themselves for battle; thyme was also used by the ancient Romans for flavouring cheese and liqueurs (Grieve, 1971).

Thyme was a symbol of action, courage and power, and in the days of chivalry, ladies embroidered a bee flying over a stem of thyme on the scarves they presented to their knights (Stahl-Biskup and Saez, 2002). During the Middle Ages, thyme was grown in monasteries in southern France, Spain and Italy and was used as a cough remedy, for digestive relief and for the treatment for intestinal parasites. Nicholas Culpeper, the famous English botanist and herbalist, wrote of thyme in 1653:

> It is a noble strengthener of the lungs, as notable a one as grows; neither is there scarce a better remedy growing for that disease in children which they commonly call the chin-cough, than it is. It purges the body of phlegm, and is an excellent remedy for shortness of breath. It kills worms in the belly, and being a notable herb of Venus, provokes the terms, gives safe and speedy delivery to women in travail, and brings away the after birth. It is so harmless you need not fear the use of it.

23.1.3 Location

Thyme is commercially grown in Austria, England, France, Germany, Greece, Italy, Poland, the Balkans, Hungary, Portugal and Spain, as well as in Morocco and the USA (Lueng and Foster, 1996). Large amounts of thyme are collected from the wild in a number of European countries, including Albania, Bulgaria, and Bosnia and Herzegovina, some of which is wild harvested under organic certification. Spanish thyme (*T. zygis*), a native of the Iberian Peninsula, is often used with *T. vulgaris* for medicinal purposes (Blumenthal *et al.*, 2000). Introduced populations of *T. vulgaris* have spread themselves across distant regions from Canada and the USA, to Chile and New Zealand (Stahl-Biskup and Saez, 2002).

23.1.4 Morphology

Thyme is a perennial subshrub with much branched, square, ascending stems with woody bases. The small, opposite, linear to elliptic, almost sessile evergreen leaves have reflexed margins. The leaves are green to grey in colour and, typically, the underside of the leaf is covered with a fine indumentum (a covering of hairs) (Panda, 2006).

The flowers grow in whorls at the tips of the branches. The calyx is tubular and is closed at the mouth by small hairs, the lip is bifurcate, and the upper lip is cut into three teeth and the lower into two. The corolla consists of a cylinder about the length of the calyx, dividing at the top into two lips of a pale purple colour; the upper lip is upright or turned back and serrated at the end, the under lip is longer and divided into three segments. The seeds are round to oval and very small.

Thyme reproduces by seed and vegetatively. Pollination is generally by bees though there can be high levels of self-pollination (up to 80%); this appears to be dependent on the genotype and the local environment. The dispersal of pollen and seeds occurs over short distances and therefore there is a high chance of population differentiation or separation in the wild (Özgüven and Tansi, 1998).

Thyme plants are generally heliophylous, i.e. they like the sun. They are often found on rocks or stones and prefer a well-drained substrate. Different thyme species require very different soil types; however, *T. vulgaris* usually lives on alkaline soils. Thyme plants are very hardy, which allows them to grow under often extreme climatic environments (Stahl-Biskup and Saez, 2002). Dense hairs and small, often narrow, leaves allow some species to tolerate very arid conditions. The high production of essential oils by thyme can also be an adaptive characteristic for arid climates, because the volatile substances evaporate and produce a moist atmosphere surrounding the plant, which helps to reduce water loss through evaporation (Stahl-Biskup and Saez, 2002).

The productive parts of the plants in terms of essential oil production are the leaves and the stems. Many members of the Lamiaceae family secrete volatile oils manufactured by glandular hairs located on the surface of the leaves and stems; these hairs are usually peltate (rounded with the hair stem beneath) or capitate (swollen at the top) in shape. The glandular hairs start to produce essential oils as soon as they are completely formed, and the oil passes through the glandular walls to the cuticle covering the hairs, where the oil is eventually released. The oil-producing glandular hairs are numerous on both young and mature leaves; the hairs are also abundant at the top of the plant but become less frequent towards the base (Boz *et al.*, 2009).

23.2 Chemistry

Thyme has many industrial, medicinal and economic applications that are attributed to its chemical composition. The continuous development of new techniques capable of identifying the plant constituents to their smallest quantities has contributed to new interest in thyme chemistry. The two main secondary chemical products that characterize the pharmacological activities of thyme are the volatile essential oils and polyphenols, mainly flavonoids (Van den Broucke and Lemli, 1981; Van den Broucke and Lemli, 1983).

23.2.1 Chemical composition/ phytochemistry

Like the rest of the Lamiaceae, thyme is rich in essential oil that has been known for its medicinal properties since early civilization. Some 360 volatile components have been reported in 162 taxa of *Thymus*. The most abundant type of compound are the terpenes, comprising 75%, of which 43% are monoterpenes and 32% are sesquiterpenes (Stahl-Biskup, 2002). The remaining components of thyme essential oil are 17% non-terpenoid aliphatic compounds (aliphates), 6% benzene derivatives and 2% phenylpropanoids.

The chemical profile of thyme shows the presence of many constituents (see Table 23.1). The most abundant essential oil constituent is thymol (2-isopropyl-5-methylphenol, $C_{10}H_{14}O$), the structure of which is shown in Fig. 23.2. Thymol makes up 20–54% of the essential oil of thyme, and is an antiseptic and antispasmodic compound that is used in traditional and modern medicinal remedies. Other compounds present in significant quantities include γ-terpinene, carvacrol, *p*-cymene, myrcene, limonene and borneol. These compounds exhibit antioxidant, anti-inflammatory, immunomodulatory, antibacterial and antifungal properties (Stahl-Biskup, 2002; Braga *et al.*, 2006; Porte and Godoy, 2008).

Table 23.1. Characteristic chemical composition of essential oils of thyme leaves. From Porte and Godoy, 2008.

Compound	Content (%)
Borneol	0.5
δ-Cadinene	0.1
Calamenene	<0.1
Camphene	0.3
Carvacrol	2.4
Carvacrol acetate	<0.1
β-Caryophyllene	0.8
ρ-Cimenene	0.1
ρ-Cymene	18.6
trans-Dihydrocarvone	0.2
2,4-Dimethyl-2,4-heptadiene	1.5
Limonene	0.8
Mentha-3,8-diene	0.4
p-Menthene-1	1.8
p-Menthene-3	0.1
Myrcene	2.4
(*E*)-β-Ocimene	0.1
(*Z*)-β-Ocimene	0.1
1,3-Octadiene	0.3
1,7-Octadiene	0.1
α-Phellandrene	0.3
α-Pinene	0.8
Sabinene	0.1
α-Terpinene	1.8
γ-Terpinene	16.5
α-Terpinolene	0.2
Thymol	44.7
Thymol methyl ether	0.1
Total	**95.1**

Synthesis of Thymus *essential oils*

Research on the essential oils of *Thymus* also focused on the synthesis of the aromatic and aliphatic compounds that it contains. Thyme oil contains volatile terpenoids that are aliphatic except for a few major aromatic monoterpenes, such as thymol, *p*-cymene and carvacrol. Little and Croteau (1999) reviewed two pathways. The first is the classical (acetate/mevalonate) pathway, whereby three molecules of acetyl-Coenzyme A (acetyl-CoA) are fused by the action of acetyl-CoA acyltransferase and hydroxymethylglutaryl-CoA (HMG-CoA) synthase to produce HMG-CoA. Mevalonic acid is then produced by the reduction of HMG-CoA, which is converted to form a C_5 compound, isopentenyl diphosphate (IPP), as the terpenoid precursor. The second pathway is the pyruvate/glyceraldehyde-3-phosphate pathway. In this pathway, a transketolase reaction of glyceraldehyde-3-phosphate with carbons 2 and 3 of pyruvate produces the C_5 intermediate, 1-deoxy-D-xylulose, followed by a series of reduction and dehydration steps and a phosphorylation, resulting in IPP or dimethylallyl diphosphate (DMAPP) as end products (Rohmer *et al.*, 1996). The acetate/mevalonate pathway dominates in the cytosol to form sesquiterpenes and triterpenes, whereas the pyruvate/glyceraldehyde-3-phosphate pathway (also known as the deoxyxylulose pathway) occurs in the plastids to form monoterpenes, diterpenes and tetraterpenes. Stahl-Biskup (2002) has provided in-depth discussion of the biosynthesis of thymus terpenoids.

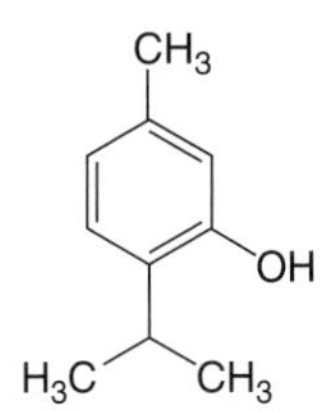

Fig. 23.2. The structure of thymol (2-isopropyl-5-methylphenol).

The biosynthesis and chemical structure of flavonoids in the genus *Thymus*, focusing primarily on flavonoid aglycones, have been described in detail by Vila (2002). Among the aglycones that were reported to occur in thyme were 32 flavones, four flavanones, two flavonols and two dihydroflavonols. The most frequently found compounds are the flavones luteolin, apigenin and scutellarein. No isoflavonoids have been reported in *Thymus* taxa (Vila, 2002).

The composition of the volatile compounds of *Thymus* varies greatly depending on several factors, including variations among *Thymus* species, sexual variations among female and hermaphrodite plants, seasonal, climatic, geographical and other environmental factors, growing conditions, and the extraction and analytical techniques used (Stahl-Biskup, 2002).

23.3 Postharvest Technology

T. vulgaris is the most cultivated commercial species of thyme, while many of the other species are harvested manually from the wild. Naturally occurring thyme is processed minimally, and this typically includes thyme for fresh use, for storage in olive oil and for drying of the leaves and flowers. However, the commercial production of thyme involves mechanical harvesting and processing, oil extraction and the production of by-products that have many cosmetic, pharmaceutical and food applications.

23.3.1 Processing

Harvesting date and time is critical for the maximum retention of thyme essential oil and aroma. The maximum aroma is obtained just before or at full bloom, thus harvesting at this growth stage is recommended (Venskutonis, 2002a). Fresh processing requires cleaning, sorting and storage of the freshly harvested leaves at freezing temperatures. Venskutonis (2002b) outlined the process used for the production of freshly harvested herbs, including thyme (Fig. 23.3),

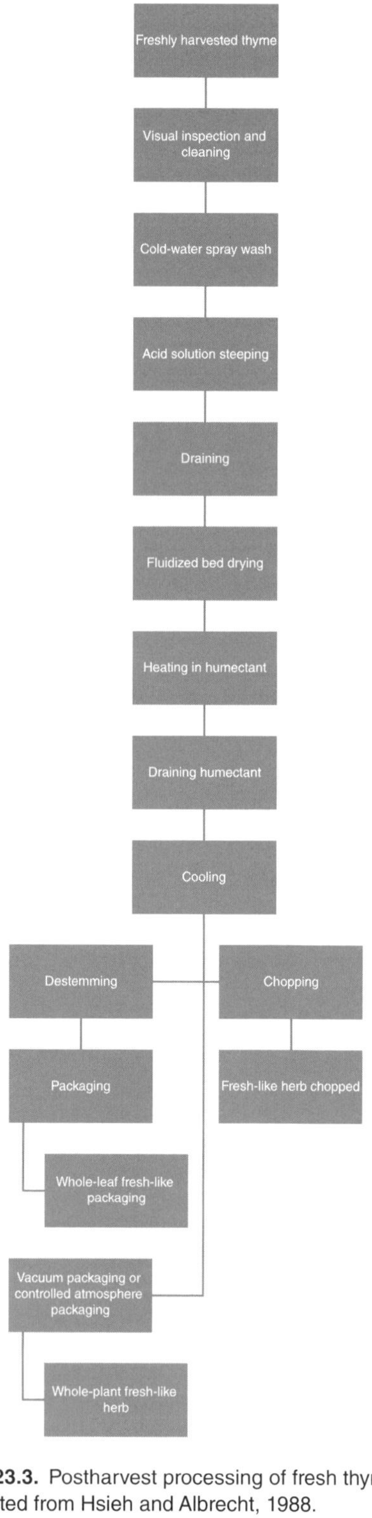

Fig. 23.3. Postharvest processing of fresh thyme. Adapted from Hsieh and Albrecht, 1988.

and highlighted the processes that were first patented by Hsieh and Albrecht (1988).

Thyme destined for production in dry form is mechanically harvested and then transported to a cooled or shaded sorting facility. Cleaning and sorting is done, followed by drying and packing of the product as whole or ground leaves (Fig. 23.4). Postharvest processing includes physical and chemical treatment to control insect infestation and fungal or other pathogenic infection of the processed thyme. Treatments may include one or more of the following, depending on the scale of production: heat and steam sterilization, ultraviolet radiation, ionization treatment, infrared irradiation and chemical fumigation (Tainter and Grenis, 1993).

Drying is carried out by thermal energy methods, such as solar, vacuum oven, microwave oven and hot air drying, or by non-thermal methods using moisture-absorbing agents such as silicon and electrolytes, and freeze-drying. Sun drying, despite being time-consuming and subject to contamination, continues to be the most commonly used method even in larger production facilities, mainly because of its low cost and minimal effects on the quantity of final product. However, it needs to be borne in mind that the drying method used directly influences the quality of aroma and essential oil content (Calín-Sánchez *et al.*, 2013).

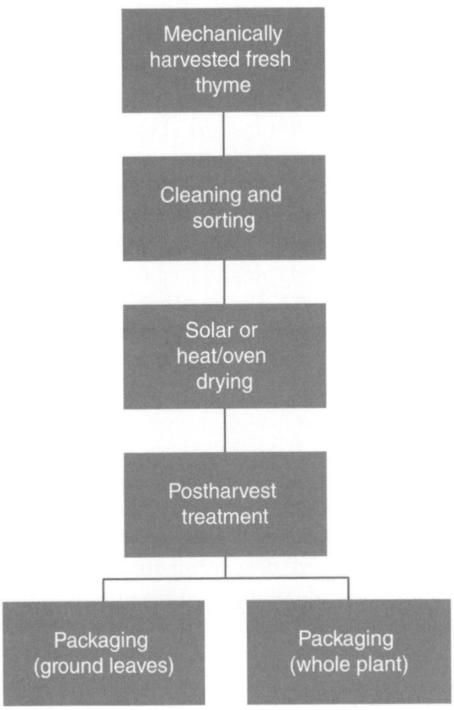

Fig. 23.4. Postharvest processing of dried thyme.

Dried thyme (Fig. 23.5) can be stored for extended periods depending on the storage temperature. Freezing temperatures (around −18°C) offer multiple years of storage, whereas at higher temperatures (5–7°C), dried thyme can last for a year. At room temperature, the shelf life of thyme drops drastically (Venskutonis, 2002a). Shelf life is highly dependent on product moisture content, the quantity stored, the permeability of the packaging material and storage conditions (i.e. temperature, humidity and light).

23.3.2 Value addition

Extracted thyme essential oil is a product of significant importance in the herbal industries. The essential oil is generally extracted from fresh or dried leaves using a variety of methods, such as hydrodistillation and steam distillation, or cold-press expression using a series of extraction solvents. Solvent extraction is commonly used for the manufacture of oleoresin (Venskutonis, 2002b).

More recently, CO_2 has been used under high-pressure extraction to obtain oleoresin,

Fig. 23.5. Dried thyme leaves.

which is later separated into essential oil and resin, and methods such as pressurized liquid extraction (PLE) have been developed. Villanueva Bermejo *et al.* (2014) used PLE with three green solvents (ethanol, limonene and ethyl lactate) at different extraction temperatures (60, 130 and 200°C) to extract thymol. They found PLE to yield more and with a lower solvent consumption than steam distillation.

The development of gas chromatography/mass spectrometry (GC/MS) analytical techniques has also provided additional details on the composition of thymus essential oils, and Stahl-Biskup (2002) has reviewed these techniques in relation to the essential oil composition of thyme.

Thyme essential oils, mainly thymol and carvacrol have antiseptic, insecticidal and antimicrobial properties that have made them suitable alternatives to traditional pesticides for crop pest and disease control (Clemente *et al.*, 2003; Hama-salih *et al.*, 2014; Nikolić *et al.*, 2014). They have also been used in aromatherapy and homeopathy, and against oxidative stress because of the antioxidant contents of thyme essential oil and flavonoids (Dorman *et al.*, 1995; Roby *et al.*, 2013).

Thyme has been used as a food preservative as well owing to its antimicrobial and insecticidal properties. In addition, it has been extensively used by the cosmetic industry as an ingredient in 'natural' cosmetics, deodorants, soaps, perfumes, lotions, toothpaste and mouthwash (Zarzuelo and Crespo, 2002).

23.4 Uses

23.4.1 General uses

Thyme has been used throughout history, and its chemical variability is well known (Zarzuelo and Crespo, 2002). In traditional medicine, some thyme species are used for their antiviral, soothing, and diaphoretic properties, and are typically consumed as an infusion. The treatment of rheumatic and skin disorders is usually undertaken by bathing in thyme-infused water (Soliman and Badeaa, 2002). The volatile oils of thyme are used as preservatives in cosmetics and as antioxidants (Zarzuelo and Crespo, 2002). Essential oils extracted from the plant are used to add fragrance to food and pharmaceuticals (Senatore, 1996).

23.4.2 Pharmacological uses

Thyme has a wide range of effects as an antiseptic, carminative, antimicrobial and antioxidant (Baranauskiene *et al.*, 2003). Warm infusions of thyme (teas) have a relieving effect on intestinal gas and parasites. Thyme also relieves coughs and has antimicrobial and astringent qualities, and it has been used to relieve heartburn, pertussis, asthma, bronchitis, chest congestion, laryngitis, gastritis and to encourage the production of saliva (Lueng and Foster, 1996; Blumenthal *et al.*, 2000, Bown, 2001; Barnes *et al.*, 2002). It is also known to promote menstrual flow and to aid occasional childhood diarrhoea and bedwetting (Blumenthal *et al.*, 2000, Bown, 2001).

Cough drops, liniments, mouthwashes, detergents, toothpastes and perfumes contain essential oils extracted from *T. vulgaris* and *T. zygis* (Lueng and Foster, 1996). Red thyme oil is preferred, as white thyme oil is distilled from the red oil. Thymol and carvacrol are the two main ingredients in thyme oil, although thymol is considered to have the greater therapeutic properties (Lueng and Foster, 1996). Coughs, bronchitis and indigestion are typically treated with thyme oil (Grieve, 1971; Blumenthal *et al.*, 2000). The whole plant and the oil are used to treat arthritis, gum disease and tonsillitis (Bown, 2001).

Thyme has a wide range of uses in the cosmetics and food industries, and in modern and traditional medicine. The medicinal effects and benefits of thyme have been recognized in the last 20–30 years and thyme has been awarded government certification in a number of countries. These permissions have been summarized by Engels (2008), as follows. In 1984, the German Commission approved thyme for oral use (as a tea infusion or fluid extract) or topical use (as a tea infusion for compresses) for bronchitis and whooping cough and for mucus-producing problems (catarrhs) of the upper respiratory tracts. The European Medicines Evaluation Agency (EMEA) published a final monograph on thyme in 2007, and approved thyme (as a dry extract, fluid extract, tea and tincture) as an expectorant in coughs associated with colds. Health Canada published its final monograph for thyme natural health products in 1988; this permitted the traditional uses of thyme for rinsing the throat to help alleviate laryngitis, tonsillitis and mucous membrane inflammations of the mouth and/or throat, and also its topical use as an antiseptic and/or antimicrobial to help manage minor wounds and sores. Thyme can also be used as an expectorant to ease the signs of bronchitis and mucous build-up of the upper respiratory tract, to relieve coughs and to help alleviate flatulent dyspepsia and indigestion.

23.5 Summary

Thyme is a very important herb that has been cultivated and used in food and medicine since the ancient Egyptian and the Greek periods. Numerous nutritional and pharmaceutical benefits are being explored today as extraction methods improve. New chemical ingredients are being unearthed, and some of the more recent include antioxidants, anticancer and immunity-enhancing compounds. Thyme is also used in numerous industrial applications, from hygiene products and natural cosmetics to food preservatives and biopesticides, and its multiple uses and benefits continue to be highly valued across the world. The potential for the discovery and development of new and exciting thyme extracts is very positive and will no doubt continue to be, long into the future.

References

Baranauskiene, R., Venskutonis, P.R., Viskelis, P. and Dambrauskiene, E. (2003) Influence of nitrogen fertilizers on the yield and composition of thyme (*Thymus vulgaris*). *Journal of Agricultural and Food Chemistry* 5, 7751–7758.

Barnes, J., Anderson, L.A. and Phillipson, J.D. (2002) *Herbal Medicines: A Guide for Healthcare*, 2nd edn. Pharmaceutical Press, London.

Blumenthal, M., Goldberg, A. and Brinckmann, J. (2000) *Herbal Medicine.* Expanded Commission E Monographs, American Botanical Council, Austin, Texas. Published by Integrative Medicine Communications, Newton, Massachusetts.

Bown, D. (2001) *The Herb Society of America: New Encyclopedia of Herbs and Their Uses.* Dorling Kindersley, New York.

Boz, I., Navarro, L., Galesm, R. and Padurariu, C. (2009) Morphological and structure of glandular hairs in development of *Thymus vulgaris* (L.). *Scientific Annals of Alexandru Ioan Cuza University of Iasi, New Series, Section 2. Vegetal Biology* 55, 81–86.

Braga, P.C., Dal Sasso, M., Culici, M., Bianchi, T., Bordoni, L. and Marabini, L. (2006) Anti-inflammatory activity of thymol: inhibitory effect on the release of human neutrophil elastase. *Pharmacology* 77, 130–136.

Calín-Sánchez, Á., Figiel, A., Lech, K., Szumny, A. and Carbonell-Barrachina, Á. (2013) Effects of drying methods on the composition of thyme (*Thymus vulgaris* L.) essential oil. *Drying Technology* 31, 224–235.

Clemente, S., Mareggiani, G., Broussalis, A., Martino, V. and Ferraro, G. (2003) Insecticidal effects of Lamiaceae species against stored products insects. *Boletín de Sanidad Vegetal Plagas* 29, 421–426.

Culpeper, N. (1653) *Culpeper's Complete Herbal: A Book of Natural Remedies of Ancient Ills.* Published in 1995 in the Wordsworth Reference Series, Wordsworth Editions, Ware, UK.

Dorman, H.J., Deans, S.G., Noble, R.C. and Sera, H. (1995) Evaluation *in vitro* of plant essential oils as natural antioxidants. *Journal of Essential Oil Research* 7, 645–650.

Engels, G. (2008) Herb profile: thyme. *HerbalGram* 80, 1–2.

Gonçalves, G.M.S., Botaro, M. and Nilson, A.C. (2011) Effect of the *Thymus vulgaris* essential oil on the growth of *Streptococcus mutans. Journal of Basic and Applied Pharmaceutical Sciences* 32, 375–380.

Grieve, M.A. (1971) *Modern Herbal, Vol.* 2. Dover Books, New York.

Hama-salih, F.M., Raoof, A.M., Rashed, R.J., Hamid, J.S. and Qdir, A.F. (2014) Effect of foliar spray with thyme extract on codling moth (*Cydia pomonella*) control and some fruit quality of pear (*Pyrus communis* L) Al-zafaraniyah selectee. *Journal of Zankoy Sulaimani, Part A* 16, 125–129.

Hsieh, R.C. and Albrecht, J.J. (1988) Method and apparatus for treating fresh vegetable products. *European Patent Application* EP 0285235 A1. Available at: http://www.google.co.uk/patents/EP0285235A1?cl=en (accessed 4 February 2016).

Little, D.B. and Croteau, R.B. (1999) Biochemistry of essential oil terpenes – a thirty-year overview. In: Teranishi, R., Wick, E.L. and Hornstein, I. (eds) *Flavor Chemistry: 30 Years of Progress.* Kluwer/Plenum, New York, pp. 239–253.

Lueng, A.Y. and Foster, S. (1996) *Encyclopedia of Common Natural Ingredients Used in Food, Drugs, and Cosmetics*, 2nd edn. Wiley, New York.

Nikolić, M., Glamočlija, J., Ferreira, I.C.F.R., Calhelha, R.C., Fernandes, Â., Marković, T., Marković, D., Giweli, A. and Soković, M. (2014) Chemical composition, antimicrobial, antioxidant and antitumor activity of *Thymus serpyllum* L., *Thymus algeriensis* Boiss. and Reut and *Thymus vulgaris* L. essential oils. *Industrial Crops and Products* 52, 183–190.

Ozcan, M. and Chalchat, J.C. (2004) Aroma profile of *Thymus vulgaris* L. growing wild in Turkey. *Bulgarian Journal of Plant Physiology* 30, 68–73.

Özgüven, M. and Tansi, S. (1998) Drug yield and essential oil of *Thymus vulgaris* L. as in [sic] influenced by ecological and ontogenetical variation. *Turkish Journal of Agriculture and Forestry* 22, 537–542.

Panda, H. (2006) *Compendium of Herbal Plants.* Asia Pacific Business Inc., New Delhi.

Porte, A. and Ronoel Godoy, L.O. (2008) Chemical composition of *Thymus vulgaris* L. (thyme) essential oil from the Rio de Janeiro State, Brazil. *Journal of the Serbian Chemical Society* 73, 307–310.

Roby, M.H.H., Sarhan, M.A., Selim, K.A.H. and Khalel, K.I. (2013) Evaluation of antioxidant activity, total phenols and phenolic compounds in thyme (*Thymus vulgaris* L.), sage (*Salvia officinalis* L.), and marjoram (*Origanum majorana* L.) extracts. *Industrial Crops Products* 43, 827–831.

Rohmer, M., Seemann, M., Horbach, S., Bringer-Meyer and Sahm, H. (1996) Glyceraldehyde 3-phosphate and pyruvate as precursors of isoprenic units in an alternative non-mevalonate pathway for terpenoid biosynthesis. *Journal of the American Chemistry Society* 118, 2564–2566.

Royal Horticultural Society (2014) *RHS-Plant Finder.* Available at: http://www.rhs.org.uk/plants/search-form (accessed 16 July 2014).

Senatore, F. (1996) Influence of harvesting time on yield and composition of the essential oil of thyme *(Thymus pulegioides* L*)* growing wild in Campania. *Journal of Agricultural and Food Chemistry* 44, 1327–1332.

Soliman, K.M. and Badeaa, R.I. (2002) Effect of oil extracted from some medicinal plants on different mycotoxigenic fungi. *Food and Chemical Toxicology* 40, 1669–1675.

Stahl-Biskup, E. (2002) Essential oil chemistry of the genus *Thymus* – a global view. In: Stahl-Biskup, E. and Sáez, F. (eds) *Thyme: The Genus* Thymus. Medicinal and Aromatic Plants – Industrial Profiles, Taylor & Francis. New York, pp. 75–124.

Stahl-Biskup, E. and Saez, F. (2002) *The Genus* Thymus. Medicinal and Aromatic Plants – Industrial Profiles, Taylor & Francis, New York.

Tainter, D.R. and Grenis, A.T. (1993) *Spices and Seasonings: Food Technology Handbook (Food Science and Technology).* VCH Publishers, New York.

Van den Broucke, C.O. and Lemli, J.A. (1981) Pharmacological and chemical investigation of thyme liquid extract. *Planta Medica* 41, 129–135.

Van den Broucke, C.O. and Lemli, J.A. (1983) Spasmolytic activity of the flavonoids from *Thymus vulgaris. Pharmaceutisch Weekblad* 5, 9–14.

Venskutonis, P.R. (2002a) Harvesting and post-harvest handling in the genus *Thymus.* In: Stahl-Biskup, E. and Sáez, F. *Thyme: The Genus* Thymus. Medicinal and Aromatic Plants – Industrial Profiles, Taylor & Francis, New York, pp. 197–223.

Venskutonis, P.R. (2002b) Thyme – processing of raw plant material. In: Stahl-Biskup, E. and Sáez, F. (eds) *Thyme: The Genus* Thymus. Medicinal and Aromatic Plants – Industrial Profiles, Taylor & Francis, New York, pp. 224–251.

Vila, R. (2002) Flavonoids and further polyphenols in the genus thymus. In: Stahl-Biskup, E. and Sáez, F. (eds) *Thyme: The Genus* Thymus. Medicinal and Aromatic Plants – Industrial Profiles, Taylor & Francis, New York, pp. 142–176.

Villanueva Bermejo, D., Angelov, I., Vicente, G., Stateva, R.P., Rodriguez García-Risco, M., Reglero, G., Ibañez, E. and Fornari, T. (2014) Extraction of thymol from different varieties of thyme plants using green solvents. *Journal of the Science of Food and Agriculture* 95, 2901–2907.

Zarzuelo, A. and Crespo, E. (2002) The medicinal and non-medicinal uses of thyme. In: Stahl-Biskup, E. and Sáez, F. (eds) *Thyme: The Genus* Thymus. Medicinal and Aromatic Plants – Industrial Profiles, Taylor & Francis, New York, pp. 263–292.

Index

Note: Page numbers in **bold** type refer to **figures**; page numbers in *italic* type refer to *tables*.